EMULSIONS AND EMULSION TECHNOLOGY

SURFACTANT SCIENCE SERIES

CONSULTING EDITORS

MARTIN J. SCHICK
*Diamond Shamrock Chemical
   Company
Nopco Chemical Division
Morristown, New Jersey*

FREDERICK M. FOWKES
*Chairman of the Department
   of Chemistry
Lehigh University
Bethlehem, Pennsylvania*

Volume 1: NONIONIC SURFACTANTS, edited by Martin J. Schick

Volume 2: SOLVENT PROPERTIES OF SURFACTANT SOLUTIONS, edited by Kozo Shinoda *(out of print)*

Volume 3: SURFACTANT BIODEGRADATION, by Robert D. Swisher *(out of print)*

Volume 4: CATIONIC SURFACTANTS, edited by Eric Jungermann

Volume 5: DETERGENCY: THEORY AND TEST METHODS *(in three parts)*, edited by W. G. Cutler and R. C. Davis

Volume 6: EMULSIONS AND EMULSION TECHNOLOGY *(in two parts)*, edited by Kenneth J. Lissant

OTHER VOLUMES IN PREPARATION

AN IMPORTANT MESSAGE TO READERS...

A Marcel Dekker, Inc. Facsimile Edition contains the exact contents of an original hard cover MDI published work but in a new soft sturdy cover.

Reprinting scholarly works in an economical format assures readers that important information they need remains accessible. Facsimile Editions provide a viable alternative to books that could go "out of print." Utilizing a contemporary printing process for Facsimile Editions, scientific and technical books are reproduced in limited quantities to meet demand.

Marcel Dekker, Inc. is pleased to offer this specialized service to its readers in the academic and scientific communities.

about the book...

This volume, in two parts, presents an introduction to emulsions and their most important applications in industry. It brings together up-to-date concepts in emulsion technology and summarizes the state of the art.

The initial chapters of the book cover the theoretical aspects of emulsions and their preparation, from a physical and geometric viewpoint. The terminology used in various sections of the industry is defined and unified in these chapters. The remainder of the book deals with the applications of emulsion technology to ten commercial areas. Each of these chapters explains why emulsions are used, how they are made, and how they behave. Numerous references are included so that the reader can have access to more detailed literature in the field.

Emulsions and Emulsion Technology is of utmost interest to colloid, surface, and polymer chemists in a wide variety of industries. All those concerned with research, development, and manufacture of agricultural and food products, drugs, paints, asphalt, paper, ink, hydraulic fluid and cutting oil, and cosmetics will find this book a valuable reference.

about the editor...

DR. KENNETH J. LISSANT is Director of Advanced Research at Petrolite Corporation in St. Louis, Missouri, where he has worked since 1944. Dr. Lissant received his B.A. in Chemistry (1941) from Ottawa University in Kansas, his M.S. in Physical Chemistry (1943) from Washington University, St. Louis, and his Ph.D. in Colloid Chemistry (1947) from Stanford University. From 1946 to 1947 he was a Bristol-Myers Fellow at Stanford.

Dr. Lissant has pioneered many new developments in the technology of high-internal-phase-ratio emulsions. He has also been active in recent years in the application of this technology to commercial products, including rocket and jet and automotive fuels. Dr. Lissant has published numerous papers on these and other topics and holds more than twenty U.S. patents. He is a member of Sigma Xi, Alpha Chi Sigma, the ACS, and the AAAS, and is listed in *American Men of Science*.

EMULSIONS AND EMULSION TECHNOLOGY

(IN TWO PARTS)

PART I

edited by Kenneth J. Lissant
Director of Advanced Research
Petrolite Corporation
St. Louis, Missouri

MARCEL DEKKER, INC.   New York and Basel

COPYRIGHT © 1974 by MARCEL DEKKER, INC.

ALL RIGHTS RESERVED

Neither this book nor any part may be reproduced or transmitted in any form or by any means, electronic or mechanical, including photocopying, microfilming, and recording, or by any information storage and retrieval system, without permission in writing from the publisher.

MARCEL DEKKER, INC.
270 Madison Avenue, New York, New York 10017

LIBRARY OF CONGRESS CATALOG CARD NUMBER: 73-82192

ISBN: 0-8247-1891-7

Current printing (last digit):
10 9 8 7 6

Printed in the United States of America

CONTENTS OF PART I

Contributors to Part I　　　　　　　　　　　　vii
Contents of Part II　　　　　　　　　　　　　ix
Preface　　　　　　　　　　　　　　　　　　　xi

Chapter 1　BASIC THEORY　　　　　　　　　　　1
　　　　　　K. J. Lissant
　　　　I.　Introduction　　　　　　　　　　　1
　　　II.　Theoretical Considerations　　　　4
　　　III.　Basic Geometry　　　　　　　　　　10
　　　IV.　Low-Internal-Phase-Ratio Emulsions　34
　　　V.　Medium-Internal-Phase-Ratio Emulsions　42
　　　VI.　High-Internal-Phase-Ratio Emulsions　49
　　　VII.　Summary　　　　　　　　　　　　　　66
　　　　　　References　　　　　　　　　　　　68

Chapter 2　MAKING AND BREAKING EMULSIONS　　71
　　　　　　K. J. Lissant
　　　　I.　Introduction　　　　　　　　　　　72
　　　II.　Reasons for Using Emulsions　　　　72
　　　III.　Selection of Emulsion Type　　　　74
　　　IV.　Selection of Emulsifier　　　　　　75
　　　V.　Techniques of Emulsification　　　　103
　　　VI.　Techniques for Breaking Emulsions　111
　　　　　　References　　　　　　　　　　　　123

Chapter 3 MICROEMULSIONS 125
Leon M. Prince

I. Introduction 126
II. Identification 127
III. Theory 145
IV. Practical Applications 161
References 175

Chapter 4 AGRICULTURAL EMULSIONS 179
Paul L. Lindner

I. Introduction 181
II. Calculations of Formulations 183
III. Solvents for Agricultural Formulations 185
IV. Surfactants for Agricultural Formulations 187
V. Emulsion Stability 190
VI. Blending Emulsifiers and Economics 200
VII. Water and Agricultural Formulations 201
VIII. Testing 208
IX. Suggested Optimization of Emulsifiable Concentrates 210
X. Selection of Emulsion Type 214
XI. Some Special Formulations 219
XII. Soil Penetration, Adsorption, and Mobility 226
XIII. Application Techniques 227
References 236

Chapter 5 FOOD EMULSIONS 249
Matthew W. Lynch and William C. Griffin

I. Introduction 250
II. Emulsion Properties 253
III. Emulsification 258
IV. Product Development 268
V. Equipment 274
VI. Food Emulsions 279

	VII.	Other Uses for Surfactants	281
	VIII.	Multiple Emulsifier Effects	286
		References	289

Chapter 6 MEDICINAL EMULSIONS 291
B. A. Mulley

	I.	Introduction	292
	II.	Demulsifying Agents and Other Emulsion Adjuvants Used in Medicine	295
	III.	Some Technical Considerations in the Design and Preparation of Medicinal Emulsions	306
	IV.	Product Review	322
		References	343

Chapter 7 EMULSION PAINTS 351
Gerould Allyn

	I.	Introduction	352
	II.	Latex-Paint Vehicles	360
	III.	Paint Formulation	367
	IV.	Trade-Sales Coatings	374
	V.	Latex Maintenance Paints	379
	VI.	Industrial Latex Paints	380
		References	384

Chapter 8 ASPHALT EMULSIONS 387
R. L. Ferm

	I.	Introduction	389
	II.	Advantages of Asphalt Emulsions	390
	III.	Product Types and Nomenclature	391
	IV.	Emulsifiers for Asphalt	392
	V.	Improvement of Asphalt Emulsifiability	399
	VI.	Additives in Emulsions	399
	VII.	Manufacture of Emulsions	403
	VIII.	Emulsion Properties and Testing	407

IX.	The Technology of Paving with Emulsions	412
X.	Industrial Emulsions	419
XI.	Agricultural and Water-Conservation Uses	425
	References	430

Cumulative Indexes appear in Part II.

CONTRIBUTORS TO PART I

Numbers in parentheses indicate the pages on which the authors' contributions begin.

G. ALLYN (351), Rohm and Haas Company, Philadelphia, Pennsylvania

R. L. FERM (387), Chevron Research Company, Richmond, California

W. C. GRIFFIN (249), ICI America Inc., Specialty Chemicals Division, Wilmington, Delaware

P. L. LINDNER (179), Witco Chemical Corporation, Chicago, Illinois

K. J. LISSANT (1, 71), Research Laboratories, Petrolite Corporation, Saint Louis, Missouri

M. J. LYNCH (249), ICI America Inc., Specialty Chemicals Division, Wilmington, Delaware

B. A. MULLEY (291), Postgraduate School of Studies in Pharmacy, University of Bradford, Bradford, Yorkshire, England

L. M. PRINCE (125), Research Center, Lever Brothers Company, Edgewater, New Jersey

CONTENTS OF PART II

EMULSION POLYMERIZATION, *T. Matsumoto*, The Faculty of Engineering, Kobe University, Rokko Nada Kobe, Japan

EMULSIONS IN THE PAPER MAKING INDUSTRY, *F. Vaurio*, 1313 Palisades Drive, Appleton, Wisconsin

EMULSIONS IN PRINTING AND THE GRAPHIC ARTS, *J. Bulloff*, State University of New York at Buffalo, Albany, New York

HYDRAULIC FLUID EMULSIONS, *R. T. Holzmann*, Quaker Chemical Corporation, Conhohocken, Pennsylvania

EMULSIONS IN THE COSMETIC INDUSTRY, *C. Fox*, Warner-Lambert Research Institute, Morris Plains, New Jersey

PREFACE

Emulsion theory and emulsion technology have always been stepchildren of the more precise theoretical aspects of physical chemistry. Because of the complexity of emulsion chemistry, many approaches have been empirical rather than theoretical and emulsion practices have been considered "art" rather than science. There has also been a feeling that all the theoretical work in this area has been done. W. D. Bancroft, in 1921, says in the preface to his "Applied Colloid Chemistry": "While we do not know much about gelatinous precipitates and jellies, the theory of the rest of the subject is in fairly good shape." He then devotes fourteen pages to the subject of emulsions and foams. In 1939 H. B. Wiser, in his "Colloid Chemistry," devotes thirteen pages to emulsions.

The heyday of theoretical emulsion studies probably came between 1930 and 1950. During this time, Professors J. W. McBain and W. D. Harkins each directed a number of graduate students in emulsion studies. w. Clayton published a landmark book, "The Theory of Emulsions and Their Technical Treatment," in 1923; the third edition appeared in 1935.

After about 1950, the emphasis on theoretical studies of emulsion behavior dwindled to a mere trickle. Paul Becker in 1957 published an excellent book, "Emulsions, Theory and Practice," the second edition of which appeared in 1965.

Emulsions, however, occur throughout industry and the natural world, and refuse to be ignored. Industry has, therefore, been forced to develop techniques for making and breaking emulsions, but most of this material is proprietary and, except for the patent literature, is unavailable to investigators. One of the purposes of writing this book was to attempt to bring the status of emulsion technology up to date; and, while this in itself might be considered a sufficient reason, we have also felt that a book was needed to introduce the concept of emulsions to a wider group of people. With the recent interest in the environment, many industries have been required to resolve their waste emulsions and recover the pollutant materials from them; and, therefore, engineers, supervisors, and architects have been forced to try to learn something about emulsions.

It is hoped that the first three chapters of this book will lay a simple but sound theoretical basis for further studies. We have deliberately avoided many of the classical mathematical derivations which can be found in books such as those mentioned above, or in the reference lists, and we have concentrated on a geometrical approach which we have found to be extremely useful in our many years of practical work with emulsions. If the general reader will study the first three chapters and then turn to the particular chapter which deals with applications in his area of interest, he should have an excellent basis on which to build the expertise which he needs to work with emulsions.

Each of the chapters which deals with the application of emulsions to a particular industry uses the terminology which is common to that industry. It will be noticed that often the terminology varies from one area to another. We have not attempted to resolve this terminology, since it was felt that anyone working in the field will have to learn to live with this problem.

In selecting authors for the chapters on practical applications, we have tried to pick people who have worked in the area for many years and who are recognized as experts in their field. I wish to express my thanks to each of the authors for the cooperation they have given in putting this book together.

In selecting the areas for discussion, we tried first to pick those areas where emulsions are most important. We realize that a number of areas which could have been discussed have been neglected. In some instances, it will be found that the technology discussed in other chapters is directly applicable to areas which we omitted. Any attempt to include all the areas where emulsions are important would have required several more volumes.

Kenneth J. Lissant

EMULSIONS AND EMULSION TECHNOLOGY

Chapter 1

BASIC THEORY

K. J. Lissant

Research Laboratories
Petrolite Corporation
Saint Louis, Missouri

I.	INTRODUCTION.	1
II.	THEORETICAL CONSIDERATIONS.	4
III.	BASIC GEOMETRY.	10
IV.	LOW-INTERNAL-PHASE-RATIO EMULSIONS.	34
V.	MEDIUM-INTERNAL-PHASE-RATIO EMULSIONS	42
VI.	HIGH-INTERNAL-PHASE-RATIO EMULSIONS	49
VII.	SUMMARY	66
	REFERENCES.	68

I. INTRODUCTION

Ecology is one of the popular words today, and it would seem therefore expedient to attempt to ensure the success of any treatise by relating it to topics that have the attention of the public. Perhaps, then, our first job is to establish at least some tenuous relationship between ecology and emulsions. It turns out that this is not at all difficult to do and that, when one does so, some rather useful ideas emerge. The first thing that becomes apparent is that the subject of emulsion, like any other topic, cannot be usefully considered detached from its normal environment. To be able to handle and manipulate any type of phenomenon, one has to be able to do so under conditions that are routinely encountered in the real world.

Most disciplines run through comparable evolutionary developments. They begin as a collection of disjointed observations. These are then collected into encyclopedic arrangements, at which point there emerge certain correlations that permit simplification of the concepts and prediction of much of the data, thus making the encyclopedic collections less necessary. Unfortunately emulsion technology is still to a large extent in the encyclopedic stage. Some 20 to 30 years ago a number of general theories were evolved and some specific mathematical treatments were developed. They have not, however, proved as useful in practical fields as could be wished. Currently relatively few academic institutions are seriously engaged in the studies of emulsions, and most of the emulsion technology is held in proprietary files around the world.

However, emulsions occur everywhere in our daily life, and as a result investigators in other disciplines have been forced to deal with the emulsions they encounter. For this reason a portion of the new work in the area of emulsions is being done by meteorologists, microbiologists, physicists, and others.

It would be presumptuous to hope that this volume would constitute the generalizations that would allow simplification of the area of emulsion technology. It is hoped, however, that it will serve as a guide for the development of more useful concepts.

Since this volume is likely to be read by investigators in a variety of fields, many of whom have not been trained in the "classical" emulsion theory, a little time will be spent in discussing the meaning of certain terms, even though these terms may be familiar to practitioners in the field. It has become apparent

1. BASIC THEORY

that the precise meaning of some terms is not universally accepted and that a good deal of looseness exists in the use of terminology. A discussion and definition of terms will therefore be undertaken.

In high school or even grade school, today, we are told that there are three states of matter: solid, liquid, and gaseous. A grade school student could probably calculate that nine possible two-way combinations of these states exist. However, since by definition all gases are miscible, the mixture of "gases in gases" always results in a single-phase homogeneous system. At least that is what the physics books say. Meteorologists may disagree. We are then left with possible mixtures of

Solids in solids	Conglomerate rocks
Solids in gases	Dust clouds, smoke
Solids in liquids	Muds, colloidal dispersions
Gases in solids	Pumice, "foam" insulation
Gases in liquids	Foams
Liquids in solids	Butter, certain lavas
Liquids in gases	Fog
Liquids in liquids	Emulsions

Classically the area concerned with mixtures of "liquids in liquid" constitutes the domain of emulsions and is the concern of this volume. However, in practice the picture is not quite this simple. Whenever we have two immiscible liquids in contact with each other there is a tendency for one of the liquids to become dispersed in the other, and as long as some mechanical agitation is present, this dispersion may persist. When allowed to remain in a quiescent state, pure liquids will separate into two bulk layers. In actual practice pure liquids do not remain pure for long, and almost any two mixtures of impure, immiscible liquids are likely to form a dispersion. Dispersions of one liquid in another are not usually called emulsions unless they have sufficient stability to persist for a reasonable period of time. This period may vary from a few seconds to several years, depending on the purpose for which the system is being used. Obviously food emulsions and other similar products are required to remain stable in containers for extended periods of time. However, if one is conducting a liquid-liquid extraction process, a dispersion that persists for even a few seconds in the dynamic process may cause trouble and be considered an "emulsion problem". It should also be remembered that solids or gases can be dispersed or dissolved in either or both of the liquid phases, and in actual practice this is often the case. So-called emulsion paints are usually dispersions of a paint vehicle in water, but the system also contains a variety of solid pigments and a number of components dissolved in either the oil or water phase. Wax-polish "emulsions" are technically dispersions of very finely divided solid wax particles in water. However, since they are usually manufactured by mixing a liquid wax formulation with water above the melting point of the wax, they are produced as emulsions and considered in the area of emulsion technology.

It becomes apparent, then, that emulsions pervade much of nature, and many of the products produced and used by man either are emulsions or involve emulsion technology. It is also apparent that very few of these "emulsions" are simple dispersions of one liquid in another and that we therefore cannot study them effectively without considering them in the context of their environment.

II. THEORETICAL CONSIDERATIONS

Textbooks on physical or colloid chemistry are not very helpful on the subject of emulsions. There is usually a short chapter at the back of the book, which is skipped or glossed over at the end of a college course. Even the most recent books are not completely up to date on the subject. For instance, a well-known physical chemistry text of over 1300 pages contains no mention of

1. BASIC THEORY

emulsions. Most of the really pertinent information is to be found in patents or in the proprietary files of private industry.

A number of books and articles have been written on the theoretical aspects of emulsions (<u>1-8</u>). Many systems have been studied, and large amounts of data have been collected. Unfortunately most of the information is on systems that are not usually encountered in actual practice. Many excellent mathematical treatments have been presented, but here again they are on idealized systems and of limited use in practice. Rather than repeat these calculations in detail, this discussion will attempt to develop a rationale of emulsion formation and formulation, and show how the theoretical material can be used as a guide for practical work.

In the past a great deal of effort has been devoted toward mathematical formulations for the energy relationships at the interfaces of emulsion systems, and a number of ingenious and fruitful approaches have been devised to explain certain emulsion phenomena. Much of this explanation has been ex post facto. In approximately 30 years of experience with emulsion systems, I have become more and more convinced that the most useful approach in attacking an entirely new system is to regard it as a mechanical "construct." In other words, many emulsion systems behave as though their performance properties were more dependent on their physical and topological configuration than on the chemical properties of the constituents. In many instances a number of functionally equivalent emulsions can be made, all of which are suitable for the practical purpose, but each of which differs substantially in chemical composition. For instance, one emulsion may be made with a cationic emulsifier and another may be made with an anionic emulsifier, and yet both may function equally well for a given purpose. However, in investigating the geometric or topological properties of the two emulsions, one finds that they are essentially identical and that the properties that make them useful for a particular purpose are more a function of their physical state than of their chemical composition. It is for this reason that we have chosen to discuss emulsions under the low-, medium-, and high-internal-phase-ratio type of classification rather than to classify them according to the type of emulsifier employed.

The classic definition of an emulsion as a continuous liquid phase in which is dispersed a second, discontinuous, immiscible liquid phase is far from complete. One very convenient way to classify emulsion is first to divide them into two large groups on the basis of the nature of the external phase. The two groups are usually called *oil-in-water* (o/w) and *water-in-oil* (w/o) emulsions. The terms "oil" and "water" are very general; almost any highly polar, hydrophilic liquid falls into the "water" category in this definition, while hydrophobic, nonpolar liquids are considered "oils."

Once the type of internal and external phase has been specified, the next step is to divide the emulsions into three classes based on the volume percentage of the internal phase, or the *internal-phase ratio*. Emulsions with less than about 30% internal phase form the first category. Emulsions with about 30 to about 70% internal phase fall into the second category, and those with more than 74% internal phase fall into the third category. We can thus present a diagram showing the six major classes (Table 1).

Although it must be admitted that the boundaries of these groups are somewhat arbitrarily drawn, there are very sound reasons for dividing all emulsions into these six groups. The volume

TABLE 1

The Six Major Classes of Emulsions

Group	Internal-phase ratio (IPR)		
	<30%	30 to 74%	>74%
Water in oil	Low-IPR w/o	Medium-IPR w/o	High-IPR w/o
Oil in water	Low-IPR o/w	Medium-IPR o/w	High-IPR o/w

1. BASIC THEORY

percentage of internal phase has a profound influence on the properties of the emulsion. This can be illustrated by considering a container partially filled with a liquid. Into this container add a small amount of a second, immiscible, liquid. It will coalesce and form a layer with an interface between the two liquids. If, however, we had added to the first liquid a material that would prevent the coalescence of the second phase, the second liquid would break up into droplets as it was poured in and form a loose emulsion. Each of these droplets would be essentially spherical, and the system can be idealized as a collection of spheres dispersed in a continuous liquid phase. Unless the individual drops are small enough to be kept in suspension by thermal forces, they will eventually settle out or rise to the top and form a layer of droplets. This is known as *creaming*.

When the internal phase constitutes less than about 30% of the volume, the individual droplets do not interfere with each other appreciably, and the physical properties of the whole system are determined primarily by the nature of the external, continuous, phase. As more internal phase is introduced, the droplets begin to collide more frequently and to interfere with each other. This causes an apparent increase in the viscosity of the whole system.

The viscosity of the system then increases slowly until the internal-phase ratio reaches about 50 to 52% by volume. This is the volume ratio of uniform spheres stacked in a cubic array. At ratios higher than this the particles of the internal phase are forced into close contact, and unless the emulsion is polydisperse, the formulation shows quite high viscosity and non-Newtonian flow behavior. This geometrical situation is also responsible for the fact that slurries of nondeformable solids in this range become too thick to flow.

At an internal volume ratio of about 68% another packing point is reached. Uniform spheres packed into a tetrakaidecahedral array occupy about 68% of the volume. Emulsions in this range are not usually stable to shear unless special emulsifier systems are used. Most of the conventional and traditional emulsifier systems lose effectiveness at or above this range, and the emulsions invert, with the greater phase becoming external.

When the internal phase of the system nears 74% by volume, the droplets are so closely crowded that they have no room to settle, and the whole system takes on the appearance of the layer of creamed material seen in the earlier examples that had stratified. If the system contains more than 75% internal phase, the droplets are so closely crowded that they can no longer be spheres unless the emulsion is polydisperse. At this point the apparent viscosity of the system increases rapidly and non-Newtonian behavior becomes marked. Emulsions of this third type--that is, high-internal-phase-ratio emulsions--cannot cream because they are already in what may be called a "supercreamed" state. As long as the emulsifiers prevent coalescence, the emulsion will be stable.

At this point it may be well to clarify some terminology. Let V_e represent the volume of external phase and V_i the volume of internal phase. Then the total volume equals

$$V_t = V_e + V_i$$

The volume percentage of internal phase then equals

$$V_{\%i} = \frac{V_i \times 100}{V_t}$$

It is this term that we shall use as the "internal-phase ratio" in our discussions. Obviously

$V_i \times d_i$ is the weight of internal phase

$V_e \times d_e$ is the weight of external phase

$$\text{weight percentage of internal phase} = \frac{V_i \times d_i \times 100}{V_i d_i + V_e d_e} = W_{\%i}$$

This term is of practical importance since one is often required to

1. BASIC THEORY

prepare emulsions containing "5# per gallon of active ingredient." It does not, however, have a direct, obvious relationship to the emulsion properties.

Sometimes in theoretical papers the term "phase ratio" is described as ϕ

$$\phi = \frac{V_i}{V_e}$$

For comparison purposes

$$V_{\%i} = \frac{100\phi}{\phi + 1} \quad \text{or} \quad \phi = \frac{V_{\%i}}{100 - V_{\%i}}$$

Similar relationships hold for $W_{\%i}$.

Although ϕ seems to be the term to use from a theoretical standpoint, experience has shown that $V_{\%i}$ is easier to visualize. More frequently ϕ is defined as

$$\phi = \frac{V_i}{V_i + V_e}$$

in which case

$$V_{\%i} = 100\phi$$

All this is simple, obvious arithmetic, but great care must be taken in reading the literature to be sure what particular term is meant. For example, in plant production the internal- and external-phase feed pumps may be interconnected to control ϕ where

$$\phi = \frac{V_i}{V_e}$$

and in this case ϕ varies from less than 0.1 to more than 100. On the other hand, if

$$\phi = \frac{V_i}{V_i + V_e} = \frac{V_i}{V_t}$$

ϕ varies from less than 0.01 to just less than 1.00.

As a further example, if equal volumes of two liquids are used in forming an emulsion, $V_{\%i}$ = 50 and ϕ = 0.50 or ϕ = 1.00, depending on the definition of ϕ.

In this volume "internal-phase ratio" will mean volume percentage of internal-phase.

Obviously, to prepare an emulsion, one must have two immiscible liquid phases. It is doubtful that any two liquids are completely mutually insoluble. For most purposes the small solubility of the liquids in each other can be ignored. It should always be remembered, however, that in any emulsion both phases eventually become saturated with respect to the other phase. Attainment of solubility equilibrium between the two phases may not be as rapid as one would think. Some aspects of emulsion storage stability may involve slow attainment of solubility equilibrium.

III. BASIC GEOMETRY

Before beginning a discussion of the geometry of emulsions, it is worthwhile to review a few fundamental geometric principles. Strangely enough, although geometry is taught at the high school level and is considered relatively simple, the geometric principles involved in space-filling--that is, tessellating--configurations have not been studied in as much detail as one would expect. probably the most complete treatment of this subject is *Regular Figures* by Tóth (9). On page 263 at the beginning of the chapter on 3-space Tóth says, "Unfortunately very little is known in this field."

We shall first consider patterns that completely fill a plane and restrict ourselves to regular figures. Only three regular figures tessellate on a plane. These are depicted in Fig. 1.

1. BASIC THEORY

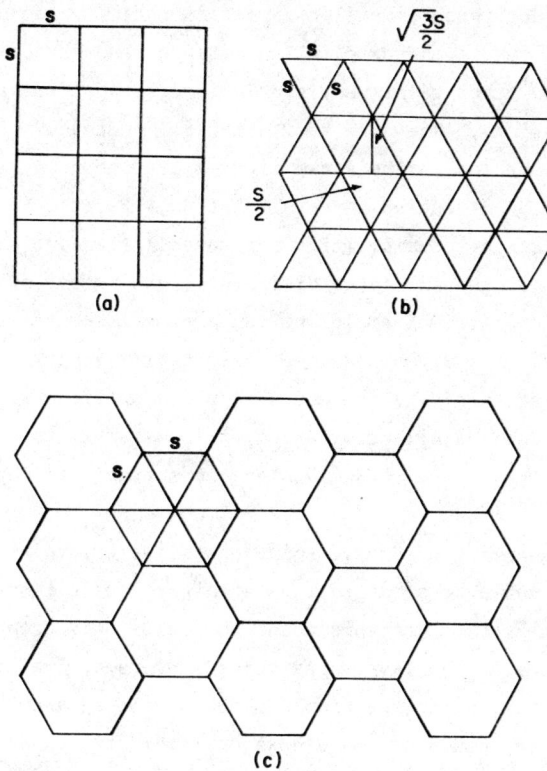

FIG. 1. The three regular figures that tessellate on a plane: (a) the square, (b) the equilateral triangle, (c) the equilateral hexagon.

They are the square, the equilateral triangle, and the equilateral hexagon. If S equals the length of the side, then

area of the square: $A = S^2$

For the equilateral triangle

$$A = \frac{\sqrt{S}}{4}$$

and for the equilateral hexagon

$$A = \frac{\sqrt{3}\,S^2}{4}$$

If we consider three dimensions, we find that, analogous to the square, the cube is a space-filling figure, and here again there are only two other relatively simple figures that tessellate in three dimensions. They are the rhomboidal dodecahedron and the tetrakaidecahedron. (These two figures are shown in Figs. 10 and 11 and will be discussed in detail later.) Because of the quite limited number of possibilities for tessellations of a simple nature in both two and three dimensions, the number of configurations that droplets of emulsions can assume is quite restricted. Let us consider a situation where essentially spherical, monodisperse droplets are settling out on a plane surface, without coalescence, to from a single droplet layer. At first, the droplets will distribute themselves randomly over the surface, as depicted in Fig. 2.

As the number of droplets increases, we arrive at the point where they could be arranged in a straight square tessellation, as depicted in Fig. 3. At this point they will only occupy $\pi/6$, 52.36% of the volume of the layer. As more droplets arrive at the interface, they will force the droplets out of the square arrangement and ultimately could assume a hexagonal pattern, as depicted in Fig. 4. Here they will occupy square $\sqrt{3}/9$ π, 60.4599% of the volume of the layer. This is hexagonal prism packing. From this point on, as more droplets arrive at the surface, a second layer will have

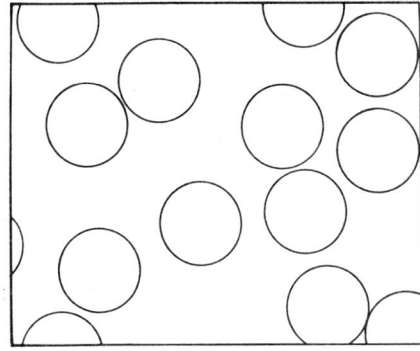

FIG. 2. Random spheres on a surface.

1. BASIC THEORY

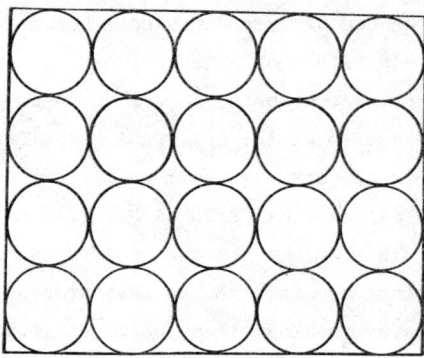

FIG. 3. Square packing on a plane.

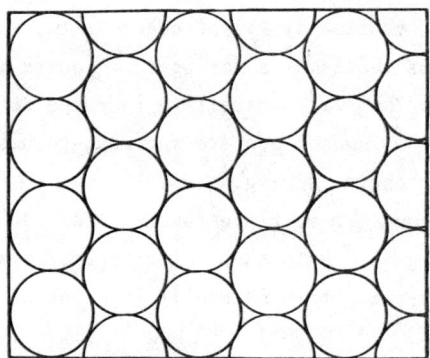

FIG. 4. Hexagonal packing on a plane.

to be started. Of course, in actual practice the weight of the additional droplets will deform the first layer, causing flattening of the spheres. However, for our initial discussion we shall assume that the droplets behave like rigid spheres.

Consider for a moment the transition of the square array to the hexagonal array. In Fig. 5(a) we depicted four of the droplets in the square array and connected the centers of the four droplets to produce a square. Alternatively we could consider the square as depicted in Fig. 6(a) so that a droplet is inscribed in the unit cube. In order for this arrangement to transform itself into the hexagonal array, we can picture the square becoming distorted into a diamond, as in Fig. 5(b), or alternatively, as in Fig. 6(b), we can consider the droplet as being inscribed in an irregular hexagon. When the droplets arrive at the hexagonal tessellation as depicted in Fig. 5(c), their centers will form equilateral triangles, while in Fig. 6(c) they can be considered as being inscribed in an equilateral hexagon. Thus we can see that all three of the simple primary tessellations are involved in this process.

For single-layer packing, then, we expect transition points possibly at about 52 and 60%. If the succeeding layers were stacked with the centers of the spheres exactly on top of those in the first layer, we would have shear planes in the packing and the whole array could be expected to flow by slippage across the shear planes. Geometrically this is the same as saying that square-, triangular-, or hexagonal-based prisms will tessellate in three dimensions, but that these configurations cannot achieve internal-volume ratios of much more than 60% as sphere-packing modes.

Let us now consider multiple layers. Consider first the square array shown in Fig. 3. Note that, if a sphere is allowed to drop onto such an array, it falls naturally into one of the gaps between

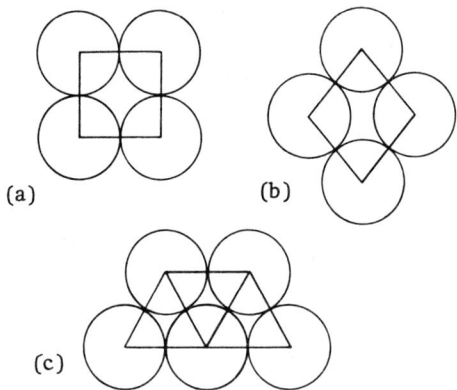

FIG. 5. Transition from square to hexagon.

1. BASIC THEORY

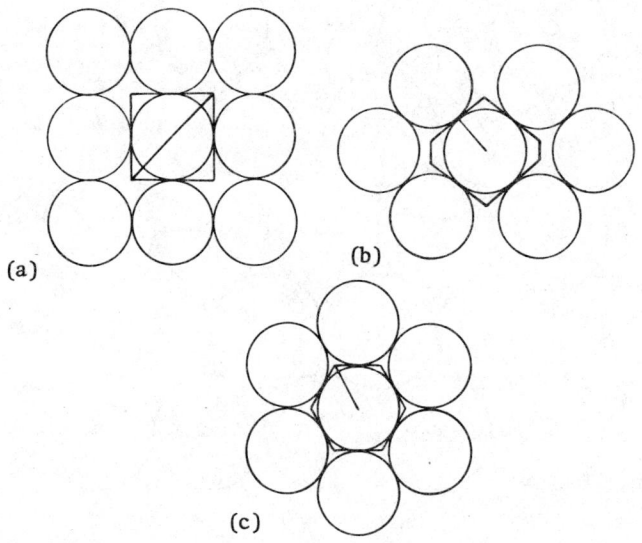

FIG. 6. Unit cells.

four spheres and rests so as to be touching each of the four spheres. This process can be continued until all gaps are filled. The result is an array like that shown in Fig. 11b. Note that there is only one way this array can be put together, and if pyramidal structures are built on this plan, the sides of the pyramids show hexagonal packing.

Next let us consider layers built up from a hexagonal base, like that shown in Fig. 4. Note that the "gaps" in this system are among three spheres instead of four. If a sphere is placed in any particular gap, we find that the next three closest adjacent gaps are blocked, and therefore only half the gaps can be occupied in any one layer. Once the first sphere is placed, the pattern is set, but two equivalent patterns are possible.

Consider Fig. 7, where we have labeled the centers of the spheres in the first layer as c and the center of the spheres in the second layer as a. Now when we go to place the third layer, we can place the centers of the third layer of spheres either over the gaps (the b position) or over the spheres (the c position).

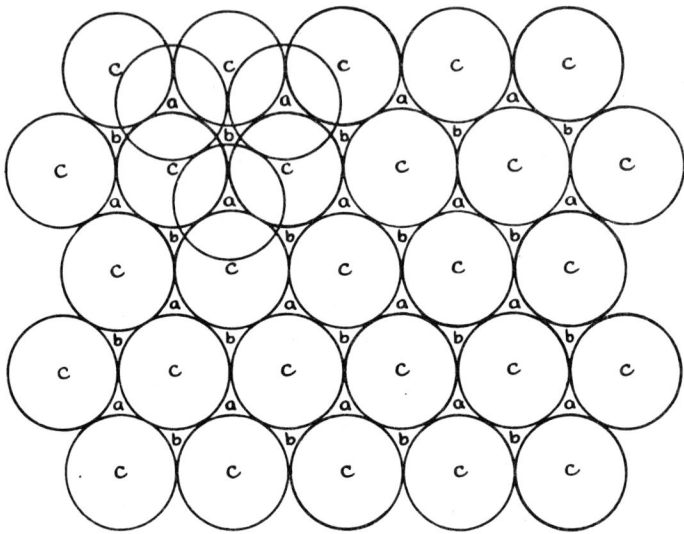

FIG. 7. Multilayer close packing.

Placing the third layer in the b position results in the same pattern as we obtained from the square array, usually called "cubic close packing." Placing the third row over the c position results in a different pattern (depicted in Fig. 12), usually referred to as "hexagonal close packing." It is interesting to note that both cubic and hexagonal close packing can arise from an initial hexagonal layer, but only the cubic can arise from an initial square layer. It should also be noted that from the standpoint of surface area and volume occupied by the spheres the two configurations are equivalent.

By now the reader will have realized that even with fairly good illustrations it is difficult to become really familiar with three-dimensional patterns without actually constructing the configurations. We have found that the simplest and least expensive way of preparing three-dimensional models in this particular area is to purchase a large number of foamed-polystyrene spheres. The 1.5-in.-diameter size can be purchased in lots of 1000 for approximately 3 cents each. They can be fastened together with toothpicks

1. BASIC THEORY

and glued with model cement and can be painted or dyed with appropriate materials.

The two-dimensional nets for the construction of such polyhedral figures as the rhomboidal dodecahedron and the tetraidecahedron are shown in Fig. 8. If this network is drawn to a suitable size, it can be reproduced on light-weight board or card stock using a Xerox or similar copier. The net can then be cut out, folded, and fastened together with plastic tape. The time spent in constructing such models both from the plastic balls and the cardboard will be found well spent in that it makes the interrelationships among the configurations much more readily observable.

We have so far considered straight cubic packing, cubic close packing, and hexagonal close packing. We still have to consider

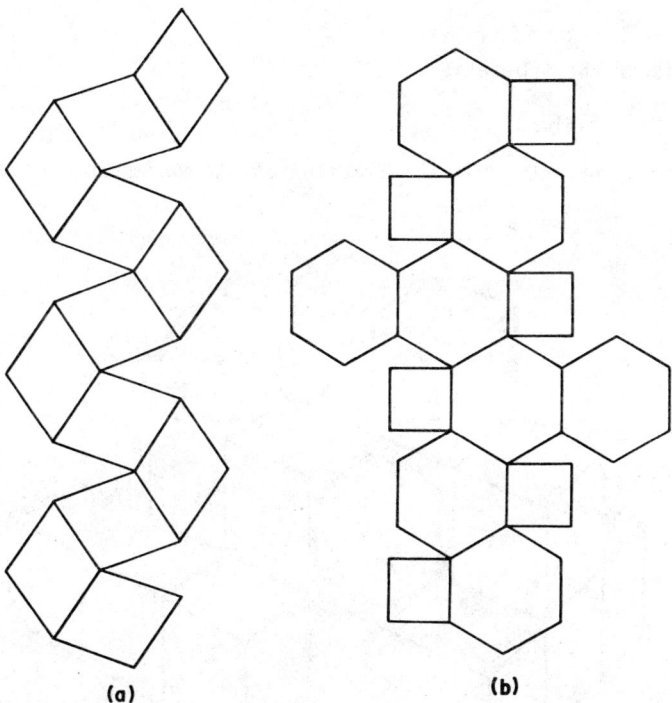

FIG. 8. Two-dimensional nets for the construction of (a) the rhomboidal dodecahedron and (b) the tetrakaidecahedron.

the tetrakaidecahedral (TKDH) pattern. This cannot be built up directly from either the square or the hexagonal first layer. If, however, the squared configuration is distorted into a diamond of the proportions found in the one face of the rhomboidal dodecahedron, we get a pattern like that described in Fig. 6(b). Note that there are again two positions that the spheres of the next layer can occupy. Strangely enough, it turns out that in the TKDH configuration the spheres of the second layers occupy a position intermediate between the a and b configurations of Fig. 7.

We now have four possible sphere packings: the cubic, the TKDH, and the cubic and hexagonal close packings. Let us now see what the unit cell is in each of these four configurations. In the case of the straight cubic packing the unit cell is a cube and the sphere is inscribed in it (Fig. 9). The volume of the cube is

$$V = S^3$$

and that of the sphere is

$$V = (4/3)\pi(S/2)^3$$

Thus it can be seen that the fraction of the volume of the unit

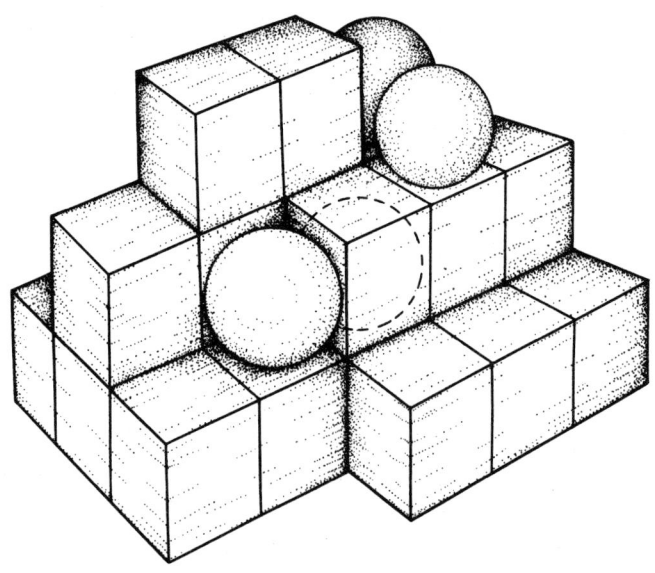

FIG. 9. Cubic packing.

1. BASIC THEORY

cell occupied by the sphere is

$$\frac{V_s}{V_c} = \frac{4/3\pi (S/2)^3}{S^3} = \frac{\pi}{6}$$

Thus in the straight cubic packing the sphere occupies 52.36% of the total volume.

For TKDH, picture the TKDH inscribed in a cube as in Fig. 10(a).

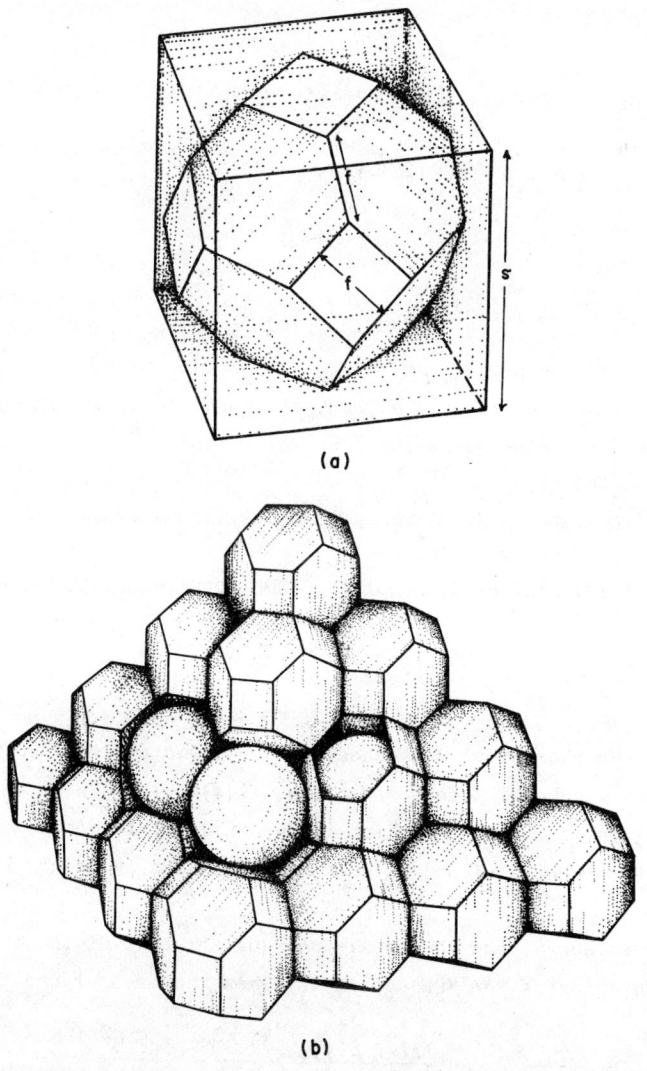

FIG. 10. The basic tetrakaidecahedron (a) and tetrakaidecahedral packing (b).

It has six square faces and eight hexagonal faces. All the edges f are of equal length.

If we call the side of the cube S, then

$$f = \frac{\sqrt{2}\,S}{4}$$

The volume of the TKDH is equal to half that of the cube:

$$V = \frac{S^3}{2}$$

The area of the TKDH is

$$A = 6f^2 + 8\,\frac{3\sqrt{3}}{2}\,f^2 = (6 + 12\sqrt{3})\,f^2$$

or

$$A = (6 \times 12\sqrt{3})\,\frac{(\sqrt{2}S)^2}{4} = \frac{(3 + 6\sqrt{3})S^2}{4}$$

If a sphere is inscribed in a TKDH, it touches only the hexagonal faces, and the diameter is equal to $\sqrt{6}f$. Thus

$$V = \frac{4}{3}\pi\,\frac{(\sqrt{6}f)^3}{2} = \frac{4}{3}\pi\,\frac{(\sqrt{3}S)^3}{4} = \frac{\sqrt{3}\,\pi S^3}{16}$$

The percentage of the volume of the TKDH occupied by the sphere is

$$\frac{(\sqrt{3}\,\pi S^3)/16}{S^3/2} = \frac{\sqrt{3}\,\pi}{8} = 68.02\%$$

For the rhomboidal dodecahedron (RDH), picture the RDH inscribed in a cube of side S as in Fig. 11(a). The RDH is a 12-sided figure, each face of which is an equilateral rhombus. The major axis of the rhombus, a, equals $\sqrt{2}\,S/2$, and the minor axis, b, equals $S/2$.

The volume of the RDH equals one-quarter that of the cube, or $S^3/4$. The area of the RDH equals 12 times the area of one face, or

$$A_{RDH} = \frac{12\sqrt{2}\,S}{2}\,\frac{S}{4} = \frac{12(\sqrt{2}\,S^2)}{8} = \frac{3\sqrt{2}\,S^2}{2} = \frac{3\sqrt{2}\,S^2}{2}$$

1. BASIC THEORY

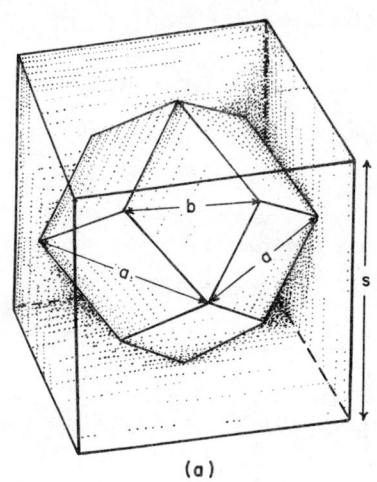

(a)

(b)

FIG. 11. The basic rhomboidal dodecahedron (a) and rhomboidal dodecahedral packing (b).

If a sphere is inscribed in the RDH, it will have a diameter of "a" and a volume of $4.3\pi(a/2)^3$, or

$$V = \frac{4}{3}\pi \frac{(\sqrt{2}\ S^2)^3}{4} = \frac{\sqrt{2}\ \pi S^3}{24}$$

Since the RDH has a volume of $S^3/4$, the percentage of the volume of the RDH occupied by the sphere is

$$\frac{(\sqrt{2}\ \pi S^3)/24}{S^3/4} = \frac{\sqrt{2}\ \pi}{6} = 74.048\%$$

This is simply another way of calculating the closest packing of spheres.

The RDH, when closely examined, will be found to have a hexagonal cross section when cut through a plane of symmetry. If such a cut is made and the pieces rotated one-sixth of a turn and reassembled, a double bee cell results. This is shown in Fig. 12(a). It is the form of the cell in a honeycomb.

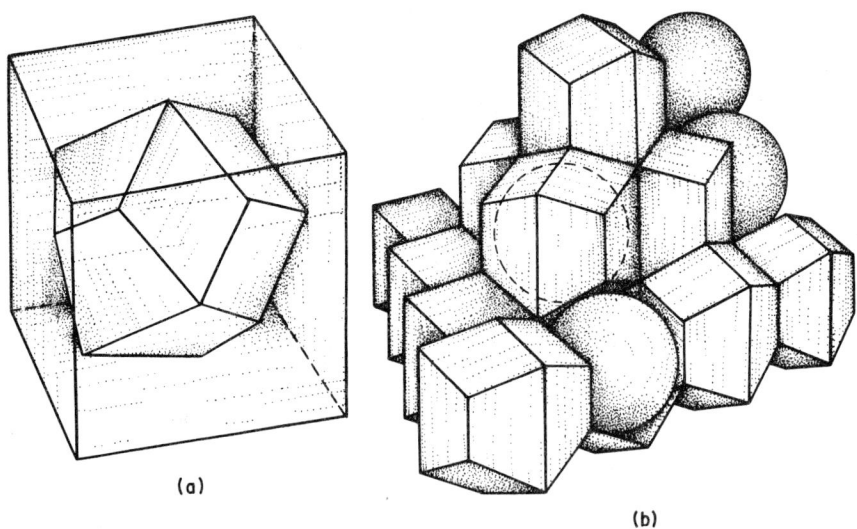

(a)

(b)

FIG. 12. The basic double bee cell (a) and bee-cell packing (b).

1. BASIC THEORY

The double bee cell has six trapezoidal and six rhomboidal faces. Obviously the area and volume are the same as in the RDH.

In retrospect, then, we can see that, since there are two ways of stacking spheres in the closest-packing mode, the unit cells have to be equivalent solid shapes.

We have now shown that the regular cubic packing occupies 52.36% of the volume, the hexagonal prismatic packing occupies 60.46%, the TKDH packing occupies 68.02%, and either the hexagonal or the cubic close packing occupies 74.048% of the total volume. A number of studies have been made concerning the actual density of packing obtainable by randomly pouring spheres into a container. Two packings of this kind are recognized, the so-called random loose packing, which occupies about 60% of the volume, and random close packing, which occupies 64% of the volume. Mathematical models of such packings have been worked out (10). It would seem, then, that as rigid spheres are suspended in a relatively non-viscous medium, they will flow fairly readily until there are enough of them to occupy approximately 50% of the total volume. Under conditions where the spheres occupy some 50 to approximately 75% of the total volume they will assume approximations of one or more of the configuration we have described. Monodisperse rigid spheres cannot occupy more than 74.05% of the total volume. Since emulsions consist of particles that are neither monodisperse nor rigid, we expect some deviation from this classical behavior.

Having dealt with the problem of close packings, let us now consider the loosest, or least dense, packings that can form a rigid configuration. This problem is discussed briefly in Gardener's *New Mathematical Divsersions* (11). It is also discussed in some detail by Tóth (9). We can approach this problem by studying ways in which spheres can be arranged in a single layer on a plane. Beginning at about page 47 of his book, Tóth (9) discusses circle coverings of a plane and portrays 31 possible single-layer packings. It is not worth our while to picture all of these packings here, although the reader is strongly advised to study them.

Most of the packings shown in Tóth's book can be eliminated from our consideration for one reason or another. The first packing he illustrates is the hexagonal packing of our Fig. 5(c). He then shows 11 gradual transitions from this packing to the square packing of our Fig. 5(a). We have already shown that these packings cover the range from 52% to approximately 60% and that, therefore, they do not represent less dense packings than the ones already discussed. Three of the packings portrayed by Tóth are not rigid in that some of the spheres are free to move, and hence the configuration would collapse into a more dense pattern. Six more of the packings are unstable in that certain groups of spheres can be rotated into positions where the pattern would collapse into a denser configuration. This leaves us nine possible configurations for more detailed study.

Of the nine possible configurations, closer study shows that they fall essentially into three categories, the first category consisting of stable patterns obtained by removing spheres from the hexagonal configuration. Figure 13(a) shows the hexagonal packing of Fig. 5(c).

Let us consider what happens if we remove single spheres like the shaded sphere in Fig. 13(a). Note that the removal of a single sphere does not allow any of the adjacent spheres to move since each of them is still in contact with at least three other spheres and no single hemisphere is without a contact. We have labeled the six spheres immediately adjacent to the hole we have produced with an a. Notice that, if any of the "a" spheres are removed, the configuration becomes unstable and adjacent spheres can then fall into the hole.

Let us consider the next ring of spheres and observe that there are two different kinds of position in the next ring. Half the spheres are in contact with two "a" spheres, and half the spheres are in contact with only one "a" sphere. We have distinguished these two classes by labeling them b and c, respectively. We now study the next ring of spheres, and here we have labeled r and p (right

1. BASIC THEORY

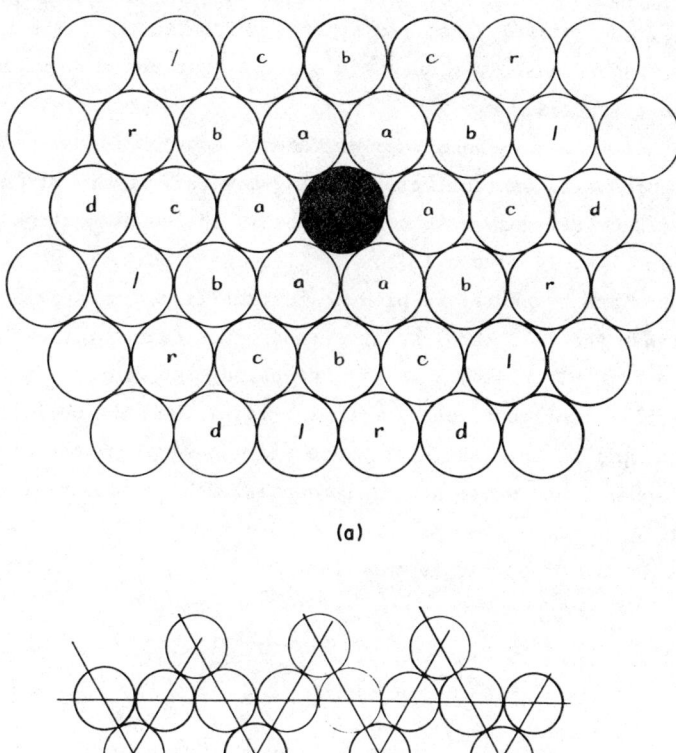

(a)

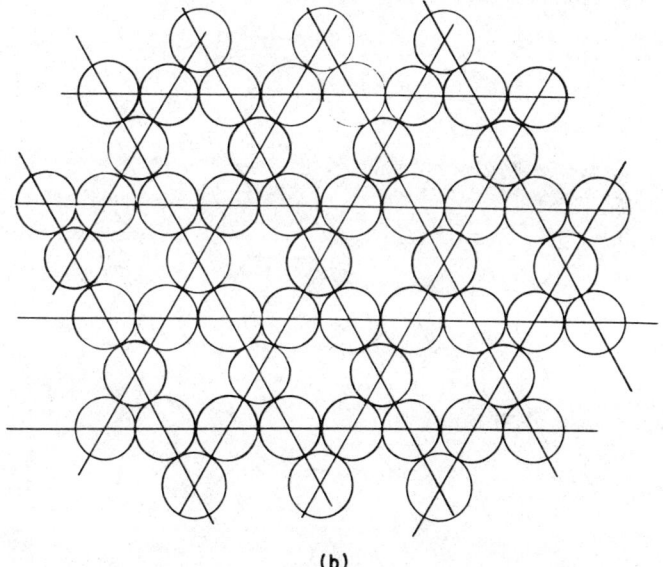

(b)

FIG. 13. Stable holes in hexagonal packing.

and left, respectively) the two spheres in contact with the b spheres, and we have labeled d the spheres that are only in contact with the c spheres.

We have already shown that we cannot remove category a spheres without producing instability. We can, however, remove either category b or category c spheres to obtain the configurations depicted in Figs. 14(a) and 13(b). Notice again that, if we remove both the c and b sphere, we produce instability, We can also remove either the r or the l spheres to produce configurations of the type shown in Fig. 14(b). We will now study the density of packing of these configurations. For Fig. 14(a) the unit cell is a hexagonal prism, the length of whose side is equal to the diameter of the spheres and whose height is equal to the diameter of the spheres.

It is thus possible to show that

$$V_t = 12\sqrt{3}\ r^3$$

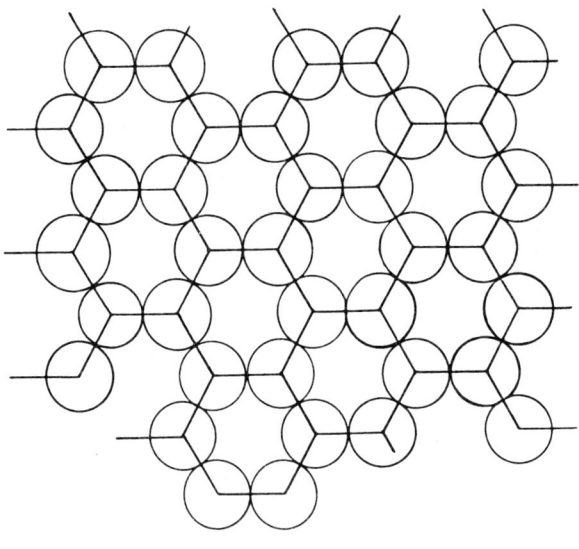

(a)

FIG. 14(a). Stable planar packings.

1. BASIC THEORY

This unit cell contains two complete spheres. Therefore

$$V_s = \frac{8}{3}\pi r^3$$

The packing density is therefore V_s divided by V_t

$$\frac{V_s}{V_t} = \frac{8/3\ \pi r^3}{12\sqrt{3}\ r^3} = \frac{2\sqrt{3}\ \pi}{27} = 0.403$$

Similar calculations show that the density of the configurations shown in Figs. 13(b) and 14(b) is approximately 0.454 and 0.519, respectively. We can thus see that occasionally spheres can be removed from the closest-packing configurations without disturbing

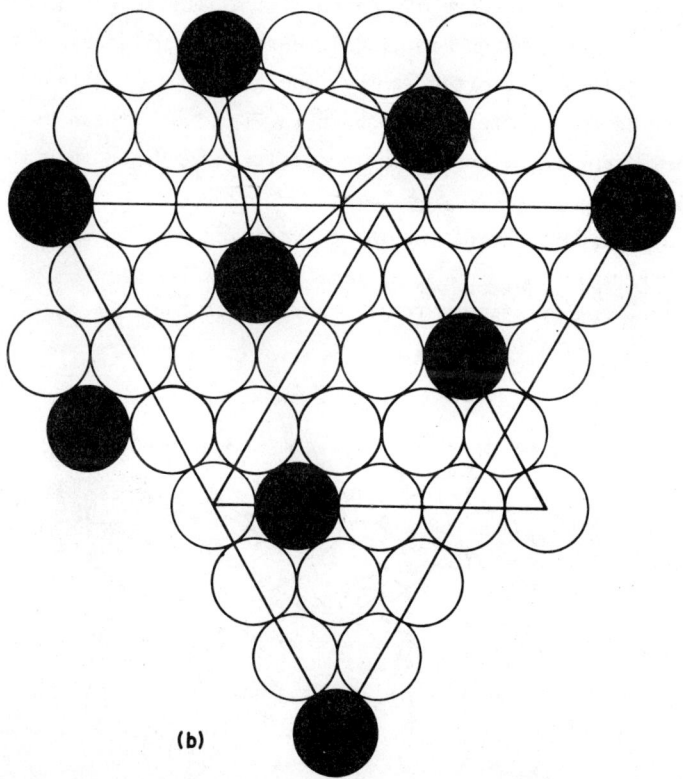

(b)

FIG. 14(b). Stable planar packings.

the stability of the whole pattern. It should be noted at this
point that we have caluclated the packing density for a single
layer and that, if successive layers are placed exactly on top of
the layers in prismatic configurations, the same density will obtain. However, the configuration will not be completely stable.
If successive layers are placed as in either cubic or hexagonal
close packing, the density has to be calculated by somewhat different processes, which will be discussed later in this section.

We have shown that, of the nine configurations left for study,
three can be considered as being derived from the hexagonal configuration by the removal of spheres from appropriate portions and
that the least dense of these three configurations is the one shown
in Fig. 14(a). Three more of the configurations shown in Tóth's book
consist of variations on Fig. 14(a) where the equilateral right
hexagonal cell is distorted so that the hexagon is flattened or
canted. The cell patterns are shown in Fig. 15. It can be shown
that these configurations are of necessity more dense than that of
Fig. 14(b).

The cell patterns of the three remaining configurations are
shown in Fig. 16.

These configurations are also more dense than the configuration of Fig. 14(b).

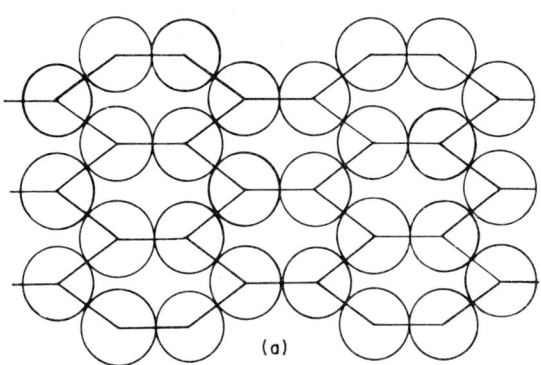

FIG. 15(a).

1. BASIC THEORY

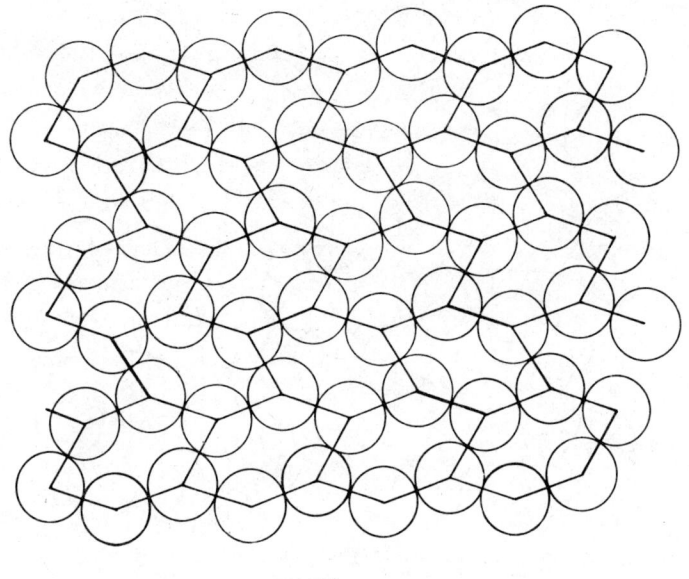

(b)

(c)

FIG. 15(b-c).

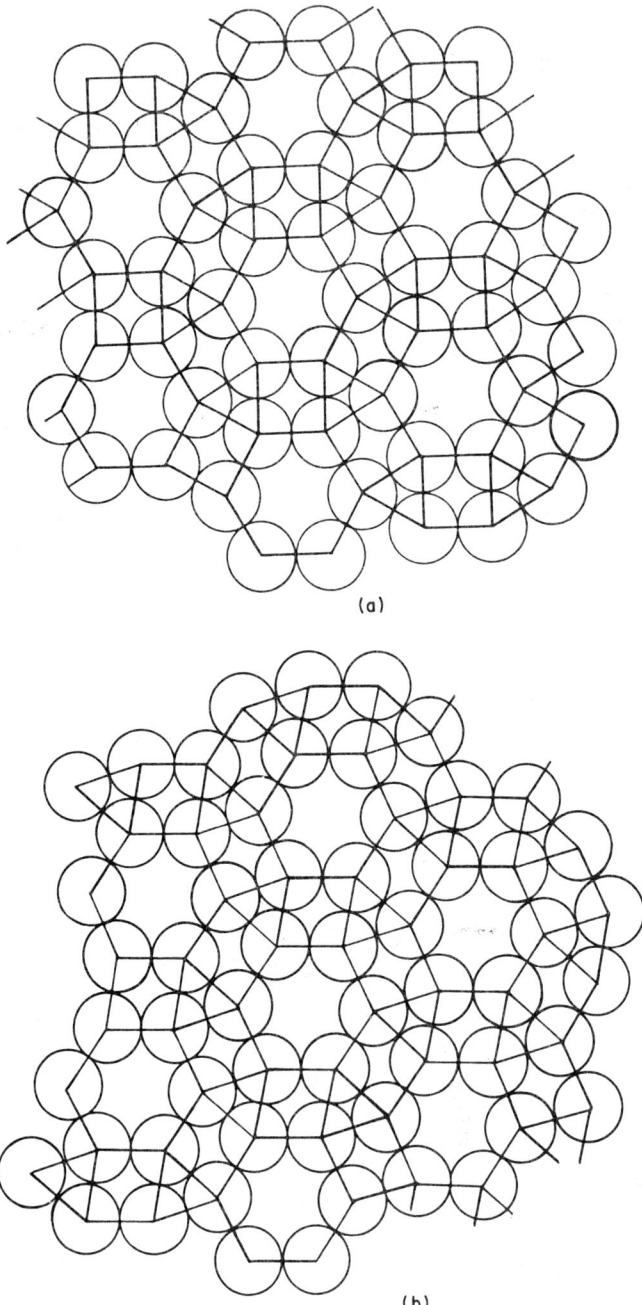

FIG. 16(a-b).

1. BASIC THEORY   31

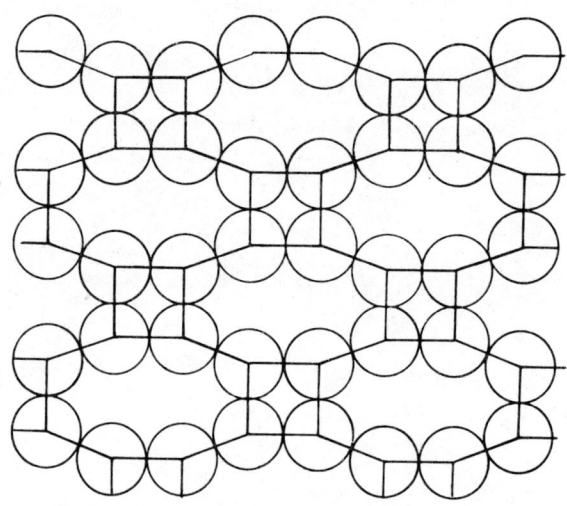

(c)
FIG. 16(c).

We have thus shown that stacking simple sphere-packing configurations may give us stable densities of not less than about 40%. Is it possible to get rigid packings less dense than this? Let us take a somewhat different approach. What are the requirements for a sphere to be held rigidly in a packing? It turns out that the requirement is quite simple. To be fixed in space a sphere must have at least four points of contact with other spheres or with rigid surfaces, and not more than three of these points may be in any hemisphere or more than two on a great circle. If a regular tetrahedron is inscribed in a sphere, the points at which the tetrahedron contacts the sphere meet these requirements.

At this point it might be well to refer back to Figs. 7, 8, and 10 and note that in the cubic array each sphere is in contact with six other spheres. The TKDH packing has 8 points of contact, and the RDH packing has 12 points of contact. In general, the smaller the number of points of contact, the lower the density. Can we find a pattern that requires less than six points of contact?

Look at Fig. 17(a). This is another view of the TKDH pattern. Suppose we place spheres of diameter f so that the center of each sphere is located at an apex of a TKDH. This gives Fig. 17(b) with the unit cell shown in Fig. 17(c). The volume of a TKDH of side S is

$$V_{TKDH} = 8\sqrt{2}\ S^3$$

Since the unit cell contains six equivalent spheres,

$$V_s = 8\pi \left(\frac{S}{2}\right)^3 = \pi S^3$$

The density of packing is thus

$$\frac{V_s}{V_{TKDH}} = \frac{\pi S^3}{8\sqrt{2}\ S^3} = \frac{\pi\sqrt{2}}{16} = 0.27767$$

Thus we have a stable packing with a density of only 0.27767. Even less dense packings are possible, but they are not always stable to the rotation of groups of spheres. It should be pointed out that the packing of Fig. 17(b) can be obtained by remomoving selected spheres from the cubic close packing. This is another way of saying that "holes" of this size and shape are stable in the cubic close packing, which may account for the difficulty of obtaining random packings more dense than 0.64.

Having established the behavior to be expected from "clouds of spheres," we can now apply these concepts to emulsions.

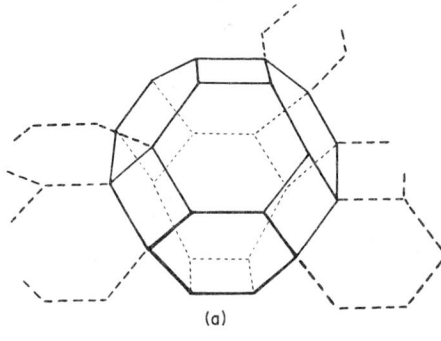

(a)

FIG. 17(a).

1. BASIC THEORY

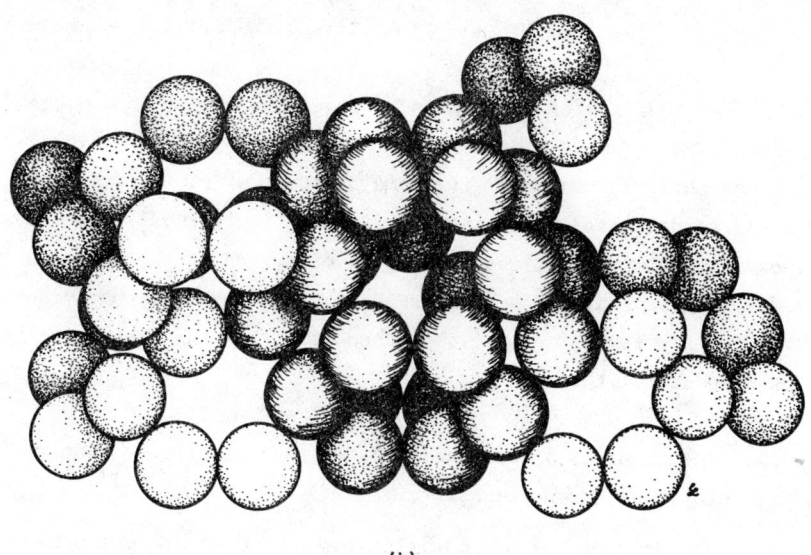

(b)

(c)

FIG. 17(b-c). One stable loose packing.

IV. LOW-INTERNAL-PHASE-RATIO EMULSIONS

It has been stated that two pure liquids will not form a stable emulsion (12). This has been demonstrated in the case of deionized water and purified benzene. If a small amount of purified benzene is added to deionized water and the mixture is vigorously agitated out of contact with air, the benzene is dispersed in the water in the form of small droplets. Observation of the behavior of such a system reveals several significant phenomena. If the dispersion is allowed to stand, three things occur:

1. The droplets of benzene collide and coalesce, so that the average droplet size increases.

2. The droplets of benzene rise to the top of the container.

3. The layer of droplets coalesces into a single layer of benzene.

We shall consider each of these factors separately. This particular experiment is so chosen that we are almost sure *not* to get an emulsion. All of the factors that lead to stable emulsions are absent, and all of the factors that destroy emulsions are emphasized. Immediately after the agitation is finished we have a system of spheres dispersed in a liquid medium. This system has a vastly greater interfacial area than would the two bulk phases. Energy has been expended to produce this interface, and the system is thermodynamically unstable. Given the opportunity, it will revert to two bulk phases with a minimum interface.

In order for the system to reduce the interfacial area, the droplets must come into contact and then coalesce. In dilute dispersions the droplets are far enough apart for the statistical chance for collision to be low. However, when two droplets do collide, the likelihood of coalescence is great since the interfaces are "clean." There is, therefore, a continual increase in droplet size and a reduction in the number of droplets. This process would

1. BASIC THEORY

eventually slow down as the number of droplets becomes small and the resulting spacing great. Before this effect can be noticed, however, the phenomenon of creaming takes over. Since the droplets of benzene are lighter than water, they rise toward the top and thus the distance between droplets is reduced. All droplets reaching the air interface coalesce, and finally two bulk phases result.

Suppose we now introduce an ionized substance that will be attracted to the oil-water interface and impart to each droplet a positive or negative charge. We shall consider this material as being so chosen that it does not interfere with the coalescence of the droplets if they actually come into contact. The material will, however, reduce the coalescence rate since the electrical repulsive forces will inhibit contact of the droplets.

In such a situation the coalescence of the droplets will be slowed because the chance of collision will be reduced. The creaming phenomenon will, however, still occur. As the droplets are forced into proximity by the action of gravity, contact will occur and coalescence will begin. Eventually two bulk phases will result. Some reasonably stable emulsions can be prepared by this technique alone.

Now suppose instead we add a substance that can be adsorbed at the interface to form a "skin" that keeps the droplets from actual contact and effectively prevents coalescence. In this case the emulsion will eventually cream, but two bulk phases will not result because the droplets cannot coalesce. It will be found that the creamed layer becomes more and more compact as the water is squeezed out from between the benzene droplets and that eventually the concentration of benzene in the cream layer will approach the closest packing fraction, 74.05%.

From this example we can see that there are two basic techniques for preparing stable dilute emulsions:

1. Attainment of small droplet size
2. Use of an "emulsifier" to prevent coalescence

The creaming of an emulsion is governed primarily by Stokes' law (13), which states that

$$u = \frac{2Gr^2(d_1 - d_2)}{9\mu}$$

where u is the rate of sedimentation of a spherical particle of radius r and density d_1 in a liquid of density d_2 and viscosity μ.

It is clear from the equation that the particle will rise if d_1 is less than d_2 and will fall if d_2 is larger than d_1. It is also clear that the rate of sedimentation increases as the square of the radius of the particle and inversely as the first power of the viscosity of the liquid. To put it simply, large drops rise (or fall) faster than small ones, and drops move faster in a less viscous liquid.

Stokes' law is strictly true only for the motion of one rigid sphere. However, for practical purposes it can be easily modified to apply (14, 15).

In general, droplets smaller than 1 to 2 μ in diameter cream only very slowly.

In actual emulsion preparation increase of the viscosity of the external phase has not been found particularly effective in increasing emulsion stability (16, 17). This may be due to other secondary effects of the thickening agents, however.

Besides creaming there is another effect that is not observed in the pure benzene-water system, but does occur in many low-internal-phase-ratio emulsions. This is flocculation, or aggregation. In its pure form it consists of droplets "sticking together," without coalescing, to form clumps or chains of larger effective size. This usually increases the settling rate, but not in a simple manner, since the clumps do not behave like spherical particles, as called for by the simple Stokes' law model.

1. BASIC THEORY

Frequently flocculation, or clumping, results in an initial increase in the apparent viscosity of the emulsion. In some instances the rheological properties become highly non-Newtonian and the formulation seems to gel. Explanations have been offered for this effect based on the fact that the clumps are irregular and therefore do not flow past each other easily.

Before proceeding further, let us review briefly the basic principles of rheology so that we can establish a vocabulary for discussing the flow properties of emulsions. For a thorough treatment see the series *Rheology, Theory and Applications*, edited by Erick ([18]), particularly the chapter by Okis ([19]).

Begin by visualizing a cube of liquid of side x with the lower face stationary and the upper face moving parallel to it at a constant velocity v produced by a force F. The rate of movement is slow enough for the flow to be laminar. The force exerted per unit area is called the shearing stress S

$$S = F/A$$

Since the bottom of the cube is stationary, the velocity goes from zero to v in a distance x, and the rate of shear is the change of velocity with distance, dv/dx.

The coefficient of viscosity η is then defined by the equation

$$S = \eta(dv/dx)$$

When S is expressed in dynes per square centimeter and the rate of shear in reciprocal seconds, viscosity is expressed in poises.

If we plot shear stress versus shear rate, we get a flow curve where the slope of the curve at any point is the viscosity coefficient of the liquid under those conditions. For simple liquids the plot is a straight line going through the origin, η is constant, and since this was the case studied by Newton, these liquids are called Newtonian.

Several non-Newtonian types of behavior are known. If the plot of shear stress versus shear rate is a straight line cutting the shear-stress axis, the behavior is called plastic. A fluid possessing such a curve will not flow under low shearing stress, but will deform like an elastic body. At the point where the flow curve cuts the shear-stress axis the fluid begins to flow. From that point the visocity remains constant until turbulent flow begins.

These two flow curves plus two others are shown in Fig. 18. Dilatant materials show an increase in apparent viscosity with shear rate, whereas pseudoplastic materials show a decrease. Either of these types may exhibit yield values or not in specific cases. (For methods of measuring viscosity see Ref. 19.) Thixotropic materials show a decrease in apparent viscosity with increase in shear rate, but also a decrease in viscosity with time as the shear rate is held constant. Such materials will thin out if they are agitated, but will thicken again on standing. Rheopectic materials thicken when stirred and thin on standing. Such materials "climb" the stirrer.

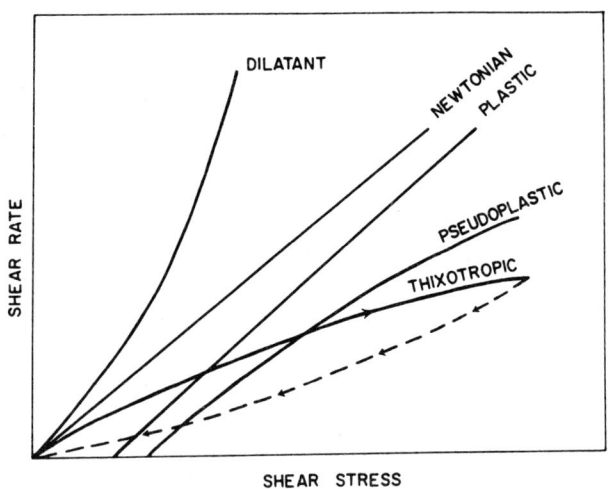

FIG. 18. Plots of shear stress versus shear rate for several types of flow behavior.

1. BASIC THEORY

Low-internal-phase-ratio emulsions show Newtonian behavior if very dilute. However, they may be mildly pseudoplastic or dilatant at concentrations of 10 to 20% internal phase.

A large number of papers have been published on the flow properties of emulsions and of dispersions of spheres in liquids (20-23). The viscosity is found to vary with the following:

1. The viscosity of the external phase
2. The rheology of the external phase
3. The percentage of internal phase
4. The viscosity of the internal phase
5. The nature of the interface
6. The nature and charge of the surfactant
7. The particle size and size distribution of the internal-phase droplets
8. The degree of clumping, or flocculation

We now return to our model of a low-internal-phase-ratio emulsion as an array of noninteracting spheres in a liquid medium. As long as the droplets do indeed show little interaction, the model works well, and in many cases the rheology of low-internal-phase-ratio emulsions is governed mainly by the flow properties of the external phase. When the percentage of external phase exceeds about 20%, the viscosity of the emulsion begins to increase and the flow properties become non-Newtonian. Deviation from Newtonian behavior is generally in the direction of thixotropic or pseudoplastic behavior.

However, when the droplets interact, non-Newtonian behavior begins at quite low phase ratios. This is particularly true if the droplets tend to form chains rather than clumps. This is not surprising since gels can attain quite high viscosity at gelling-agent concentrations of 0.5 to 3.0%. These materials are believed to form long threadlike structures that become "tangled" and result in the gelling effect. By the same reasoning, if droplets "chain out" to form long aggregates, these can also restrict the flow and produce thixotropic formulations.

It has been shown that small emulsion droplets behave in many ways as if they were rigid spheres (24).

As noted in Section III, rigid spheres can form stable packings at very low volume fractions. It is entirely possible that such packings can be formed and cause some low-internal-phase-ratio emulsions to seem to gel on standing.

Having developed the concept of an emulsion as a cloud of spheres and having discussed the general applications of this concept to low-internal-phase-ratio emulsions, it is now appropriate to consider how such emulsions are formed, where they occur naturally, and what uses have been found for them. Why are emulsions used at all? There are several reasons. To mention a few:

1. To dilute an expensive or concentrated ingredient with an inexpensive, but immiscible, diluent. This may be desirable to reduce the cost, to obtain even application of a small amount of material, to control dosage, or to reduce the viscosity of the formulation.

2. To introduce an insoluble ingredient into a system.

3. To control the odor, taste, toxicity, or reactivity of a material.

4. To control the physical properties of a formulation.

Why does one consider using low-internal-phase-ratio emulsions?

1. They maximize the dilution effect.

2. They are generally low-viscosity, fluid systems.

3. They are relatively easy to manufacture and inexpensive to produce.

1. BASIC THEORY

How are these emulsions made? Usually by brute force. The most common methods consist of adding the material to be dispersed to the external phase, containing emulsifiers, while subjecting the system to high shear. Shear can be obtained by impellers, pumps, orifices, ultrasonics, and other means. The first general rule seems to be to add the ingredients and mix well.

In some instances where severe agitation is not available or is otherwise prohibited, the emulsifier may be dissolved in the internal phase to make an "emulsifiable concentrate." This material, when poured into the external phase with mild agitation, "blooms" and is dispersed by the surface activity of the emulsifying agents.

We have already mentioned that low-internal-phase-ratio emulsions are obtained and stabilized by two primary factors: the attainment of small droplet size and the use of emulsifiers. A study of the devices used to produce such emulsions shows clearly that their primary purpose is to achieve small droplet size. Essentially they consist of mechanical devices that subject the emulsion ingredients to high shear.

The choice of emulsifiers for the production of low-internal-phase-ratio emulsions has been the subject of a voluminous literature. A few basic points apply. Usually the emulsifiers employed in low-internal-phase-ratio applications are ionic emulsifiers, which will induce a charge on the dispersed droplets. However, nonionic emulsifiers find wide use, and in almost all practical instances a mixture of emulsifiers is used.

Usually the formulator of a particular emulsion has a considerable range of choice with respect to emulsifiers that will produce a satisfactory emulsion. And the actual choice of an emulsifier system for the particular application is usually dictated by secondary considerations, such as the relative cost of the emulsifier, toxicity, color or odor, or conformance to specifications. Particularly in the area of food, pharmaceutical, and agricultural emulsions the choice of emulsifiers is dictated more by legal considerations than anything else since emulsifiers used in these areas must be approved by government agencies and the cost of obtaining such approval is considerable. It is therefore, difficult to introduce new emulsifiers into this area since the expenditure of time and money required to obtain their certification is not justified unless they offer exceptional advantages.

Although it is difficult to document, it is quite probable that most emulsions used commercially are of the low-internal-phase-ratio, oil-in-water, type. Several areas of application for this type of emulsion will be mentioned in succeeding chapters, where techniques for the selection of particular emulsifiers will be discussed in more detail.

Low-internal-phase-ratio, water-in-oil, type are not very common as articles of commerce. They are, however, encountered fairly frequently in actual practice and usually constitute an undesirable situation. For example, crude oil when produced from the wells frequently contains significant amounts of water or brine emulsified in the oil. This water must be removed before the oil can be refined, and a substantial industry has developed solely to accomplish this objective. Similarly small amounts of water may become emulsified in lubricating oils, hydraulic oils, and heat-exchange systems. Here again the removal of the emulsified water must be accomplished in order to prevent corrosion and other undesirable effects. Some ways in which this is done will be discussed in succeeding application chapters.

V. MEDIUM-INTERNAL-PHASE-RATIO EMULSIONS

We have set the range of medium-internal-phase-ratio emulsions as between 30 and approximately 70% internal phase. Again, it should be noted that these designations are somewhat arbitrary and that there is substantial overlap at each end of the range.

Medium-internal-phase-ratio emulsions are characterized by substantially higher viscosities, non-Newtonian rheological

1. BASIC THEORY

behavior, and usually some difficulty in achieving long-term stability.

If we start with the simplifying assumption that the internal-phase droplets do not tend to stick together, or flocculate, and if we apply the previously detailed geometric concepts to our cloud-of-spheres model, we find that many of the properties of these emulsions are readily explainable. As previously noted, the packing fraction for a cubic array is approximately 0.52; thus, if the internal phase approaches approximately 50%, we can expect the emulsion droplets to begin to interfere with each other rather drastically, and hence the flow properties will be non-Newtonian and the apparent viscosity of the formulation will be high. A number of experimenters have called attention to the fact that it is often most difficult to produce an emulsion with an internal-phase ratio of approximately 50% by volume. They have cited instances where an oil-in-water emulsion can be made up to approximately 50% oil and similarly the same two fluids with appropriate emulsifying agents can be made into water-in-oil emulsions of up to approximately 50% internal phase, but in this region difficulty is encountered in producing either type of emulsion with extended stability.

Further study of the experimental work leading to these conclusions usually shows that the emulsifiers being used are ionic in type and function primarily by a charge-repulsion mechanism. It is easy to see that, if the internal-phase ratio approaches 50%, the droplets will inevitably be forced into contact with each other in spite of the charge repulsion produced by the emulsifiers, and if the emulsifier is not effective in preventing coalescence when contact does occur, the emulsion will be unstable. This seems to explain many instances of emulsion instability in the regions where equal volumes of internal and external phase are used.

We can say, therefore, that in the lower portion of the medium range small particle size and ionic emulsifiers can be employed to produce reasonably stable emulsions using the same rationale and stabilization mechanisms commonly employed for low-internal-phase-ratio emulsions.. However, if the phase ratio approaches or exceeds 50% external phase, it is necessary to use emulsifiers that function as though they formed a "skin" over the droplets and thus prevented coalescence. The incorporation of a certain amount of ionic emulsifier may still be useful to prevent flocculation of droplets, but the presence of at least some film-forming emulsifier is necessary. The film-forming emulsifier may be either ionic or nonionic. Techniques for the selection of emulsifiers or emulsifier combinations for particular applications will be discussed in later chapters.

In the preparation of medium-internal-phase-ratio emulsions, particularly in the range above 50% internal phase, the attainment of small particle size is not as necessary as it is in the low-internal-phase-ratio range. Actually the use of emulsifying equipment that produces polydisperse droplet distributions will usually result in a formulation with more fluidity than one that produces very small, monodisperse droplets.

It is important to remember that all of our geometric discussions have assumed monodisperse spherical droplets. In actual emulsion preparation complete monodispersity is seldom attained, although some methods will produce systems remarkably close to monodisperse. When systems are polydisperse, certain situations can occur that would lead to misinterpretation of experimental results.

Suppose we use, as an external phase, an aqueous solution of a nonionic emulsifier chosen to be very effective in preventing coalescence of droplets. Suppose we add an oil, while stirring, until we have a 60 vol % oil-in-water emulsion that is reasonably monodisperse, with droplets of about 10 µ in diameter. We then take a 100-ml sample and store it in a stoppered graduated cylinder, labeling it sample A. Now we increase the stirring rate and stir until five parts of the oil volume are still in 10-µ size and one part is of 1-µ size.

Again we take a sample and label it "B." Now we stir again

1. BASIC THEORY 45

until four parts of the oil droplets are still 10-μ size and two parts are of 1-μ size. We now take sample C. Observation of the samples shows that sample A clears to a 81-ml layer, leaving 19 ml of clear fluid. This is because 60 ml of internal-phase droplets cream until they occupy approximately 75% of an 81-ml cream layer. In sample B the 50 ml of large droplets occupy 74% of a 67.5-ml cream layer, which also contains 17.5 ml of external phase, and the small droplets, if by themselves, would occupy 74% of a 13.5-ml layer. However, in the mixture the small droplets can be dispersed in the 17.5 ml of interstitial external phase, driving out some of the external phase without increasing the total volume of the cream layer. We, therefore, observe that there is about 32.5 ml of clear fluid below the cream layer.

In sample C the large drops pack into 54 ml with 14 ml of free space. The small droplets cream into 27 ml; 14 ml of this can fit between the large droplets, so that the whole system could cream, leaving 33 ml of clear fluid below the cream.

Now suppose we stir the remaining emulsion until it is all in 1-μ droplets. Call this sample D. This again would cream leaving a free space of 19 ml, but it would cream much more slowly than sample A.

We now have before us four emulsion samples. The first one has creamed showing approximately 19 ml of clear fluid at the bottom. The second and third graduates show approximately 33 ml of clear fluid, and the fourth graduate is creaming very slowly and ultimately would show again approximately 19 ml of clear fluid. We will now consider the line of reasoning that might be applied to these results by an experimenter unaware of the geometric effects encountered in emulsions. Looking at the first graduate, he would probably correctly interpret this as creaming and consider that the system was close to what he wanted, but not quite well enough dispersed. Looking at the second sample, he could interpret this as almost complete breaking of the emulsion, since the cream layer occupies only 7 ml more than the original bulk internal phase. The third sample could also be interpreted as almost complete breaking, and the experimenter could be led to the conclusion that the additional agitation was destroying the emulsion, and he could conceivably abandon the experiment at this point without ever obtaining sample D. On the other hand, if he were to continue the agitation and achieve sample D, he might very well obtain an emulsion that creamed sufficiently slowly to meet his requirements and therefore was entirely satisfactory. Having completed the full series of experiments and obtained what he considers to be a satisfactory emulsion, he might then conclude that this particular emulsifier combination required severe agitation in order to produce a stable emulsion. Actually none of the emulsions were really breaking, and the effects were purely the result of the inherent geometry.

From the preceding discussion it can be seen that, if an emulsion contains some fairly large droplets mixed with very small droplets, all the droplets will be less crowded and therefore the emulsion will appear to be less viscous. However, it must be remembered that the larger droplets will settle more rapidly and that this emulsion may be therefore less stable.

A number of studies seem to indicate that emulsions with particles of very small diameters are, at any given phase ratio, more viscous than those containing coarse droplets. This phenomenon has been explained in terms of the increased surface area and hence greater surface energy, and also in terms of the adsorption of emulsifiers from the surface of the droplets to produce droplets with larger apparent diameters and hence an effective higher phase ratio. Although in some instances these explanations may have validity, it is not necessary always to invoke these more complex arguments since the geometric situation explains the phenomenon quite well. Most emulsions are made by subjecting a mixture of the two liquids to a high-shear environment that breaks the liquid, which would be the internal phase, into droplets. Studies of the mechanism of droplet formation have shown that, when large drops are formed, a number of much smaller "satellite" drops are formed at the same time. Thus the first droplets that are formed consist

1. BASIC THEORY

of a group of fairly large droplets and a group of quite small droplets, or a highly polydisperse form of bimodal system. Since the small droplets can fit inside the coarse-droplet-packing pattern, a fairly fluid system results.

Subjecting an emulsion of this kind to additional shear results in breaking the larger droplets into a number of smaller droplets without reducing the size of the smaller ones. The smaller the droplets become, the more difficult it is for the shear mechanism to further reduce them in size. This means that the system gradually becomes more and more monodisperse as the particle sizes are reduced, and it then approaches one of the classical packing patterns. If the phase ratio is in excess of 50% internal phase, the final emulsion will be quite viscous simply because the particles are nearly all the same size and hence cannot be fitted in between each other as was the case with the polydisperse system.

It is true that these systems with small particle sizes will have a much greater surface area and hence may have an increased tendency to clump, or flocculate. Since the total surface area may be greatly increased over that of a coarse emulsion, more emulsifier may be required to completely fill the surface, and if enough emulsifier has not been provided, the emulsion may become unstable and either coalesce until a stable state can be reached or break completely. It should be noted also that the monodisperse system forces the droplets into much closer contact than does the polydisperse one and that therefore a film-forming emulsifier may be required to achieve stability.

Medium-internal-phase-ratio emulsions have been used for many years to produce thick, creamy formulations. Products such as mayonnaise and cold cream fall into this category. Medium-internal-phase-ratio emulsions would have received considerably more acceptance long ago if they could have been stabilized more effectively. Prior to about 1940 the variety of commercially available emulsifiers was considerably restricted, and most of the more common emulsifiers were ionic in character. We have already shown that those ionic emulsifiers that function purely by a charge-repulsion mechanism are unable successfully to stabilize emulsions where the droplets are forced into relatively close proximity. It is for this reason that we find statement in the older chemistry books and even in some recent books to the effect that phase ratios above approximately 75% cannot be attained. With the advent of a wide variety of synthetic emulsifiers, particularly of the nonionic type, the emulsion formulator is able to chose film-forming emulsifiers that will stabilize emulsions at much higher phase ratios.

Of what use are medium-internal-phase-ratio emulsions and of what importance are they in actual practice? We have already pointed out that the most common use for medium-internal-phase-ratio emulsions is as a technique for producing a viscous, creamy formulation. Products such as salad dressing, mayonnaise, cold cream, vanishing cream, and certain pharmaceutical formulations fall into this category. Many of these formulations are of the oil-internal type and are used because the aqueous external phase minimizes the greasy feel or, in the case of formulations for internal use, improves the palatability. However, a number of medium-internal-phase-ratio, water-internal, formulations are also in common use. Examples of both kinds will be found in succeeding applications chapters.

Medium-internal-phase-ratio emulsions also occur naturally or are produced as undesirable effects in commercial processes. Butter falls in this category as a commonly occurring useful oroduct. Certain crude-oil emulsions are medium-internal-phase-ratio emulsion of water in oil. These emulsions are quite viscous and in some instances very difficult to pump and to resolve. "Slop oils" are produced in many commercial processes. These usually result from small amounts of oily products, lubricants, and solvents escaping from the equipment and being flushed into the sewers. These water-immiscible materials cannot be discharged to the environment and are therefore collected in separators, skimmers, and the like. Almost invariably this slop oil

1. BASIC THEORY

contains substantial amounts of water emulsified in it, and this water must be removed before the oily material can be reclaimed. Oil recovered from accidental spills, such as the *Torrey Canyon* or Santa Barbara incidents, almost always contains some water emulsified in it, and not infrequently these emulsions reach the medium-internal-phase-ratio range. This seriously increases the bulk of material that must be handled and makes it more difficult to devise suitable disposal procedures.

Industry is increasingly being required to clean up effluent water before discharging it to the environment. Frequently these water streams contain small amounts of nonmiscible oily materials. These materials must be concentrated and removed before the water can be discharged. The concentration is achieved by settling techniques, centrifugation, filtration, or air flotation. As the oily material is concentrated, it often forms a medium-internal-phase-ratio emulsion, either water or oil external, and frequently containing substantial amounts of finely divided solids. These products usually are referred to as froths or scums and often are very difficult to dispose of. They frequently contain enough water not to burn well and are therefore difficult to incinerate. Where the oily material has economic value, removal of the entrained water by distillation is often economical. Much attention is presently being devoted to means of dealing with these materials. Further discussion will be found in later chapters.

VI. HIGH-INTERNAL-PHASE-RATIO EMULSIONS

In the period from 1890 to 1920 a number of investigators claimed to be able to prepare emulsions with high internal-phase ratios. As early as 1881, Du Motay et al. (25) received a patent for a "jellified" petroleum that some people have claimed was a high-internal-phase-ratio emulsion. Similarly Pickering [26] reported that he was able to prepare an emulsion of benzene in water with 98 parts of benzene in 2 parts of sodium oleate solution. Such claims were always met with considerable skepticism by the scientific community. Frequently it was found difficult to repeat the work of the investigators or, as in the case of Du Motay and Pickering, the work was repeatable, but it was contended that the compositions formed were not truly emulsions.

From an actual, practical, standpoint, very few true high-internal-phase-ratio emulsions were produced until late in the 1950s, and the commercial application of this type of composition is really just beginning. The current position of most experts in the field today is as follows:

1. High-internal-phase-ratio emulsions can be prepared under certain special conditions.

2. Most of the older types of formulation were probably not true emulsions, possibly dispersions in gels.

3. In the opinion of some investigators, the distinction is not important from a practical standpoint.

What sort of composition did these early investigators produce? This can probably best be described by referring to Fig. 19. A true emulsion consists of drops of one liquid dispersed in another liquid, and usually the requirement is added that the dispersion be at least minimally stable. This stability in the case of high-internal-phase,ratio emulsions must be obtained by generating on the surface of the dispersed droplets a film that will prevent coalescence when two drops come in contact. This film may be a liquid film, a layer of solid particles, or an adsorbed, oriented, micellelike structure. In any event, it must be rigid enough to resist breaking, but flexible and reversible enough to adjust to changing conditions in the environment. However, it should be noted that the external liquid phase of an emulsion constitutes a continuous region, any part of which can be reached from any other part without passing out of the external-phase region.

1. BASIC THEORY   51

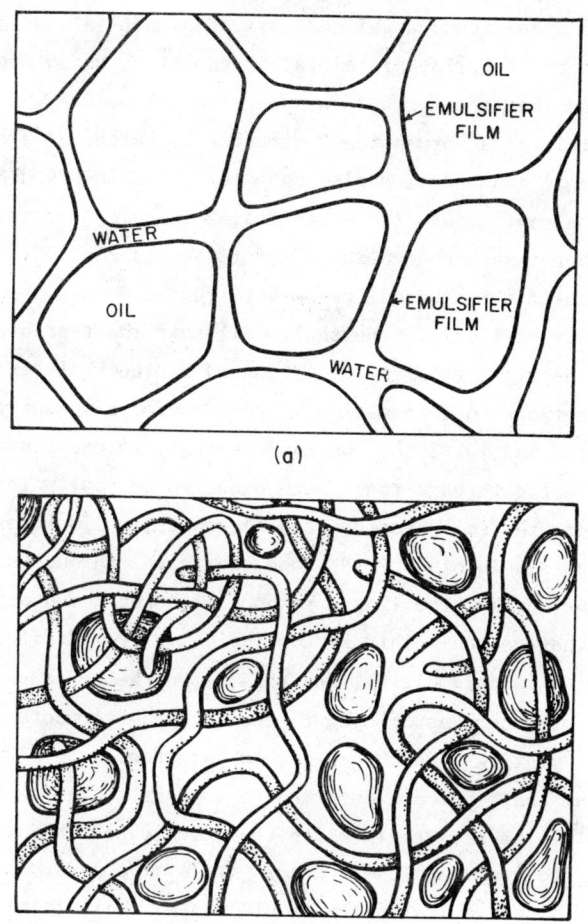

FIG. 19. The structure of a true emulsion (a) and the fiber network of a gel (b).

On the other hand, the internal phase is divided into a large number of separate and discreet regions, and one cannot travel from one of these regions to another without passing through the external-phase region. Topologically, when, we would say that a true emulsion shows no connectivity with respect to the internal phase. This situation is depicted in Fig. 19(a), and for the sake of convenience we shall refer to this as a type a composition.

On the other hand, a gel consists of a three-dimensional network of fibers, fibrils, platelets, or the like, so interconnected and entwined that they produce a spongelike structure. McBain (5) refers to this as a "brush heap" structure. Weiser (27) says: "In both inorganic and organic gels, both the solid and liquid phases appear to be continuous, the solid framework being comparable to a sponge rather than a honeycomb." Refer then to Fig. 19(b), which tries to depict this threadlike network that enmeshes a considerable quantity of liquid. (We shall call this the type b composition.) The rigidity of a gel is caused by the resistance of the solid gel network to deformation. Ignoring for a moment the droplets depicted in Fig. 19(b), we can see that, if one considers the points where the threads touch each other to be real points of contact, both the solid network and the enmeshed liquid are continuous phases. In other words, one can travel throughout either the solid network or the liquid without it being necessary to pass into the other region. This is topologically very different from the situation in an emulsion, and this is why Weiser emphasizes the sponge structure as opposed to the honeycomb structure. McBain (28) mentions that, when bubbles are formed in a gel, they are not spherical, but are deformed to conform to the shape of the region of the network in which they are trapped. Just as it is possible to trap bubbles in a gel, it is also possible to trap droplets of a second immiscible liquid phase. This is pictured in Fig. 19(b). These droplets, if they come in close contact, may coalesce to produce somewhat larger droplets, but they cannot escape from the gel network to produce a single separate liquid phase. What we have then in the case of the compositions produced by Du Motay et al., Pickering, and others is a situation of the type depicted in Fig. 19(b), where a temporary dispersion of one liquid in another is made, and this dispersion is trapped in the lattice of a gel. It has been claimed that the jellylike compositions of Du Motay et al. are emulsions because "gelatin is a well-known emulsifier." This argument can be countered in several

1. BASIC THEORY

ways. Many of the instances where gelatin-produced "emulsions" really result in composition of type b. Although gelatin can indeed function as an emulsifier or as a stabilization aid in emulsion, it is also well known that gelatin is a gel-producing substance. According to Werbin (29), "often the gums are referred to as emulsifying agents. They are considered not so much as emulsifiers but rather as emulsion protectors or stabilizers. To a large extent, the function is to increase the viscosity of the aqueous phase." He further distinguishes between thickening agents and true emulsifiers: "A true emulsifying agent is usually used to form the emulsion and for cleansing properties. Vegetable gums...are added... because of the added viscosity effect." What we have then in the case of Du Motay (25) is a viscous paste or gel in which droplets of a second immiscible liquid are trapped. This is borne out by a quote from Clayton (30), who, incidentally, does not make a distinction between type a and type b structures: "The nature of the precursor and the treatment to which it is subjected before the gelatin is extracted have a marked effect on the emulsifying power of the gelatin." He also states: "The viscosity, which depends on the previous history of the precursor, closely parallels the emulsifying power. Gelatins made from alkali-treated precursors are better emulsifiers and they also show greater viscosity in solution."

Returning to Du Motay, we find that he very carefully refrained from claiming that what he produced was an emulsion. He even disclaimed its similarity to certain known emulsions. He also stated that he could use, instead of the glue or gelatin of his preferred example, "ordinary flour paste, gluten and equivalent glutinous substances" and that he could achieve additional thickening of his external phase by the addition of plaster of paris. Furthermore, he noted that it is necessary to use more gelling agent when higher temperatures are to be encountered. All this clearly shows that what Du Motay had was a composition of type b, and not truly an emulsion. Type b compositions are stable only as long as the gel structure persists, and if the gel structure is disrupted either mechanically or thermally, or by the addition of coagulating agents, the whole system becomes unstable

What Pickering and Du Motay probably produced was a dispersion of oil droplets in a gel. This may seem to be a rather fine distinction. However, dispersions of this type, though stable on storage, tend to "bleed" rather badly when subjected to kneading or other gentle shearing action.

Beginning in the late 1940s, a wide variety of synthetic surface-active agents became commercially available. These included alkyl sulfonates, alcohol sulfonates and sulfates, and a considerable variety of nonionic materials, such as esters of polyhydric alcohols and oxyalkylated derivatives of a number of basic materials. By this time most workers in the area of practical emulsions had become convinced that stable emulsions in the upper part of the medium-phase-ratio range and in the higher-phase-ratio range were not sufficiently stable to allow commercial applications. For this reason one finds statements in textbooks and papers stating that internal-phase ratios in excess of 75% result in inversion. If one examines books of recipes for emulsions, one finds that the maximum internal-phase ratio usually mentioned is in the neighborhood of 65%.

In the early 1960s, however, an awareness of the possibility of producing stable high-internal-phase-ratio emulsions began to grow, and in 1965 we find Becher (1) citing examples of high-internal-phase-ratio emulsions. It has, however, been only quite recently that emulsions of this type have received commercial exploitation.

This attitude concerning emulsions with phase ratios higher than about 65% persists even today. Although a number of papers concerning high-internal-phase-ratio emulsions have been published, many practitioners of emulsion technology in specialized fields are not aware of the properties and potentialities of emulsions with higher phase ratios. An example is a patent (31) issued on November 23, 1971. This patent was applied for on September 27, 1968, and we can assume that at the time the application was filed,

1. BASIC THEORY

it represented the state of the art as far as at least one commercial laboratory was concerned. In reading the specifications for this patent we find that the type of emulsion under discussion is what the inventors call an "invert." By this they mean a water-in-oil emulsion. They state: "In preparing and employing such emulsions, however, several difficulties have been encountered. One of the principal difficulties arises from the fact that few emulsifiers effectively form water-in-oil emulsions that contain more water than oil. When such an emulsifier is found, it is generally due to empirical observation as there is no effective means to predict whether known emulsifiers will form invert emulsions that will contain more water than oil." In claim 1 of the same patent the inventors claim:

1. Water-in-oil invert emulsion compositions comprising about 50 to 75 wt. percent water, about 5 to 30 wt. percent oil, and... about 1 to 10 wt. percent of an emulsifier.

Actually the claims are directed toward one specific type of emulsifier, which does not concern us at the moment. What should be noted is that these inventors were able to find an emulsifier that would produce emulsions in the 50 to 75 st % internal-phase range and that this was considered sufficiently unusual for an emulsifier with this capability to be patented. Note also that the inventors state "there is no effective means to predict whether known emulsifiers will form invert emulsions that will contain more water than oil."

An example of the commercial use of high-internal-phase-ratio emulsions is that of hydrochloric acid-in-oil emulsions used to increase the productivity of oil wells. These emulsions are prepared with an amine salt as the emulsifier, and their pseudoplastic rheology makes them capable of suspending significant amounts of solids such as sand. These emulsions are injected down an oil well where the external oil phase protects the piping from the action of the acid. On reaching the rock formations, the acid reacts with the carbonate portions of the rock, and as the acid is neutralized, the amine emulsifier becomes unstable and the emulsion breaks. At the same time sufficient hydraulic pressure is applied to the formation to fracture the rock, forcing the acid and sand into the cracks. The acid widens the cracks, and when the pressure is released, the sand props the formation open, thus allowing increased production of oil. Other applications will be discussed later.

We have set the range of high-internal-phase-ratio emulsions as being above about 70% internal phase. On the basis of our discussion of geometric configurations, we would have reason to begin the class at 68.2% internal phase. At this point the spheres can be arranged in the TKDH configuration shown in Fig. 10. Suppose we took a monodisperse emulsion in this configuration and were able to withdraw the external phase slowly so as to increase the internal-phase ratio without otherwise distrubing the system. As the internal-phase ratio is increased, the configuration would have two alternatives. Either it could rearrange into one of the more compact configurations, that is, the cubic or bee-cell configurations (Fig. 11 or 12), or it could retain the TKDH form, the spheres beginning to flatten slightly. Figure 20(a) shows how a sphere is fitted into the TKDH. Note that the TKDH has six square and eight hexagonal faces, and that the edges are all of equal length. Note that the sphere touches only the hexagonal faces.

As the phase ratio increases, a small circular flattened area appears on each of the hexagonal faces. This grows until the radius of the truncated sphere equals $\sqrt{2}\, f$ [Fig. 20(c)]. From then on, flattening occurs on all 14 faces. The volume of the flattened sphere will be

$$V_{TKDH} = 4/3\pi r^3 - 8/3\pi h^2 (3r - h) - 2\pi (h')^2 (3r - h')$$

and the area will be

$$A = 4\pi r^2 - 16\pi\, rh + 8\pi (r^2 - 3/2) - 12\pi rh' + 6\pi (r^2 - 2)$$

1. BASIC THEORY

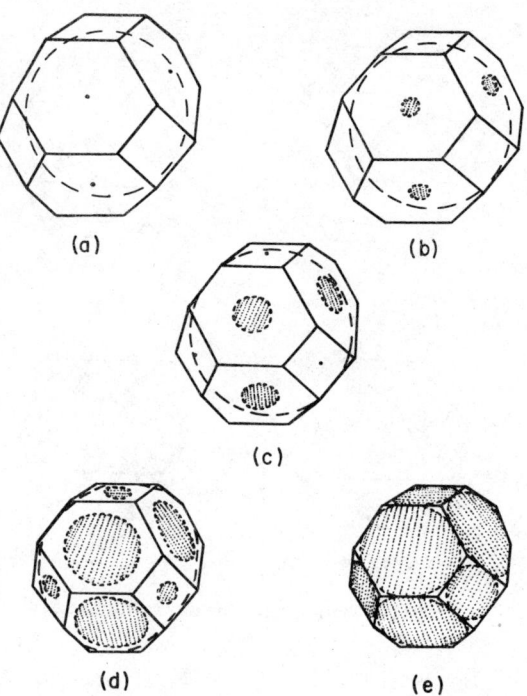

FIG. 20. Transition from sphere to tetrakaidecahedron.

where r is the radius of the truncated sphere, h is the height of the first sector and h' is the height of the second sector. Note that r ranges from $\sqrt{6}/2$ through $\sqrt{2}$ to $3/2$ and $r = \sqrt{6}/2 + h$ and $h' = r - \sqrt{2}$. Dimensions r, h, and h' are expressed in terms of f, the key dimension of the unit tessellating figure.

Similarly, if the spheres assume an RDH configuration (close cubic packing), flattening will occur according to the process illustrated in Fig. 21. If the flattening occurs in this mode, it will take place at each of the 12 rhomboidal faces, and the volume of the figure will be

volume = volume of sphere - 12 (volume of flattened sector)

volume of internal unit = $4\pi r^3/3 - 4\pi h^2(3r - h)$

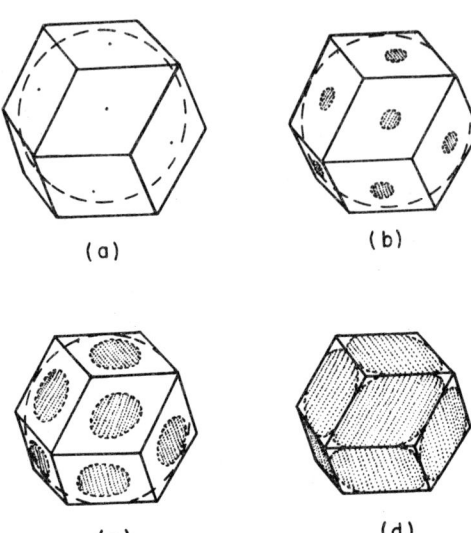

FIG. 21. Transition from sphere to RDH.

The area of the figure will be

$$\text{area of internal unit} = 4\pi r^2 - 24\pi rh + 12\pi(r^2 - 1/4)$$

where r is the radius of the truncated sphere, h is the height of sector, and $r = 0.5 + h$ and ranges from 0.5 to $\sqrt{3}/3$. Both r and h are expressed in terms of a, the key dimension of the unit tessellating figure.

This simple form of flattening will proceed [see Fig. 21(c)] until $r = \sqrt{2}/3$, at which time the flattened circles will just reach the edge of the rhombus [Fig. 21(d)]. At this point geometrical analysis becomes somewhat more complex.

We shall now proceed to calculate the volumes and areas of some of these transitional figures and then attempt to compare the various forms. In order to do this, we need some expression that will relate the volume of a particle to its area, but will be independent of the size of the particle. Such a geometrical factor is the expression $(A/V)^{2/3}$. Table 2 shows a series of calculations

1. BASIC THEORY

TABLE 2
Parameters for TKDH Flattening

r	h	h'	V	A	%V	$(A/V)^{2/3}$
$\sqrt{6}/2$	0	0	$\sqrt{6}\pi$	6π	68.02	4.386
1.30	0.0753	0	9.0209	21.092	79.73	4.871
1.35	0.1253	0	9.7898	22.505	86.53	4.918
$\sqrt{2}$	0.1895	0	10.6289	24.228	93.952	5.012
1.45	0.2253	0.0358	10.981	25.118	97.055	5.0844

TABLE 3
Parameters for RDH Flattening

r	h	V	A	%V	$(A/V)^{2/3}$
0.5	0	0.52356	3.1415	74.054	4.8360
0.52	0.02	0.58121	3.38202	82.210	4.8560
0.54	0.04	0.627772	3.6039	88.785	4.9156
0.56	0.04	0.662234	3.8050	93.666	5.008
$\sqrt{3}/3$	0.7735	0.68173	3.9639	96.425	5.117

for TKDH flattening, and Table 3 shows a similar set of calculations for the RDH mode.

Figure 22 shows a graph of the geometrical factor versus the volume percentage of internal phase. This graph brings out the previously unrecognized fact that the two curves intersect at a point that on the TKDH curve corresponds to the sphere just beginning to touch the square faces. It has always previously been assumed that the RDH packing or its equivalent, the bee-cell packing, was the most economical packing under all situations. This graph now shows that above approximately 94% internal phase the TKDH

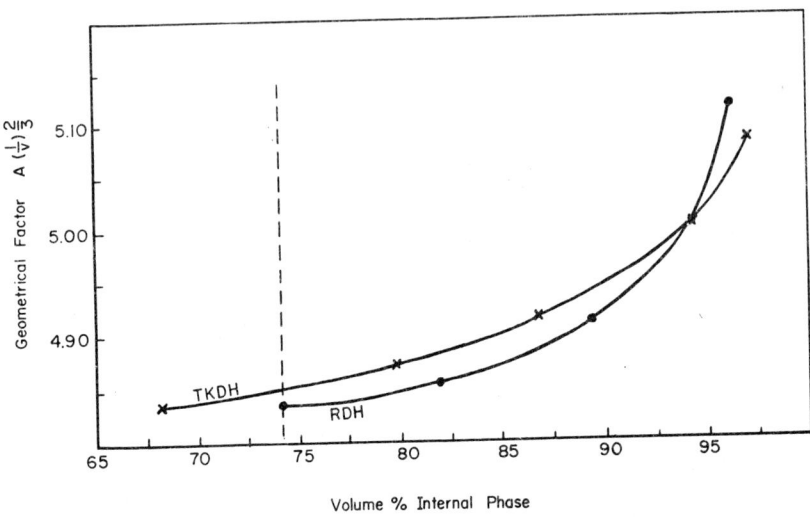

FIG. 22. Geometrical factor versus the volume percentage of internal phase.

packing is preferred. This result becomes more understandable when one considers that, at very nearly 100% internal phase, the RDH packing would require vertices with six droplets around each vertex where the TKDH packing would in no case require more than four droplets around a vertex.

Figure 22 also leads one to inquire whether the emulsions would show any unusual properties near the crossover region where either type of configuration is possible. One might also expect another change in properties in the neighborhood of 96.4% where r is equal to $\sqrt{3}/3$. This is the point where the flattened circle on the square face just reaches the edge. In 1966 Lissant (32) first pointed out the significance of Fig. 22 and postulated the transition points. He ran some simple tests which indicated that there were indeed transition points where the theory predicted. In 1969 Nixon and Beerbower (33) ran some more precise measurements on a high-internal-phase-ratio fuel emulsion and also found that the properties varied near the predicted transition point. They attributed the slight deviation from the predicted value to the fact that their emulsion was not completely monodisperse.

1. BASIC THEORY

One of the characteristic properties of high-internal-phase-ratio emulsions is that they seem to slowly "relax," or "unwind," on standing undisturbed. Yield values or viscosity-versus-shear relationships determined on freshly made emulsions are nearly always higher than those obtained on an emulsion that has remained undisturbed for a period of 24 hr or more. This phenomenon has been explained in two ways. First, that the smaller droplets in the emulsion tend to be squeezed together and to coalesce, thereby producing a more monodisperse emulsion. Actually we have shown that monodisperse emulsions would appear to be "thicker" than polydisperse ones in the medium-internal-phase-ratio region. It can be shown, however, that in the high-internal-phase-ratio region monodispersity should give lower values since the structure will be more regular and the geometric factor will be smaller. Another possible explanation for the relaxation phenomenon would be that, when freshly made, the emulsions may have regions that are in the less preferred geometric configuration and that as they stand they tend to rearrange into the preferred packing and hence are more fluid. To date no one has been able to show conclusively whether either of these explanations is the correct one.

Because the high-internal-phase-ratio emulsions are in a supercreamed state, they possess non-Newtonian rheological properties. When at rest, they may have yield values as high as 600 to 6000 dynes/cm^2. Apparent viscosities at low shear rates can run into the 100,000-cP region. However, at shear rates commonly encountered in pumping operations the apparent viscosities may get as low as 100 cP. In February of 1968 Lissant, Hopper, and Harris (34) presented a paper concerning the pumping properties of high-internal-phase-ratio fuel emulsions. They noted that common methods of determining the viscosity of these emulsions had not achieved satisfactory reproducibility and therefore attempted to pump emulsions through tubes whose diameters were comparable to those encountered in actual engine fuel lines. The three tubes were 152.4 cm long and had average radii of 0.535, 0.396, and 0.185 cm. Test gauges at each end of the tubes were used to measure the pressure drop, and the flow rate was obtained by measuring the time required to pump 10 kg of fluids. The system was calibrated with water, and then data were obtained on two fuel emulsions.

The assumption is made that in capillary rheometers the tube is very long compard to its diameter, that the flow is lamellar, and that the fluid enters and leaves the tube without turbulence. In tubes of the sizes used in this study all these assumptions are not necessarily valid. Therefore the ordinary viscosity equations were rearranged so as to incorporate any deviations from these liquids into a set of constants.

The basic equation is

$$\eta = \frac{68944\pi R^4 \, \Delta P}{8(L + nR)v/t} - \frac{m\rho(v/t)}{8\pi(L + nR)}$$

where ρ is the density and m and n are empirical constants.

If we assume R and L to be constant for any particular tube, this equation reduces to an equation of the type

$$\eta = \frac{K_1 \, \Delta P}{v/t} - K_2 \rho \frac{v}{t}$$

or

$$\eta = \frac{K_1 \, \Delta P}{10,000 \, \rho t} - K_2 \rho \frac{10,000}{\rho t}$$

where to is the time in seconds to pump 10 kg of fluid. Simplifying and combining constants, we obtain

$$\eta = K_3 \times \rho \Delta P \times t - K_4 \times \frac{1}{t}$$

where

$$K_3 = \frac{K_1}{10,000} \quad \text{and} \quad K_4 = 10,000 \, K_2$$

On rearranging, we obtain

1. BASIC THEORY

$$\frac{1}{t} = \frac{K_3}{K_4} \rho (\Delta P \times t) - \frac{\eta}{K_4}$$

This equation indicates that, if we plot $1/t$ against $\Delta P \times t$, we should obtain a straight line where

$$\text{slope} = \frac{K_3 \rho}{K_4}$$

and

$$\text{intercept} = \frac{-\eta}{K_4}$$

or

$$K_4 = \frac{-\eta}{\text{intercept}} \frac{\text{slope}}{\rho}$$

Plots of $1/t$ versus $\Delta P \times t$ were made for each of the three tubes using the data obtained in the water-pumping tests. It was found that these plots were reasonably linear. Slopes and intercepts were then determined by the method of least squares, and K_3 and K_4 were calculated.

The effective viscosity of the emulsions was then calculated by using the equation

$$\eta_{eff} = K_3 \rho \, \Delta P \times t - K_4 \frac{1}{t}$$

where

ρ is the density of the emulsion, ΔP is the pressure of drop across the test section, t is the time in seconds to pump 10kg, and K_3, K_4 are the arbitrary constants determined above.

Proceeding in this manner, the effective viscosity at various shear rates was determined, and Fig. 23 shows a typical curve obtained from the data.

The curve of Fig. 23 is the sort of curve one would expect to obtain from a thixotropic fluid, except that there is almost no

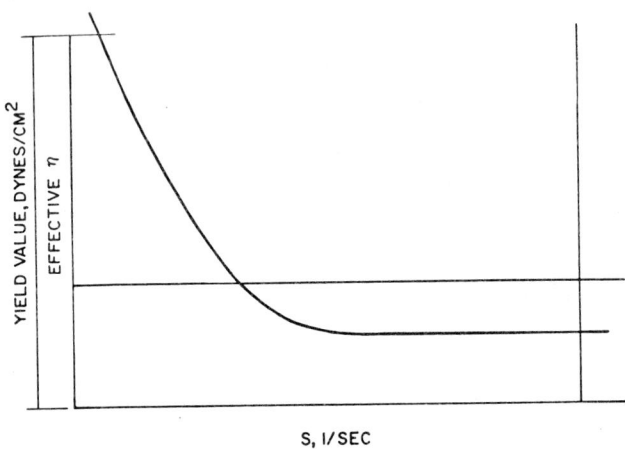

FIG. 23. Emulsion parameter relationships.

hysteresis loop in the curve. The rheological behavio of these emulsions is sufficiently unusual to make it doubtful that they really fit any of the classical categories. This point can best be illustrated by a typical case. If one make a high-internal-phase-ratio emulsion of kerosene in water using an appropriate emulsifier in the water phase, one will find that, if one subjects this emulsion to moderate shear by pumping it around a loop or through a piping system, or by subjecting it to the action of beaters or paddles, the emulsion will seem to get thicker and thicker, and if yield-value determinations are made at various stages of the mixing process, the yield value will increase and finally reach a maximum. Further moderate mixing will not increase the yield value. If the emulsion is subjected at this point to higher shear, the yield value may increase slightly, but if too high a shear is applied, the emulsion will start to "bleed." If the emulsion is allowed to stand undisturbed for a period of time, the yield value will be found to decrease with time and slowly approach a stable relaxed value. The emulsion can be wound up again by subjecting it to the previous shear regime. The relaxed and fully wound-up levels of the yield value are characteristic of the

1. BASIC THEORY

particular emulsion system and are reasonably stable and reproducible. However, determinations made at intermediate levels are found to vary considerably with the work history of the emulsion. If reproducible rheological values are to be determined on a given high-internal-phase-ratio system, it is absolutely necessary that the determinations be made at the top or bottom of the range or that the method of producing and handling the emulsion be very carefully specified.

For a given internal and external phase the yield value and the viscosity-versus-shear curve will vary with the following:

1. The particular emulsifier system
2. The concentration of the emulsifier
3. The method of manufacture
4. The shear history of the sample
5. The internal-phase ratio

The effect of the selection of emulsifier on the rheological properties will be discussed in more detail in a later chapter. In general, the higher the concentration of emulsifier in the external phase, the thicker the emulsion will be. However, in some instances, if too much emulsifier is used, the emulsion may become unstable. The emulsifier concentration usually ranges from a high of about 25 to 1% or less.

We have already discussed how the amount of shear to which an emulsion is subjected will affect its physical properties and have pointed out that the emulsion will relax when allowed to stand undisturbed. It has also been pointed out that the emulsions become thicker as the internal-phase ratio is increased. In general emulsion with a phase ratio of less than 80% will have the properties of a thick cream or mayonnaise. As the internal-phase ratio increases from 80 to 90%, the formulations will become thicker and more like custard. Above 90%, the thickening becomes even more marked, and in the 95 to 98% range the emulsions can be cut with a spatula and small cubes will stand unsupported.

It is difficult for anyone inexperienced with emulsions to realize that the yield value of an emulsion can increase without the viscosity-versus-shear curve's being affected drastically. This means that two emulsions, one of which seems to be much thicker than the other, may both show almost the same pressure drop when pumped through a piping system. The thicker emulsion, however, may require a larger pump intake or a higher pressure to keep the pump primed. Because of the rather steep slope of the viscosity-versus-shear curve, these emulsions are often easier to pump at high flow rates than at very low flow rates. Once the emulsion has been picked up by a pump, it can be pushed for considerable distances without excessive pressure drop or horsepower requirements, but unless a positive feed is provided to the pump, cavitation may occur. Also, provisions should be made to keep the emulsion in the pump reservoir from "coning" and allowing air to enter the system. This is usually achieved by having a sloping bottom on the tank or by subjecting the tank to a vibratory motion to keep the emulsion flowing to the pump inlet.

VII. SUMMARY

Having considered the various types of emulsions, we can now use our geometric model to guide us in selecting the type of emulsion for a particular purpose. Certain basic principles can be applied:

1. If we wish the final emulsion to have a low viscosity, we should
    a. Stay in the low-internal-phase-ratio range
    b. Obtain stability by using charged emulsifiers

1. BASIC THEORY 67

      c. Select emulsifiers that combat flocculation and clumping
      d. Use high-shear devices in the manufacture of the emulsion to obtain small particle size, thus avoiding creaming
      e. Keep the viscosity of the external phase low
2. If we desire high-viscosity emulsions, we should
      a. Operate in the medium- or high-internal-phase-ratio range
      b. Use film-forming emulsifiers or blends of film-forming and charged emulsifiers
      c. Use a viscous external phase
      d. Induce controlled flocculation by the addition of appropriate reagents
3. Sometimes it is necessary to keep the concentration of a particular active ingredient low, either for economic reasons or to control the application rate, or to control toxicity or dosage in medical emulsions. To introduce a small concentration of an active ingredient into an emulsion, we may
      a. Dissolve the active component in the internal phase of a low-internal-phase-ratio emulsion to obtain an emulsion with a low viscosity.
      b. Dissolve the active component in the external phase of a medium- or high-internal-phase-ratio emulsion to obtain a thick, smooth, noncreaming formulation
      c. Prepare a medium- to high-internal-phase ratio emulsion and suspend the active ingreadient as a finely divided solid in this emulsion. Obviously in this case the active ingredient must be insoluble in either phase.

In the chapters that follow, where the application of emulsion technology to specific fields is discussed, it will be noted that either consciously or unconsciously procedures similar to these have been evolved to guide in the selection of appropriate formulations. Even where the author does not specifically relate practices in a field directly to the geometric concepts, the reader will find it useful to do so for himself.

REFERENCES

1. P. Becher, *Emulsions: Theory and Practice*, 2nd ed., Reinhold, New York, 1965.
2. J. Alexander, *Colloid Chemistry*, Van Nostrand, New York, 1924.
3. W. Clayton, *Theory of Emulsions*, 4th ed., Blakiston, Philadelphia, 1943.
4. J. L. Moilliet, B. Collie, and W. Black, *Surface Activity*, Van Nostrand, New York, 1961.
5. J. W. McBain, *Colloid Science*, Heath, Boston, 1950.
6. S. Berkman and G. Egloff, *Emulsions and Foams*, Reinhold, New York, 1941.
7. W. R. Atkins et al., *Emulsion Technology*, Chemical Publishing Co., New York, 1946.
8. H. B. Weiser, *Colloid Chemistry*, Wiley, New York, 1939.
9. F. Tóth, *Regular Figures*, International Series of Monographs on Pure and Applied Mathematics, Vo. 48, Macmillan, New York, 1964.
10. R. H. Brisford, *Nature (London)*, *224*, 550 (1969).
11. M. Gardener, *New Mathematical Diversions*, Simon and Schuster, New York, 1966, p. 88.
12. W. Clayton, op. cit., p. 78.
13. G. G. Stokes, *Trans. Cambridge Phil. Soc.*, *9* (1851).
14. J. Hadamard, *Compt. Rend.*, *154*, 173 (1911).
15. P. Becher, op. cit., p. 153.
16. R. K. Neogy and B. N. Ghash, *J. Indian Chem. Soc.*, *30*, 113 (1953).

17. E. L. Knoechel and D. E. Wurster, *J. Amer. Pharm. Assoc., Sci. Ed.*, *49*, 249 (1960).
18. F. R. Erick, ed., *Rheology, Theory and Applications*, Academic Press, New York, 1962.
19. S. Okis, in *Rheology, Theory and Applications* (F. R. Erick, ed.), Vol. III, Academic Press, New York, 1962, Chap. e.
20. A. Einstein, *Ann. Physik (4)*, *19*, 1289 (1906); ibid., *34*, 591 (1911).
21. P. Sherman, *Rheol. Acta*, *2*, 74 (1962).
22. W. Bartok and S. G. Moson, *J. Colloid Sci.*, *13*, 293 (1958).
23. M. A. Nawab and S. G. Moson, *Trans. Faraday Soc.*, *54*, 1712 (1958).
24. N. N. Bond and D. A. Newton, *Phil. Mag.*, *5*, 794 (1928).
25. C. T. Du Motay, A. J. Rossi, A. Bourgaugnon, and P. Cosamajor, U.S. Pat. 241,505 (1881).
26. S. U. Pickering, *J. Chem. Soc.*, *91*, 2002 (1902).
27. H. B. Weiser, op. cit., p. 288.
28. J. W. McBain, op. cit., p. 164.
29. S. J. Werbin, *Physical Functions of Hydrocolloids*, ACS Monograph No. 25, American Chemical Society, Washington, D.C., 1956, p. 8.
30. W. Clayton, op. cit., p. 151.
31. J. G. Atherton and H. C. Nemeth (to Armour Industrial Chemical CO.), U.S. Pat. 3,622,518 (1971).
32. K. J. Lissant, *J. Colloid Interface Sci.*, *22*, 462 (1966).
33. J. Nixon and A. Beerbower, paper presented before the Division of Petroleum Chemistry, American Chemical Society meeting, Minneapolis, Minn., Apr. 13-18, 1969.
34. K. J. Lissant, L. R. Hopper, and J. L. Harris, paper presented at American Society of Mechanical Engineers meeting, Feb. 26, 1968.

Chapter 2

MAKING AND BREAKING EMULSIONS

K. J. Lissant

Research Laboratories
Petrolite Corporation
Saint Louis, Missouri

I.	INTRODUCTION.	72
II.	REASONS FOR USING EMULSIONS	72
III.	SELECTION OF EMULSION TYPE.	74
IV.	SELECTION OF EMULSIFIER	75
	A. Introduction.	75
	B. The HLB Method.	77
	C. A Method for Classifying, Testing, and Selecting Nonionic Emulsifiers.	78
	D. Micelle Formation	100
V.	TECHNIQUES OF EMULSIFICATION.	103
	A. The State of the Art and Equipment.	103
	B. Low-Internal-Phase-Ratio Emulsions.	103
	C. High-Internal-Phase-Ratio Emulsions	106
	D. The Geometric Approach.	110
VI.	TECHNIQUES FOR BREAKING EMULSIONS	111
	REFERENCES.	123

I. INTRODUCTION

In considering the application of emulsion technology to practical problems four questions should be considered:

1. Is an emulsion really needed, or would a solution or suspension do the job better?
2. If an emulsion is required, what type would be most suitable?
3. What emulsifiers should be used?
4. What equipment should be used?

The next four sections of this chapter will attempt to deal with these questions. The final section describes situations in which undesirable emulsion may occur and discusses the techniques for breaking such emulsions.

II. REASONS FOR USING EMULSIONS

Let us consider briefly why an emulsion should be used at all. The most usual instance of the application of emulsion technology is in a situation where it is necessary to incorporate into a single formulation ingredients that are not mutually soluble. For instance, in the case of the typical oil-and-vinegar salad dressing some of the flavoring ingredients are not soluble in the salad oil and some are not soluble in the vinegar, and the two phases, the salad oil and the vinegar, are not soluble in each other. Therefore one either has to shake the mixture vigorously just before using or add emulsifying agents that will keep the two immiscible liquids intimately mixed. Another reason for employing emulsion technology lies in the ability to control flavor or stability. Here again the salad-dressing example can be cited. Many people would feel that a salad dressing consisting only of, for instance, olive oil would be too oily or greasy to the taste, and it has been found by experience that the mixture of a water phase with the oil improves the taste and the mouth feel, and simplifies the incorporation of certain flavoring ingredients.

Another reason for the use of emulsion technology is to enable one to dilute a formulation with a less expensive ingredient. For instance, water-base paints consist essentially of small drops of an oil paint emulsified in water. This makes it possible to use a more concentrated oil formulation and to replace the solvents, normally incorporated into the oil paint, by water. This reduces the cost of the formulation, diminishes air pollution by eliminating the necessity of evaporating organic solvents, and simplifies cleanup of equipment since water can be used instead of a more expensive solvent. Many insecticides and other agricultural toxicants are insoluble in water. These materials must be applied to the growing crops in very small, but even, doses. The small dosage is called for both because the materials are expensive and because, if applied in doses that are too large, they may be toxic to humans or the plants themselves. Since they are not soluble in water, the materials must either be applied from an oil solution or dissolved in oil and then emulsified in water. Application of toxicants in oil solutions is not usually desirable because the oil itself may damage the crops, most suitable oily fluids are expensive, and there is an increased fire hazard. For this reason the use of emulsions in agricultural applications is quite widespread, and this area is discussed in detail in Chapter 4.

In general, it is usually preferable to obtain a particular formulation as a single phase because solutions are inherently more stable than emulsions. However, in many real-life instances it is difficult to avoid multiphase formulations.

Summarizing then, we can say that, if we are faced with the necessity of producing a homogeneous formulation from a group of mutually insoluble ingredients, the utilization of emulsion technology may be the most acceptable procedure.

Emulsion technology can also be used to control the physical properties of a formulation. Thus a viscous material may be emulsified in a fluid external phase to produce a formulation of decreased viscosity, or a fluid material may be emulsified in a viscous system to increase the viscosity. Similarly medium- and high-internal-phase-ratio emulsions can be used to produce thick, non-Newtonian formulations. In some instances medium- or high-internal-phase-ratio emulsions can be employed in the place of gelling agents to produce thick formulations. This may be a distinct advantage either from the standpoint of economics or because the gelling agent may be undesirable because it leaves a residue or is subject to bacterial attack.

III. SELECTION OF EMULSION TYPE

Once we have determined that we want to use emulsion technology, we have to decide what type of emulsion will be most suitable. We mentioned in Chapter 1 that emulsions are classified into two types, water-in-oil and oil-in-water, and we have also discussed three categories, low-, medium-, and high-internal-phase-ratio emulsions. We need then to choose from six different general emulsion types.

The choice of emulsion type will depend on a number of factors. If we desire low-viscosity, fluid emulsions, we shall have to use low-internal-phase-ratio emulsions or emulsions at the low end of the medium range. If thicker products are desired, medium- and high-internal-phase-ratio emulsions are called for. If the emulsion must be electrically conductive, a water-external type is called for. If electrical conductivity is undesirable, the oil-in-water types are indicated.

In a practical case one must decide what considerations have the highest priority. Very often economics is the overriding factor, in which case one chooses an emulsion type that produces

2. MAKING AND BREAKING EMULSIONS

the least costly formulation. On the other hand, in the area of pharmaceuticals and cosmetics the physical properties of the formulations may be of paramount importance, with economics being secondary. In such cases one chooses the emulsion type that produces the most desirable physical properties.

By far the most common emulsions are the oil-in-water, low-internal-phase-ratio ones. As mentioned in Chapter 1, these emulsions have a low viscosity and essentially Newtonian rheological characteristics. They are usually simple to produce and can be made with low-cost materials. This particular emulsion type is so common, in fact, that many discussions of emulsions assume that this type is implied and devote almost no time to the other five types. Water-in-oil emulsions of the low-internal-phase-ratio type are less common, but are occasionally encountered as undesirable situations in manufacturing processes and effluent-cleanup systems. They will be discussed in more detail in Section VI.

A number of cosmetic and food emulsions fall into the medium-internal-phase-ratio region and may be of either the oil-in-water or the water-in-oil type. In general it seems to be felt that emulsions to be taken internally should be of the water-external type if possible because this improves their flavor and mouth feel. There are, however, exceptions to this generalization. Oil-external emulsions are used for salves and creams to be applied to the skin as well as in some grease and lubricant formulations.

IV. SELECTION OF EMULSIFIER

A. Introduction

Once we have selected the type of emulsion we wish to produce and have chosen the composition of the internal and external phases, our next problem is to select an emulsifier that will produce a stable emulsion. An enormous amount of literature and thousands of patents have been devoted to the subject of suitable emulsifiers for specific applications, but in spite of all this work, many people feel that the selection of an appropriate emulsifier is still as much an art as a science. Although there is a good deal of evidence for this point of view, it is true that a number of general principles can be applied. Some of the more obvious generalizations are the following:

1. Low-internal-phase-ratio emulsions are stabilized by ionic emulsifiers and by producing very small internal droplets.

2. As the internal-phase ratio of the emulsion increases, particle charge becomes less important and the film-forming capability of the emulsifier comes to the fore.

3. A mixture of emulsifiers is often more effective than a single pure material.

4. In general the emulsifier should be more soluble in the external phase than in the internal phase, and usually the more effective emulsifiers have limited solubility in either phase.

5. Where natural products or technical grade emulsifiers are used, the performance may vary significantly from batch to batch, depending on the source and previous history of the emulsifier.

A number of books and papers have discussed the chemistry of emulsifiers and the rationale behind their selection. Becher (1) devotes a large section of his book to the subject and gives it excellent coverage. Moilliet, Collie, and Black (2) discuss types of surface-active agents in some considerable detail. McCutcheon (3) issues an annual edition listing most of the commercial detergents and emulsifiers available in the United States. This book is particularly valuable for identifying the chemical type of materials when the trade name is known.

Once the type of emulsion to be produced has been decided on and the general type of emulsifier selected, one can determine from

2. MAKING AND BREAKING EMULSIONS

McCutcheon (3) and the trade literature what materials are available in a certain category. Nearly all manufacturers offer pamphlets describing the applications of their materials, and these are extremely helpful in specific situations. Many of these trade pamphlets list patents held by the manufacturer, and often a study of these patents will supply the experimenter with additional formulations and suggestions.

B. The HLB Method

One of the most widely used methods of selecting emulsifiers was initiated by Griffin (4) in 1949 and is known as the hydrophile-lipophile balance (HLB) method. It is based on the theory that emulsifying agents contain oil-soluble and water-soluble moieties. This dual nature of the surfactant molecule tends to limit its solubility in both phases and causes it to concentrate at the interface. Thus a long-chain alcohol ester of sulfuric acid would contain the water-soluble sulfate group and the long, oil-soluble, aliphatic chain. The balance between these groups would determine whether a particular surfactant would be soluble in either oil or water and the type of emulsion it might stabilize. Becher (1) devotes several pages to the description of the HLB method. Atlas Chemical Industries, Inc., has also published a pamphlet (5) explaining how this method may be used.

The HLB system is particularly useful for certain types of nonionic surfactants. It also serves as a general rationale for selecting materials. It does, however, have serious limitations in that, though it may select an emulsifier or combination emulsifiers for solving a particular problem, it will not necessarily provide the most effective material.

One of the big problems in selecting emulsifiers for testing in the laboratory is the enormous number of possible candidates. The HLB system will help to eliminate large groups of these emulsifiers and considerably narrow the field. It still, however, leaves large areas to be tested and eventually leaves the selection up to the skill and experience of the experimenter to a considerable extent. The HLB system, however, has been widely used and accepted for many years, and an understanding of its operation is essential for anyone active in the field of emulsion technology.

C. A Method for Classifying, Testing, and Selecting Nonionic Emulsifiers

The number of distinct ionic emulsifiers available is considerably limited, and the selection of an appropriate ionic emulsifier for a particular problem can usually be accomplished with the aid of trade literature and formulary suggestions. However, in situations where no previous work has been done or where the more conventional emulsifiers are unsatisfactory, the problem becomes much more difficult. This is particularly true in selecting a nonionic emulsifier since here the number of individual species is almost limitless. So great is the potential number of nonionic surface-active materials that a method of classifying and searching this group of materials is essential before any systematic experimentation can be done in the field.

In 1963 Lissant (6, 7) attempted to classify all nonionics in a single coherent system and outlined a method for classifying, testing, and selecting emulsifiers according to this classification system. This method has proved to be an effective laboratory tool for selecting nonionic emulsifiers (8). Because it has received less attention in the literature than the HLB method, this method will be described in some detail here. In many respects it is an extension and refinement of the HLB system coupled with a graphical method of plotting results.

By far the largest group of nonionic emulsifiers are those known as the oxyalkylates. They are produced by reacting an alkylene oxide with a base material having one or more reactive sites. Alkylene oxides have the general formula

2. MAKING AND BREAKING EMULSIONS

$$R^1-\underset{\underset{O}{|}}{\underset{|}{C}}-\underset{|}{\overset{R^3}{\underset{|}{C}}}-R^4$$
$$\diagdown\diagup$$

Here R^1, R^2, R^3, and R^4 may be, for example, hydrogen, an aliphatic radical, a cycloaliphatic radical, an aryl radical, and so forth. The R groups may also be joined to form a cyclic structure. They may also contain epoxide groups, resulting in a material that is a diepoxide or polyepoxide.

Materials of this sort react with active centers of other organic or inorganic molecules to build up polyether chains of varying length. The general reaction product may be represented as follows:

$$Z\left[\left(\underset{\underset{R^1}{|}\ \underset{R^4}{|}}{\overset{\overset{R^2}{|}\ \overset{R^3}{|}}{C-C-O}}\right)_n H\right]_x$$

Here n is the number of monomer units in the chain and x is the number of reactive sites in the starting molecule. Since the relative reactivity in the original terminal groups and the terminal groups on the chains determine the position and relative length of the chains, single pure species are almost never produced, and it is understood that the reaction products are "cogeneric mixtures." Reactions of this type can be represented mathematically by breaking down the operation into several steps. These steps are described as follows (6):

1. Designation of starting material
2. Designation of reaction conditions
3. Stepwise reaction of monomer with starting material

    a. Selection of monomer composition
    b. Specification of number of monomer units to be reacted with starting material

The product resulting from step 3 can then be used as the starting material for a new family of polymers by repeating steps 2 and 3. This process can be continued indefinitely, but usually for practical purposes is not repeated more than two or three times. We will now take a closer look at each of these three steps.

First let us consider the designation of starting materials. All materials susceptible to oxyalkylation may be said to fall into a class designated S_1, S_2, S_3,..., S_x,...,S_n, where the subscript refers to the chemical identity of the starting material. Examples of such subclasses would be alcohols, amines, or phenols. This particular class of materials could be organized, for instance, according to the method outlined by Wiswisser (9). The Wiswisser form of notation has received wide use, particularly by those employing computers for information-retrieval purposes. It assigns to each chemical identity a unique line-formula notation that is topologically related to its chemical structure. It establishes a hierarchy for ordering these notations and thus arrays them in an ordered set. If all materials that can be oxyalkylated were rendered in their Wiswisser notation and arranged according to Wiswisser's order rules, we would be able to assign to each chemical identity a specific designator S_x. Alternatively the Wiswisser notation itself could serve as the designator. However, some provision would have to be made for mixtures and technical grade products.

Let us consider for a moment the designation of reaction conditions. Anyone familar with oxyalkylation procedures is well aware that the reaction conditions have a profound effect on the properties of the final product. The temperature, the pressure, rate of addition and method of mixing, type and amount of catalyst, and many other factors may affect the final product. However, as long as the reaction conditions are specified and controlled carefully, reproducible plant production is possible and is carried out by a number of companies on a routine basis. Therefore we can

2. MAKING AND BREAKING EMULSIONS

mathematically represent this aspect of the product by postulating a function $F_x(T,P,c,...)$ from a general class of all possible reaction conditions.

We can now proceed to the stepwise reaction of the monomer with the starting material S_x under reaction conditions F_x. First we select the monomer. Four 1,2-alkylene oxides are commercially available: ethylene oxide, propylene oxide, and 1,2- or 2,3-butylene oxide. Less common materials are octylene oxide, cyclohexene oxide, styrene oxide, and various diepoxides and polyepoxides. These materials may all be grouped into a class $M_1, M_2, ..., M_x ..., M_n$. In the general case any or all of the members of this class may be used by themselves or in combination as the monomer mixture. Usually a single monomer will be used. However, instances of commercial production using a mixture of monomers are known. We can therefore represent the monomer composition symbolically as follows: $O_x(aM_1, bM_2, ..., yM_x)$, where O represents an individual polymerization step; $M_1, M_2, ..., M_x$ are the particular monomers in the mixture; and $a, b, ..., y$ are the proportions of each monomer. If $a, b, ..., y$ are expressed as weight or mole percentages, then $a + b + \cdots + y = 100$, and if they are expressed as decimal fractions, then $a + b + \cdots + y = 1$. For example, if pure ethylene oxide were used as the monomer, the notation would be M(EtO), and if equal weights of ethylene and propylene oxides were mixed, the notation would be M(0.5 EtO, 0.5 PrO, w/w), where the indices are expressed as weight fractions. The designation would be M(0.5 EtO, 0.5 PrO, m/m) if equal molar amounts were used. It should be emphasized that these two mixtures are not the same, because M(0.50 EtO, 0.50 PrO, m/m) is equivalent to M(0.4314 EtO, 0.5686 PrO, w/w).

We are now ready to specigy the number of monomer units, M_x, to be reacted with S_x under the conditions F_x. The only real problem here is deciding what set of units to use. The most commonly used are weight of monomer per weight of starting material, moles of monomer per mole of starting material, or percentage of monomer in the final product. Each of these methods has been championed by various groups. Oxyalkylated phenols are usually sold with the designation of the number of moles of material added to the phenol. On the other hand, certain other polyether derivatives are designated by the molecular weight of the starting polyglycol and the percentage of another oxide added. It has been the experience of the author that either moles per mole of starting material or weight per weight of starting material is equally convenient, but pilot-plant personnel seem to prefer the weight ratios.

We are now ready to set down a general notation for the composition of any particular oxyalkylated product in terms of the starting material, reaction conditions, and kinds and amount of monomers used. Our general notation is

$$S_x, F_x, O_1(a_1M_1, b_1M_2, \ldots, y_1M_x)N_1$$

$$O_2(a_2M_1, b_2M_2, \ldots, y_2M_x)N_2$$

$$O_n(a_nM_1, b_nM_2, \ldots, y_n(M_x)N_n$$

where $O_1, O_2, \ldots, O_n$ represent successive oxyalkylation steps and $N_1, N_2, \ldots, N_n$ are the amounts of oxide added at each step.

It may be helpful to cite a few specific examples. For instance, the Wyandotte Chemical Company markets a material designated Pluronic L-64. It is stated that this material is made by adding ethylene oxide to a polypropylene glycol of molecular weight 1750 until the ethylene oxide portion represents 40% of the weight of the final molecule. In our notation this material is considered to be the two-step reaction product, where water is reacted with propylene oxide and then ethylene oxide. If one adds about 30 moles of propylene oxide to 1 mole of water, a poly(propylene glycol) of molecular weight 1750 may be obtained. This represents 60% of a total weight of about 2900. The addition of 26 moles of ethylene oxide to the poly(propylene glycol) would give this final product. Our notation would then be

H_2O, F_p, O_1(PrO 100% m) 30.17, O_2(EtO 100% m) 26.51

F_p representing the reaction conditions used by the manufacturer.

The Union Carbide Chemical Company produces a group of materials sold under the trade name Ucon. They are mixed polyglycols produced by the reaction of a mixture of ethylene and propylene oxides on water as a starting material. For instance, Ucon 50 HB 260 is made by treating 1 mole of water with about 20 moles of a mixture of ethylene and propylene oxides in equal proportions by weight. It is made by a one-stage process, and the notation is H_2O, F_u, O_1(0.5 EtO, 0.5 PrO, w/w) 20 m, where F_u refers to the reaction conditions used by the manufacturer. Here we have chosen to express the ratio of monomers as a weight ratio, but the addition as a 20-mole addition. If desired, both proportions could be expressed in weight ratios or in moles.

What we have now done is provide a method whereby any oxyalkylated derivative can be provided with an unambiguous, unique notation. What this notation really does is to provide an address for each possible chemical identity exactly analogous to the way that a computer memory is set up.

It has been found that one can take the various notations and devise composition spaces in which all types of oxyalkylated materials can be plotted. As described in Ref. 6, the procedure consists of a mapping technique wherein a suitable composition space is chosen and the properties of members of a class of polymers are so mapped in the space that their relationships are readily displayed. Mathematically this amounts to establishing a one-to-one correspondence between the individual members of a class of polymers and the individual points in an appropriately chosen composition space.

This technique has been used before for other purposes. For instance, all the possible mixtures of methanol and water can be represented by points on a line segment like Fig. 1 (7).

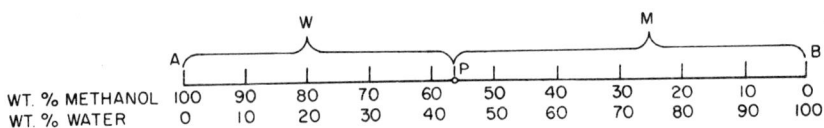

FIG. 1. Examples of a one-dimensional composition space (7).

In Fig. 1, A represents pure methanol and B represents pure water. Point P represents a mixture of methanol and water, its location being determined by the proportions of the two compounds in the mixture. As point P approaches A, the mixture becomes richer in methanol. This particular composition space is one-dimensional, unambiguous, definitive, and commutative. It represents all possible mixtures, each point represents one and only one mixture, and each mixture can be made either by adding water to methanol or methanol to water. Mathematically this is the same as saying that point P may be approached from either direction without changing its meaning.

In preparing oxyalkylates, as in most chemical syntheses, the order in which various steps are performed has a profound effect on the product obtained. Thus, if one treats a phenol first with ethylene oxide and then with propylene oxide, one obtains a different material than if the additions were performed in the reverse order. In other words, the process is noncommutative, and for this reason most of the composition spaces utilized in this procedure are also noncommutative. Referring again to Ref. 6, we find that the appropriate composition space for the polymeric species under consideration is characterized as follows:

1. The indices S_x and F_x serve to differentiate individual composition spaces that are otherwise mathematically identical.

2. The spaces are fundamentally noncommutative in that each of the dimensions making up the space must be traversed in the designated sequential order.

3. The dimensional segments of the composition space are of two kinds: selection figures, where the monomer composition is

2. MAKING AND BREAKING EMULSIONS

specified, and polymerization figures, where the amount of monomer is specified. In general the noncommutative aspects of the space require traversing alternately first through a selection figure and then through a polymerization figure.

4. The composition space contains one selection figure and one polymerization figure for each successive different oxyalkylation step required to produce the particular molecular species.

Proceeding now from the purely mathematical to concrete examples, we will illustrate a few of these composition spaces. For instance, one-component polyglycols, such as the series consisting of ethylene glycol, diethylene glycol, triethylene glycol, etc., are easily displayed in one-dimensional, noncommutative diagrams like those shown in Fig. 2 (7).

Line 1 in Fig. 2 represents the polyethylene glycols, S_1 being water and the distance from the origin moving toward the right representing moles of ethylene oxide added per mole of water; F_1 represents an appropriate set of conditions for conducting such a reaction, and each point on the line represents a possible cogeneric mixture obtained by carrying out such a reaction. If the conditions of the reaction were such that pure compounds, rather than cogeneric mixtures, were produced, a noncontinuous space such as

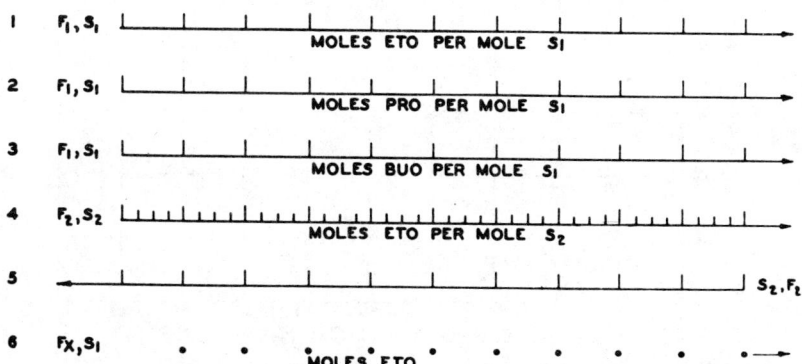

FIG. 2. One-dimensional composition spaces for one-component polyglycols (7).

that represented by line 6 would be used. Lines 2 and 3 in Fig. 2 are typical of what might be used for displaying polypropylene and polybutylene glycols. If a different starting material, for example, an amine, were used instead of water, a similar space could be employed, except that S_x would be different, such as in line 4. There is no reason why the scale has to run to the right, and therefore lines 4 and 5 are equivalent.

Proceeding now to two-step processes, we plot them in two-dimensional composition spaces like that shown in Fig. 3 (7).

This particular diagram could be used to plot the Pluronics mentioned previously.

It is important to remember that both the one- and two-dimensional spaces are noncommutative. In the one-dimensional spaces

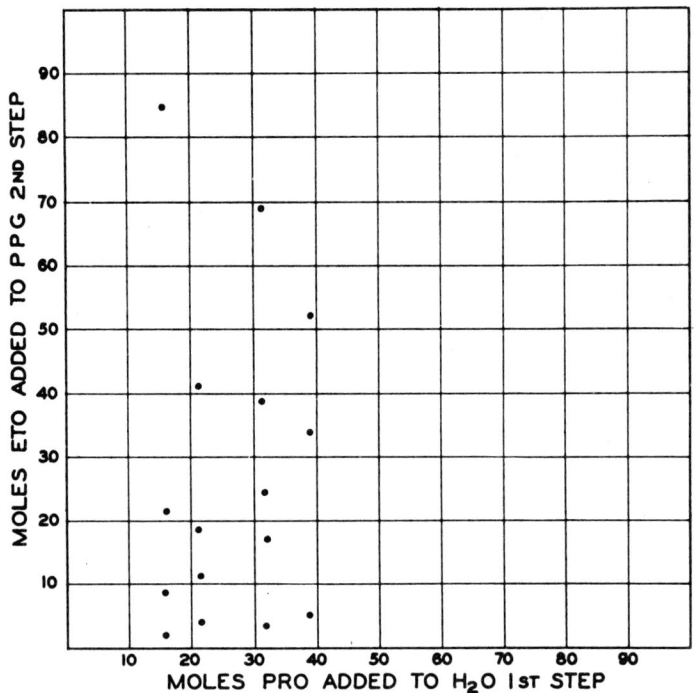

FIG. 3. A two-dimensional composition space (7).

one cannot "back up" and expect to obtain the same material. In other words, the depolymerization reaction is not the reverse of the polymerization. In the two-dimensional figure one must proceed along the O_1 direction to a particular point and then proceed in the O_2 direction. The reverse operation would not produce the same material.

Numerous other examples of types of composition space are given in Refs. 6 and 7.

Having devised a system of classifying these polymers and a method for mapping them onto an equivalent composition space, we are now ready to study methods of using this capability for selecting emulsifiers for a particular emulsification system. A detailed description of how emulsifiers may be selected is given in Ref. 10. For instance, one may set up a two-dimensional graph where the ordinate is a one-dimensional composition space and the abscissa is some measure of the effectiveness of a particular compound. A series of potential emulsifiers is then made by oxyalkylating some appropriate starting material, and a uniform test is performed to measure the emulsification capability of each member of the series. A bar graph is then prepared by erecting, at the appropriate place on the ordinate, a bar of a height corresponding to the effectiveness of a particular compound. Such a graph is shown in Fig. 4 (10).

Alternatively, if one were to prepare a series of oxyalkylates by treating a starting material first with propylene oxide and then with ethylene oxide, the members of the series could be arrayed on a two-dimensional composition space, and if the effectiveness of each member of the series were measured in a uniform test and circles of varying radius corresponding to the varying effectiveness were centered on the appropriate position in the composition space, a chart like Fig. 5 (10) might result.

Figure 5 represents a series of materials prepared by treating a phenolic resin first with propylene oxide and then with ethylene oxide. Note that, the larger the circle, the more effective the

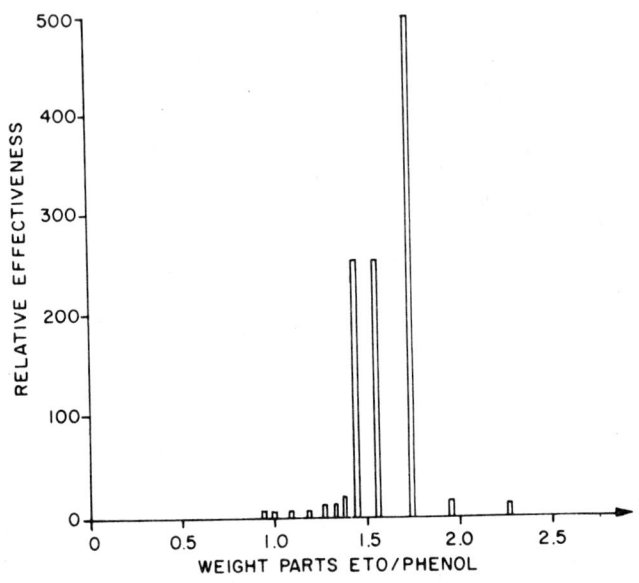

FIG. 4. Bar graph showing the effectiveness of a series of potential emulsifiers (10).

particular material was as an emulsifier agent for the system under test. It is easy to see that an area of maximum effectiveness is beginning to take shape on the diagram.

One more example of this type probably should be included to show how multidimensional systems can be diagramed.

In Fig. 6 (10) a five-dimensional composition space has been diagramed. Fourteen actual compounds were tested for effectiveness as emulsifiers. These materials were prepared by a five-step process. An amine was treated with epichlorohydrin to produce three first-stage compounds, each containing differing amounts of epichlorohydrin. Each of these precursors was then treated with propylene oxide, ethylene oxide, propylene oxide again, and finally with varying amounts of ethylene oxide to produce the 14 final products, numbered 52 through 65. The complete composition space is then a five-dimensional "cube" with five dimensions each orthogonal to each of the others. We need not concern ourselves about the "shape" or "appearance" of this five-dimensional space. All we need to do is pass a plane through this space with the appropriate orientation, and we will find that all 14 final products can be depicted on this plane since step 4 is represented by one of the axes and step 5 by the other [see Fig. 7 (10)]. Again the relative sizes of the circles show the effectiveness, and the more fruitful areas of the composition space are delineated.

Rather than retest a whole series of compounds each time a new emulsion system is studied, it would be preferable to have a single general test that would prove to be an index of the

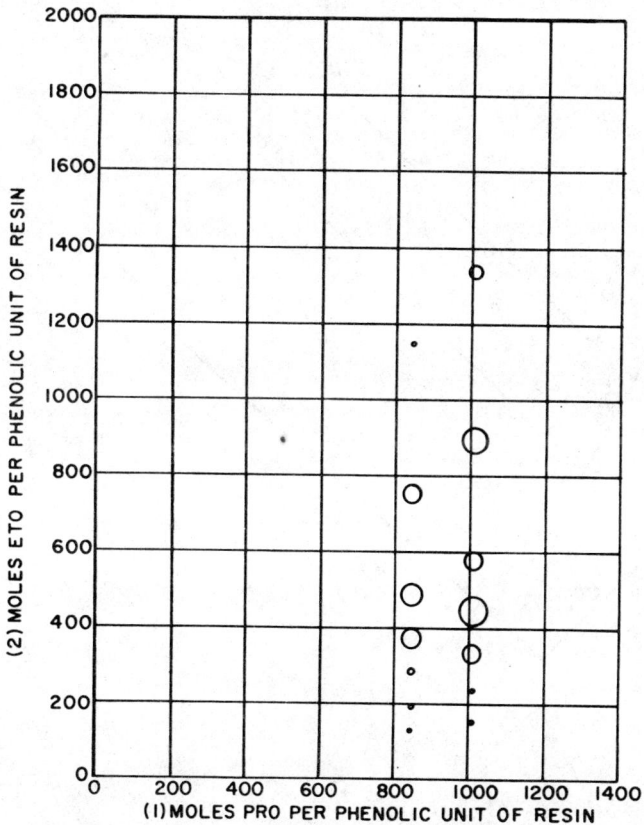

FIG. 5. Two-dimensional composition space showing the emulsifying effectiveness of a series of oxyalkylates (10).

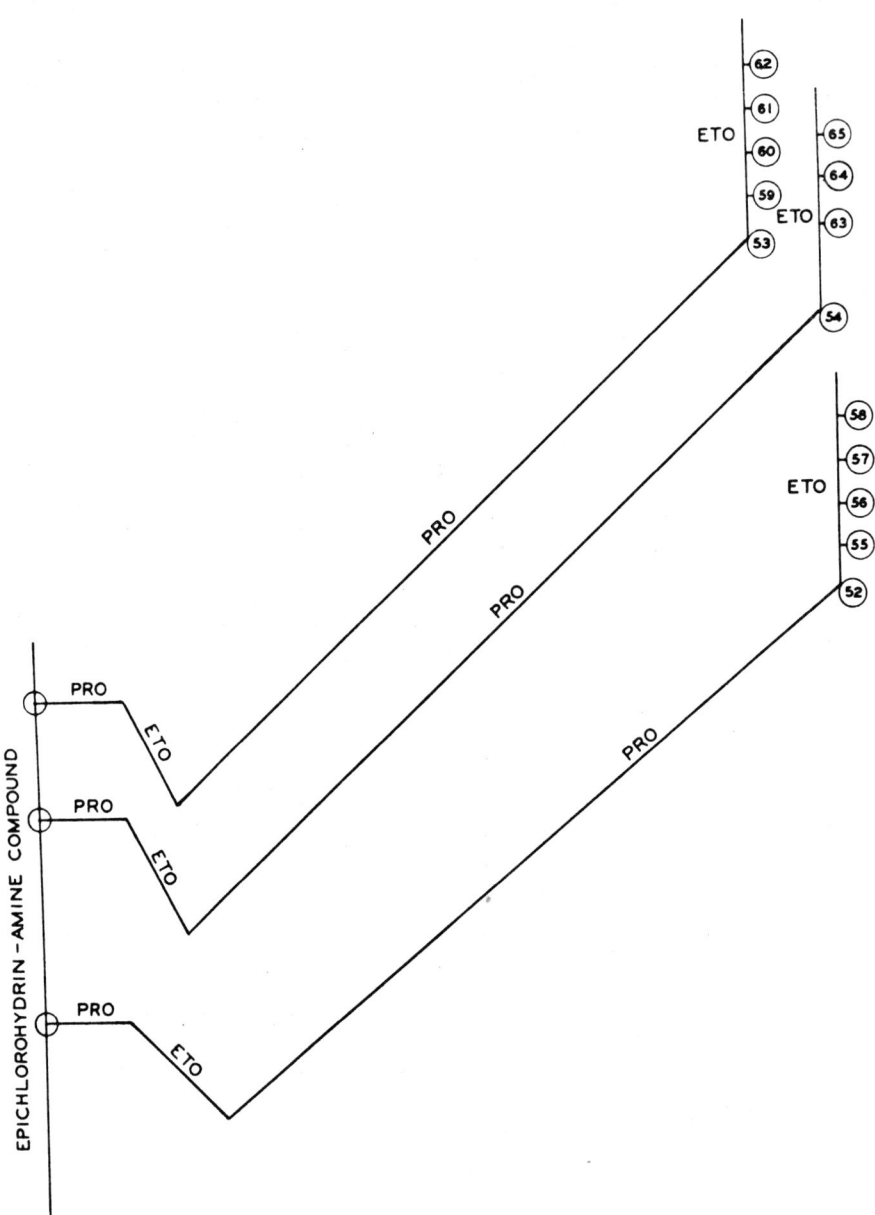

FIG. 6. Five-dimensional composition space (10).

2. MAKING AND BREAKING EMULSIONS

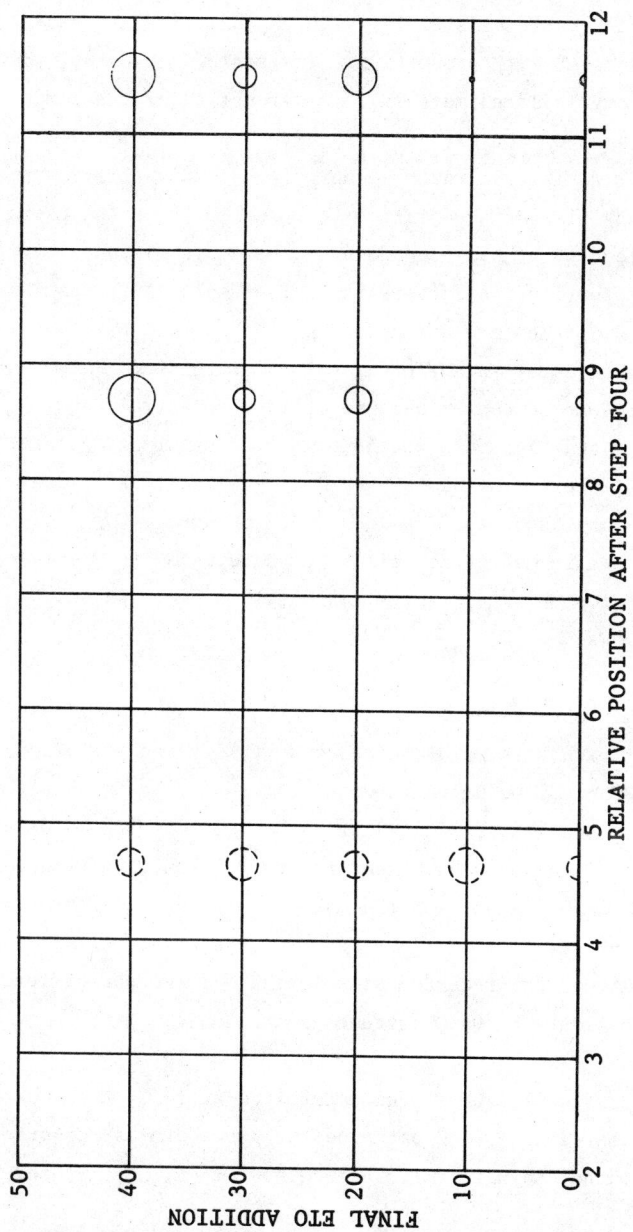

FIG. 7 Chart showing the relative emulsifying effectiveness of the 14 final products mapped in Fig. 6 (10).

compound's emulsifying potential. This can indeed be done. It is well known that a compound's emulsifying capabilities are closely related to its solubility properties. It is also known that the solubility of oxyalkylated materials is a function of the temperature and that indeed many oxyalkylates have less solubility in hot water than they do in cold water. The point at which they become insoluble at any given concentration is referred to as the cloud point. The cloud point has been found to be a useful index of emulsifying properties. A battery of uniform solubility tests and a way of recording them are discussed in Refs. 6 and 7.

In this procedure two different solvents are used, water and kerosene. Kerosene is chosen because it is a relatively poor organic solvent and therefore yields a more discriminating test. Perhaps in the interest of neatness a more precisely defined aliphatic hydrocarbon, such as n-decane, could be used.

As described in Ref. 7 solubility is determined in distilled water and in kerosene at concentrations of 1, 5, 10, and 50% by volume as follows:

> Eight test tubes were placed in a rack, and an appropriate amount of each sample, of water, and of kerosene was added to produce 20 ml of solution in each test tube. The rack with tubes was then placed in a water bath at 25°C. The solutions were allowed to reach bath temperature, shaken, and allowed to stand in the bath for 5 min more. Each tube was then inspected for solubility properties. If the solution was clear and bright, it was recorded as soluble. If two phases were clearly present, it was recorded as insoluble.

After the solubility has been determined at 25°C, the rack of test tubes is moved to a 35°C bath and the procedure is repeated. In a similar manner solubility is determined at 45, 55, 65, and 75°C.

This gives us solubility at four concentrations and six

temperatures in two different solvents, or a total of 48 separate pieces of solubility data. This set of solubility data can be referred to as the solubility profile of the compound. All 48 pieces of solubility data can be displayed on a composition-space diagram by the use of an appropriate symbology. The formulation of the solubility symbol is illustrated in Fig. 8 (7).

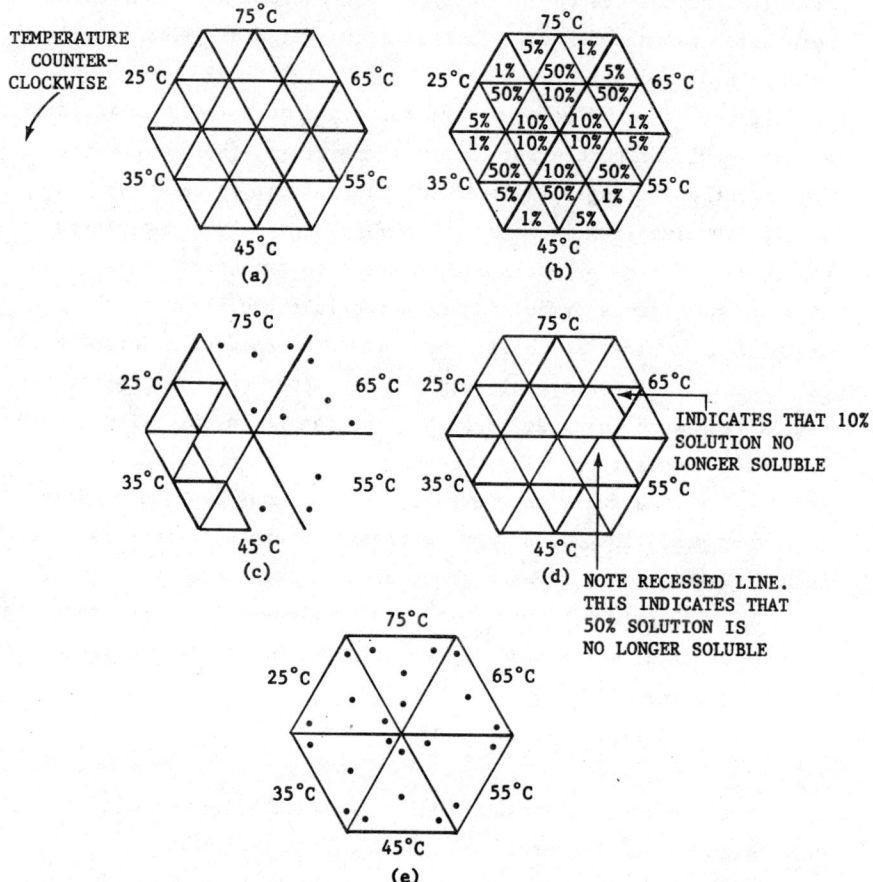

FIG. 8. Hexagonal solubility-data symbol (7): (a) representation of temperature; (b) representation of temperature and concentration; (c) typical example of a compound that is soluble in water at low temperatures and soluble in kerosene at higher temperatures; (d) example of solubility in water; (e) example of solubility in kerosene [same as example (d)]. Dots are used to indicate soluble points in kerosene, triangles are used for water.

The symbol is a hexagon formed by combining six equilateral triangles. Each triangle represents a single temperature at which the solubility was determined. The upper left-hand triangle of the hexagon represents 25°C, and proceeding thence clockwise, the remaining triangles represent 35 through 75°C in order. Each triangle is divided into four smaller triangles, and each of these triangles represents one of the four concentrations at which the solubility was determined. This is illustrated diagrammatically in Fig. 8(a) and (b).

If the compound being tested was found to be soluble in water at the specified concentration and temperature, the appropriate concentration triangle is drawn in. If not, it is omitted. Similarly solubility in kerosene is represented by a dot placed in the center of the appropriate concentration triangle. This generates an unambiguous symbol for each possible profile. Specific examples of typical solubility profiles are depicted in Figs. 8(c), (d), and (e). These symbols can now be mapped onto an appropriate composition-space diagram to display the change in solubility profile with composition. This technique makes it possible to record and display in a fully retrievable form enormous amounts of data in a very small space. A very impressive example of this is given in Ref. 7. Figure 9 shows a group of materials made by reacting water with a mixture of propylene and ethylene oxides, and then further reacting the mixed polymer with either ethylene oxide or propylene oxide (7).

Figure 10 represents a plane such as the plane EFGH of Fig. 9 (7).

In the example given 261 individual oxyalkylated materials were prepared, distributed over three of the planes of Fig. 9. It takes 12 typewritten pages just to cite the compositions of these examples and 31 more typewritten pages to record the solubility profile of each composition. All these data are displayed in a completely recoverable form on three 8½ x 11 sheets similar to Fig. 10. Not only is this a more concise way of displaying the

2. MAKING AND BREAKING EMULSIONS

data, but it makes it possible for one readily to discern the change of solubility with composition. For instance, one can connect by lines identical symbols and thereby delineate "isosolubility" lines or regions.

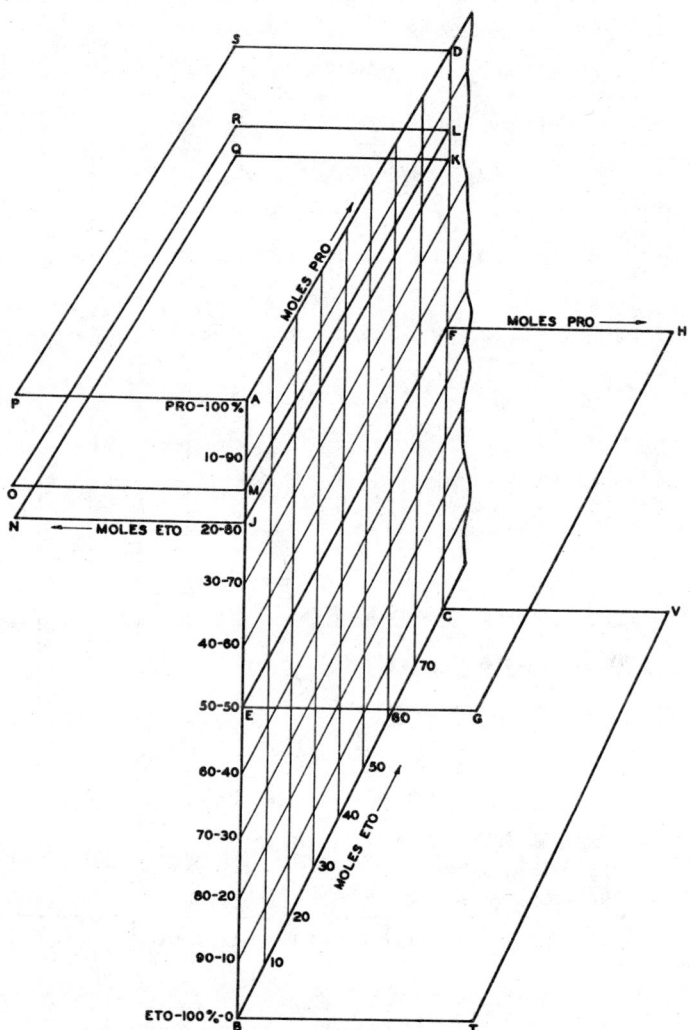

FIG. 9. Diagrammatic representation of a group of materials made by reacting water with a mixture of propylene and ethylene oxides, and then further reacting the mixed polymer with either ethylene oxide or propylene oxide (7).

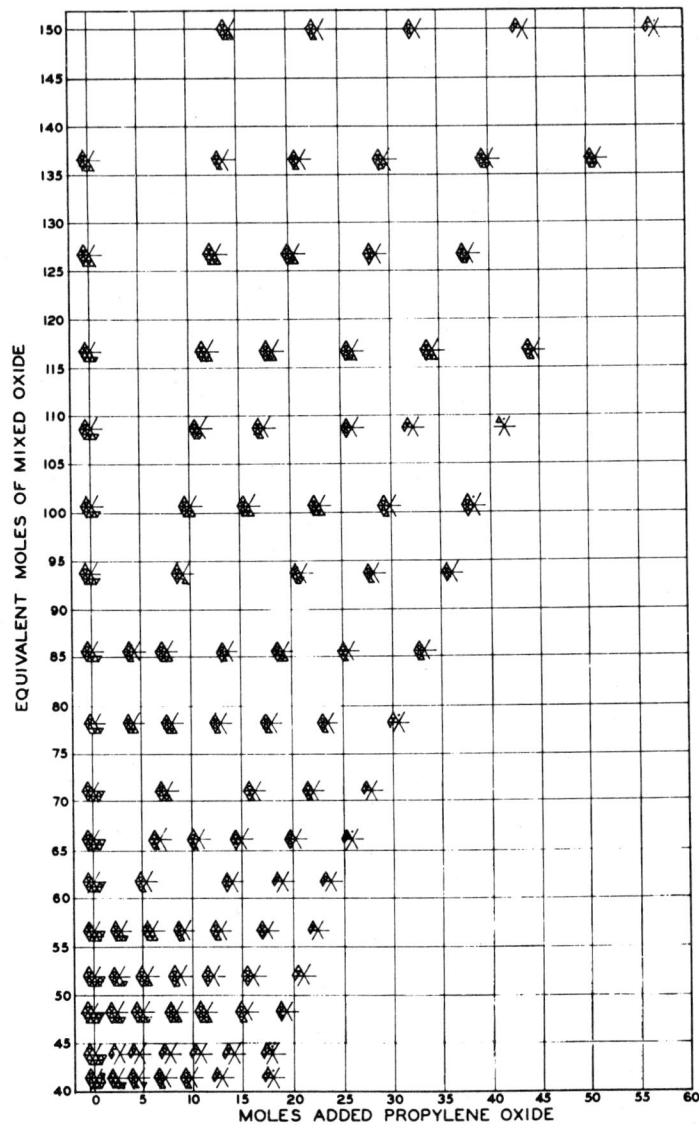

FIG. 10. A plane representing the plane EFGH of Fig. 9 (7).

2. MAKING AND BREAKING EMULSIONS

Having once determined the solubility profile for a class of compounds and plotted the data on an appropriate composition map, one can select at a glance members of the class having appropriate solubility properties for a given application, or one can predict what region should be investigated more intensively to achieve the desired solubility profile.

To give another example of the mapping technique, consider Fig. 11 (7). Here we are considering a large group of polymers that can be made by reacting a starting material with any conceivable mixture of four different alkylene compounds. The particular monomers under consideration are ethylene oxide, butylene oxide, propylene oxide, and propylene sulfide, the sulfur analog of propylene oxide. We are considering a group of materials made by reacting these monomers in any proportion in a four-step process. Obviously the number of different materials that can be made is enormously large, and the capability of designing a composition space that will hold all of them demonstrates the power of the method.

Since this is a four-step process, the composition space will be a noncommutative four-dimensional continuum. As soon as one says "four-dimensional," there is a tendency for the mind to snap shut. Actually there is no need to try to picture what this space "looks like." There is, however, a rather useful, if somewhat frivolous, analogy that can be made. Suppose we were to walk in the door of a large office building and find that the reception room was in the shape of an equilateral tetrahedron, with each of the four corners of the tetrahedron representing one of the pure monomers. Mixtures of the two monomers would be displayed on the connecting edge between the two appropriate corners, mixtures of three monomers would be displayed on the four flat walls of the tetrahedron, and mixtures of all four monomers would be represented by a point somewhere in the room, in exactly the same way that we use triangular and tetrahedral diagrams in physical chemistry.

Suppose you have the "address" of a particular compound such as that depicted in Fig. 11(a) and listed in Table 1 (7). Suppose

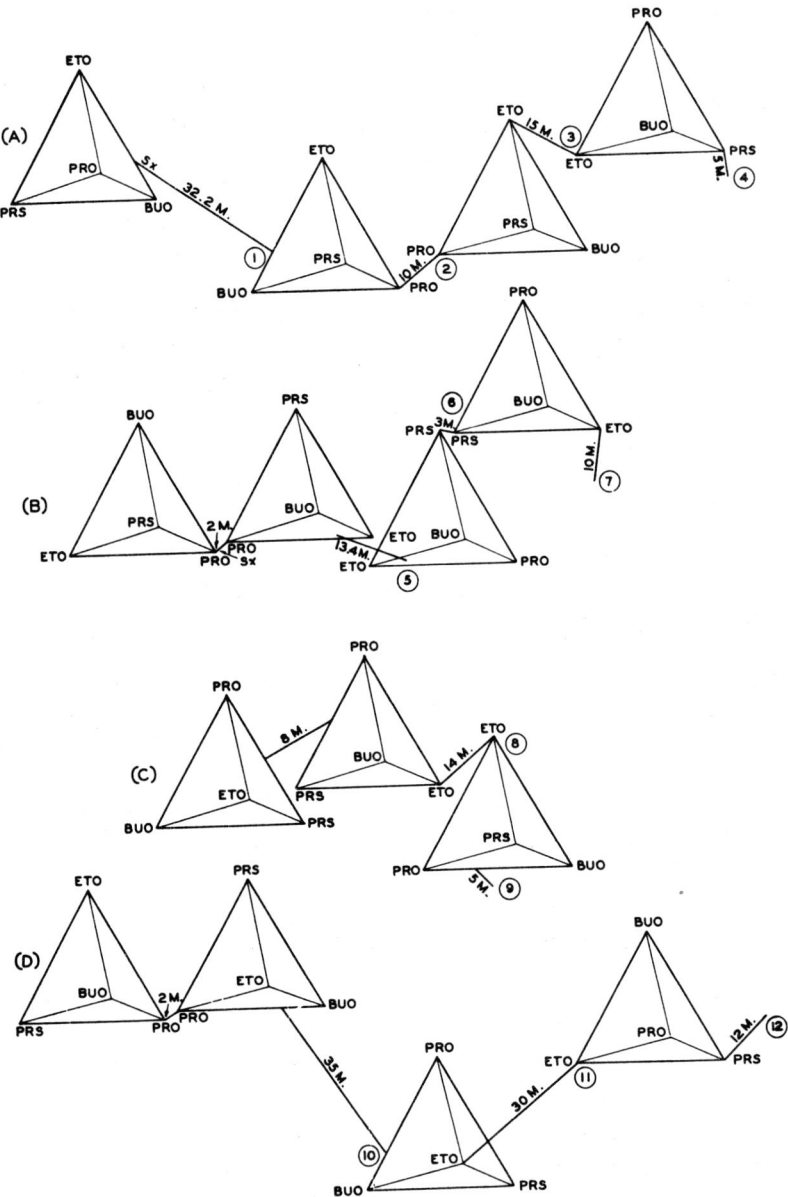

FIG. 11. The mapping of a large group of polymers that can be made by reacting a starting material with any mixture of four different alkylene compounds (ethylene oxide, butylene oxide, propylene oxide, and propylene sulfide) (7).

TABLE 1

Example	Step I Monomer composition[a,b]	Step I Moles added[c]	Step II Monomer composition[a,b]	Step II Moles added[c]	Step III Monomer composition[a,b]	Step III Moles added[c]	Step IV Monomer composition[a]	Step IV Moles added[c]
1	EtO-BuO, 7:3	32.2						
2	EtO-BuO, 7:3	32.2	PrO 100%	10				
3	EtO-BuO, 7:3	32.2	PrO 100%	10	EtO 100%	15		
4	EtO-BuO, 7:3	32.2	PrO 100%	10	EtO 100%	15	PrS 100%	5
5	PrO 100%	2	EtO-PrO-BuO, 6:2:2	13.4	PrS 100%	3		
6	PrO 100%	2	EtO-PrO-BuO, 6:2:2	13.4	PrS 100%	3	EtO 100%	10
7	PrO 100%	2	EtO-PrO-BuO, 6:2:2	13.4				
8	PrO-PrS, 1:1	8	EtO 100%	14	PrO-BuO, 2:1	5		
9	PrO-PrS, 1:1	8	EtO 100%	14				
10	PrO 100%	2	PrO-BuO, 3:7	35	EtO 100%	30		
11	PrO 100%	2	PrO-BuO, 3:7	35	Eto	30		
12	PrO 100%	2	PrO-BuO, 3:7	35	Eto	30	PrO 100%	12

[a] Abbreviations: EtO, ethylene oxide; BuO, butylene oxide; PrO, propylene oxide; PrS, propylene sulfide.
[b] All ratios given are on a weight-per-weight basis.
[c] Moles based on oxirane equivalents.

that the address were the one given in example 4 of Table 1. Picturing yourself again in the reception room, you would proceed to a point along the edge between the ethylene oxide and butylene oxide corners which would be seven-tenths of the way from the ethylene oxide corner to the butylene oxide corner. At this point, you open a door and step into an elevator, push the button labeled 32.2 moles and when the elevator stops, you get out and find yourself in another similar room. Looking again at your address, you proceed to the propylene oxide corner, enter an elevator, and proceed to the 10-mole floor. Again entering an outer room you proceed to the ethylene oxide corner and walk down a hall until you come to the 15-mole door. Again you enter a tetrahedral room, go through a door at the propylene sulfide corner, and at a 5-mole distance down the hall you will find the room holding the compound you seek. Obviously it is impossible to picture the sort of building in which one could take such a journey. It is, however, quite simple to learn to find one's way from one floor to another and from one compound to another. Figures 11(b), (c), and (d) represent the compounds of examples 7, 9, and 12 of Table 1.

To complete the analogy, the "reception rooms" are characteristic selection figures while the elevator and corridors are polymerization figures. It must be remembered that this is a deliberately complex example and that most practical cases will not require a tetrahedral selection figure, nor will they require as many successive steps.

In actual practice very few materials are prepared with as many successive additions. Most of the materials available on the open market are one- or two-stage materials and can therefore be depicted in very simple composition spaces.

D. Micelle Formation

We can summarize our techniques for choosing suitable emulsifiers by saying that, in general, emulsifiers function either by a

2. MAKING AND BREAKING EMULSIONS

charge-repulsion mechanism or by a film-forming mechanism, or in some cases by both mechanisms. The choice of the type of emulsifier to be used is, then, dictated by which mechanism is called for to produce the particular type of emulsion under consideration.

The picture is, however, somewhat complicated by the fact that many emulsifiers undergo micelle formation in solution. That is to say, if the emulsifier is dissolved in a solvent at very low concentration, it will exist in the solution as individual, discreet, emulsifier molecules. However, as the concentration of the emulsifier in the solvent is increased, the emulsifier molecules will tend to group themselves together in aggregates known as micelles. These micelles have been shown to exist in a number of forms, varying in complexity from one or two molecules held together by the mutual attraction of their hydrocarbon chains to large aggregates in a "lamellar" or "spherical" configuration. The preponderance of experimental evidence seems to indicate that at low concentrations simple head-to-head or tail-to-tail aggregates of a few molecules exist and that, as the concentration increases, the "micelle molecular weight" also increases.

It has also been shown that the same surfactant can produce different micelles when placed in different solvents. For instance, a long-straight-chain alcohol can be reacted with ethylene oxide to produce a polyether chain on the original hydroxyl group. If the length of this polyether chain with respect to the length of the hydrocarbon chain is appropriately adjusted, one can obtain a material that is soluble either in oil or in water, and the experimental evidence indicates that in the oil solution, the molecules in the micelle are oriented so that the hydrocarbon chains are on the outside of the micelle and the polyether portion is hidden from the oily solvent. Conversely, in an aqueous solution the micelles are so oriented that the polyether chains are on the outside and the hydrocarbon interior of the micelle is hidden from the solvent.

Further evidence as to this point of view can be found in the fact that if one takes an oxyethylated alkyl phenol of the right alkyl-chain length and appropriate polyether-chain length--for instance, if one treats a nonyl phenol with about 5 or 6 moles of ethylene oxide--one can produce materials that are soluble in kerosene at low concentrations, but soluble in the range from about 5 to 20% by weight. At concentrations higher then 20% by weight they are again insoluble in kerosene. Furthermore, if one adds about 4% by weight of one of these materials to a typical kerosene, the result will be a cloudy dispersion that will settle on standing. If, however, one adds a drop or two of water to this mixture and agitates it, the surfactant will dissolve to produce a bright micellar solution. This behavior can be explained by saying that below about 5 or 6% concentration in the kerosene the surfactant does not form micelles and the individual molecules are insoluble in the kerosene. In the range from about 6 to 20%, however, micelles are formed with a hydrocarbon exterior, and these are soluble in the kerosene. At concentrations higher than about 20% the micelle molecular weight increases to the point at which they will again become insoluble. The addition of water to a dilute mixture of the surfactant in kerosene results in the formation of a micellar solution apparently because the polyether chains orient themselves around small amounts of water, leaving the hydrocarbon portion of the surfactant oriented toward the kerosene. In effect the addition of a small amount of water has helped to enhance the orientation of the individual surfactant molecules into micelles and has thereby lowered the effective critical micelle concentration.

One must bear in mind, therefore, that a particular surfactant can exist in solution either as individual molecules or as a mixture of a number of different micelles and that in certain circumstances either the individual molecules or a particular species of micelle may be the effective emulsifying agent. Hence it is important that the emulsifier be given the opportunity to assume the optimum configuration in the process of preparing the emulsion. This is probably why some emulsifiers work best if added to the internal phase, whereas others perform better if added to the external phase.

2. MAKING AND BREAKING EMULSIONS

It is also why one frequently finds that below a certain concentration level a particular emulsifier suddenly loses effectiveness because below this level micelles do not form, and it is the micelles that in this case are functioning as emulsifiers, not the individual molecules.

This may also be why one frequently finds that a mixture of surfactants functions better than a single species since the mixture of surfactants provides a potentially greater variety of micelles which can adapt themselves to minor variations in the ingredients of the emulsion or to conditions to which the emulsion is subjected.

V. TECHNIQUES OF EMULSIFICATION

The choice of emulsifiers for a particular formulation will sometimes also be dictated by the method one intends to use in making the emulsion. Becher (11) cites four methods of adding one liquid to another to produce an emulsion. Other authors discuss what they refer to as the English and Continental methods of preparing emulsions. None of these approaches takes into account directly the type of emulsion being made, and often they simply assume that an oil-in-water emulsion of the low- or possibly medium-phase-ratio type is intended. We shall try to discuss techniques for making emulsions within the framework of our six emulsion types.

A. The State of the Art and Equipment

Actually in studying the literature concerning techniques of emulsification one finds very little advance in the art over an amazing period of time. For instance, in 1923 Clayton produced a classical book, *The Theory of Emulsions and Technical Treatment*, which went through successive editions for many years. In the third edition (12) he discusses stirring devices, colloid mills, and other apparatus, giving detailed diagrams and pictures of many pieces of equipment. In 1965 the second edition of *Emulsions, Theory and Practice* by Becher (1) was published. It, too, contains an excellent discussion of mechanical means for producing emulsions, particularly low-internal-phase-ratio ones.

The most striking observation that emerges from the study of these two books is that over a period of approximately 30 years very little real advance had occurred in the mechanical aspects of making emulsions. Most of the mechanical devices depend, for their effectiveness, on forcing a mixture of the two phases through small apertures to produce high-shear situations. This can be done by pumping the mixtures through small holes in orifice plates where a high pressure drop occurs or by forcing them between a tightly fitting rotor and stator. Most of the improvements seem to lie in the area of using less corrodable materials of construction and in designing the equipment so that it can be easily opened and cleaned.

1. Ultrasonic Equipment

One real innovation seems to have occurred in the period between these two books, and that is the development and commercialization of ultrasonic emulsifying equipment. In such devices the high shear is produced by passing the mixture of phases through an area where an ultrasonic field is developed. The acoustical energy may be generated either by a piezoelectric device or by a mechanical "whistle." In either event the effect is to produce rapid local variations in the pressure applied to the system, and as a result cavitation occurs. In this situation a very high local shear is produced as well as a fairly energetic shock wave. Also in some instances this method employs the principle of introducing a vapor of one phase into the other liquid. An extensive bibliography of more than 150 references is given at the end of Chapter VII in Becher's book (11).

For emulsions where heat-sensitive materials are involved the ultrasonic devices are coming into more frequent use.

2. MAKING AND BREAKING EMULSIONS

2. *The Colloid Mill*

When one gets into the area of medium- or high-internal-phase-ratio emulsions, where the "viscosity" of the formulation is much higher, it is increasingly difficult to achieve high shear levels, and in these situations the colloid mill is one of the most commonly used commercial devices. In a mill of this kind a rotor, often conically shaped and with grooves or other irregularities machined into its surface, rotates at a high speed in a conical cavity that fits very closely. Clearance between the rotor and stator usually can be adjusted and is on the order of a few thousandths of an inch. Since such devices are driven by high-horsepower engines and generate considerable heat, they must be so designed that this heat can be dissipated. The various machines differ primarily in the specific design of rotor and stator, and in the means used to dissipate the thermal energy

B. Low-Internal-Phase-Ratio Emulsions

As already mentioned in Chapter 1, most emulsion work is directed toward the production of low-internal-phase-ratio oil-in-water emulsions. We have pointed out that, for these emulsions to be stable, it is necessary to either match the density of the two phases or reduce the internal phase to very-small-size droplets. We have noted that ionic emulsifiers are usually most effective since they tend to prevent the particles from approaching each other and coalescing.

It is not very often that one has the option of matching densities between the two liquids since usually the liquids cannot be brought to the same density without the addition of undesirable components. Generally, therefore, one is required to depend on small particle size and charge repulsion to stabilize the emulsion. Most of the methods described for producing low-internal-phase-ratio oil-in-water emulsions concentrate on the production of small particle size. This can be done in a number of ways:

1. The most common method is to subject the system to very high shear. This is done by rapid stirring, by forcing the mixture through orifices, or by the use of such equipment as colloid mills.

2. Another much less frequently used technique is to introduce the internal liquid as a vapor into the water, kept at a temperature below the boiling point of the oily liquid. This condensation method can produce extremely fine emulsions and is indeed used to produce colloidal dispersions of solids, such as gold or copper.

3. Another technique consists of producing an emulsifiable concentrate where the surface-active agent is incorporated into the oil, which is then introduced into the water. The surfactant is one that has a strong affinity for the water phase and a tendency to produce oil-internal micelles. As the emulsifier attempts to pass from the oil to the water phase, the mass of oil is broken up and very small droplets of oil surrounded by hydrated emulsifier molecules are produced. Many agricultural emulsions are produced in this way. Some of the so-called soluble oils will produce droplets of such a small size that microemulsions result and the system appears to be optically clear.

4. Another way to obtain small particle size is to incorporate a water-soluble material like alcohol into the oil phase. When the oil is dispersed in the water, the alcohol will migrate out of the oil phase into the water phase, and the resultant droplets will be correspondingly reduced in volume. This technique is particularly useful when viscous oils are to be emulsified since ethyl or methyl alcohol can be used to reduce the viscosity of the oil phase as well as help produce small particle size.

5. Another way of achieving very small particle size is known as the phase-inversion method, often used for producing wax-polish emulsions. In this procedure an amine salt emulsifier is often used. A fatty acid (e.g., oleic acid) is added to the mixture of waxes and oils, and this mixture is heated to just below the boiling

point of water (if a pressure vessel is used, temperatures above 100°C may be employed). An amine, such as triethanolamine, is added to the water phase, and the water is heated to just below its boiling point. The water is then added to the oil with vigorous stirring, and at first a water-in-oil emulsion results. As it approaches 75% internal phase, this emulsion becomes quite thick and creamy. Since the emulsifier used is not effective in producing high-internal-phase-ratio emulsions, the thick water-in-oil emulsion inverts suddenly, resulting in a fluid oil-in-water emulsion with very small droplet size. The rest of the water is then added at a lower temperature to cool the wax below its solidification point.

In the low- and medium-internal-phase-ratio range the techniques required for the production of water-in-oil emulsions are much the same as those required for the oil-in-water types. However, water-in-oil emulsions are less common, and the availability of emulsifiers is somewhat more restricted. Because the oily external phases are usually more viscous and, of course, less electrically conductive, the techniques for obtaining small particle size and achieving satisfactory particle charge are somewhat different. In general, however, the same principles apply.

C. High-Internal-Phase-Ratio Emulsions

When one gets into the area of high-internal-phase-ratio emulsions, the available published art is greatly restricted, and to our knowledge, no one has extensively marketed a device specifically for producing such emulsions. The principles involved are indeed somewhat different from those encountered in lower-phase-ration emulsions.

Whereas high shear is frequently used in the production of low- and medium-phase-ratio emulsions in order to obtain small particle size, it is not required in the production of high-internal-phase-ratio types. The use of excessively high shear rates may result in either failure to prepare an emulsion or preparation of a loose, unstable emulsion of the conventional type with the reversed external phase. Mixing techniques should be used which provide thorough mixing at moderate shear rates. In particular, provision should be made for efficient mixing as the "thickness" increases.

The internal phase must be added to the external phase a little at a time. If the two phases are simply placed in a container and mixed, the result will frequently be a low-internal-phase-ratio emulsion of the reverse type. Two convenient laboratory techniques for the preparation of high-internal-phase-ratio emulsions are quoted below from a technical brochure distributed by the Petrolite Corporation (13).

Method 1

a. Equipment Required. Hobart Kitchen Aid Mixer, Model 3C or equivalent.

b. Emulsification Procedure. Place 10 ml of the appropriate external phase in the bowl of the mixer, and with a speed setting of 1 to 2 add the internal phase in 5-ml increments, mixing thoroughly between additions, until at least 50 ml of internal phase has been added. Additional internal phase may then be added continuously in a slow stream until the desired phase ratio has been obtained. With suitably chosen emulsifiers, phase ratios of 0.98 are readily obtained.

Method 2

a. Equipment Required. Tall-form (dye pot) beakers of 400-ml capacity, split-disk stirrer, and variable-speed motor.

The apparatus is so set up that the tall-form beaker is clamped securely and the split-disk stirrer is situated as close to the bottom of the beaker as possible. Provision should be made for raising the stirrer during the course of the emulsification.

2. MAKING AND BREAKING EMULSIONS

b. <u>Emulsification Procedure</u>. Place 5 to 10 ml of an appropriate external phase in the tall-form beaker and set the stirring motor to stir at a low speed. Avoid excessive splashing. Add 2 to 3 ml of the desired internal phase and allow the mixture to stir until homogeneous. With thorough mixing, add additional 2- to 3-ml increments until approximately 25 ml of internal phase has been added. At this point the emulsion should have the consistency of thick cream. Internal phase may now be added more rapidly; be careful never to add a volume of internal phase larger than the amount of emulsion already present in the beaker. Addition of internal phase is continued until the desired phase ratio is obtained. As the level of liquid increases in the beaker, the stirrer should be raised and the speed increased to ensure thorough mixing. Avoid excessive incorporation of air. When phase ratios in excess of 90% are obtained, the material will have the appearance of a stiff gel.

Both of these laboratory methods are essentially batch processes. Attempts have been made to produce emulsions of this kind in commercial quantities. Several laboratories have been successful in using essentially method 1 with larger commercial dough-processing equipment. These processes have been successful except that great care must be taken not to lose the more volatile portions of either phase, and precautions must be made to prevent contamination of the batches by airborne solid particles.

In an attempt to avoid the deficiencies of batch processing, some experimenters have attempted to prepare high-internal-phase-ratio emulsions in a stage-wise process. In such a system the external phase is metered into a mixing chamber and slightly less than an equal amount of the internal phase is introduced at the same time. A moderate-shear, paddle-type mixer or a series of screens and baffles have been proposed as mixing devices. It has also been suggested that the two metered streams could be mixed and fed into the intake of a circulating-pump loop. The result is a medium-internal-phase-ratio emulsion of approximately 50% internal phase. This emulsion passes from the first mixing stage into a second stage, where again approximately an equal amount of internal phase is added. The roughly 66% internal-phase emulsion passes to a third stage and as many more stages as required to achieve the desired phase ratio. Equipment of this type has been constructed and operated, but the instrumentation required to meter liquids to the various stages results in an excessively complicated process.

Another method of continuously producing high-internal-phase-ratio emulsions is described in a patent (14). In this process a mixing chamber is filled with a previously prepared emulsion of approximately the desired composition. Suitable stirring means are provided so that this emulsion is subjected to shear rates that will reduce its effective viscosity close to that of the separate phases to facilitate mixing. The shear level is kept below the inherent shear-stability point of the particular emulsion being produced. The internal and external phases are simultaneously admitted in the appropriate ratios to the mixing chamber, and the finished emulsion is displaced thereby. Pilot-plant models of equipment employing this principle have been prepared and operated at production rates of approximately 25 gal/min. Other devices of this type have been field tested at rates of 50 to 200 gal/min. Descriptions of such devices are given in both Ref. 14 and in a British patent (15).

D. The Geometric Approach

It has been said that the manufacture of emulsions is as much an art as a science, and indeed one might expect this to be partially true because, as we have pointed out in Chapter 1, emulsions are "constructs." The properties of an emulsion are determined as much or more by the structure and the way it is put together as by the actual constituents. The geometric approach emphasized in Chapter 1 clearly points out that often the structure of the emulsion has more effect on the properties than any other single factor.

2. MAKING AND BREAKING EMULSIONS

This means that, as a particular individual becomes familiar with the idiosyncrasies of the materials and equipment available to him, he will develop a subtle skill in combining these factors to produce the exact results he wants. There is nothing really mysterious about this process. It occurs in all fields. One of the advantages of the geometric approach to the study of emulsions is that it allows one to rationalize some of these skills and hence to acquire them more readily and transmit them to others.

VI. TECHNIQUES FOR BREAKING EMULSIONS

Having considered the reasons for wanting to prepare emulsions and the techniques whereby emulsions can be prepared, we shall now consider undesirable emulsions that occur in certain situations and the techniques for breaking them.

Whenever two immiscible liquids come together in nature, there is a tendency for emulsions to form. In certain instances this is desirable and even fortunate. For instance, fats and oils are emulsified in the digestive tract, and small globules of water-immiscible materials are transported in the bloodstream in an essentially emulsified form. It has been shown that oily substances eventually become emulsified and that the emulsification process speeds up their degradation by microorganisms.

There are, however, many instances in which emulsions are troublesome. When a crude-oil deposit is found in the ground, it is usually trapped in a porous formation that is sealed at the top by a more impervious stratum. The porous formation usually has both oil and water in it. Frequently the water phase contains dissolved salts and is a brine. When petroleum is produced from an oil well, often both oil and water flow into the gathering system. If this mixture of oil in water passes through a partially open valve or a choke, it may become emulsified into a very stable, usually water-in-oil, emulsion. In the early days of the petroleum industry this was called "roily" oil. A number of detailed accounts of methods of breaking petroleum emulsions have been written (16-20). Several books also contain chapters on this subject (21, 22). There is little to be gained by repeating most of this material in detail here. However, a brief discussion of the basic principles may be useful since these principles also apply in areas other than the petroleum industry.

As already stated, oil-filed emulsions are usually of the water-in-oil type and are usually of the low-internal-phase-ratio type. Common oil-field emulsions most frequently contain some 5 to 30% water. However, emulsions containing considerably more water than this are known. Less frequently the emulsions are of the oil-in-water type, again usually of low internal-phase ratio. These are known as "reverse emulsions" in the petroleum industry. It should be noted that what the petroleum industry calls "reverse emulsions" most of the rest of the world considers to be normal emulsions, and the water-in-oil emulsions common in the petroleum industry are considered "inverts" by many other people.

Oil-external emulsions can be broken by electrical as well as mechanical and chemical procedures. The electrical processes are similar to the well-known Cottrell precipitators used in zinc smelters and other applications to remove fine dust particles from a gas stream. The oil emulsion is passed between a series of charged electrodes, and the electric field induces charges on the water particles. These then tend to coalesce and settle out.

Many oil-field emulsions can be broken by simply subjecting the emulsion to heat to reduce the viscosity of the crude oil and then letting the water settle out. Sometimes the emulsion is passed through a chamber containing a large amount of porous, water-wetted solids, originally hay or straw. The water droplets tend to adhere to the water-wetted surfaces and coalesce, and the bulk water drains to the bottom of the vessel. Combinations of "hay tanks" and "heater treaters" are common in the oil fields.

2. MAKING AND BREAKING EMULSIONS

With the recent emphasis on water pollution, a number of emulsion problems have come to light, and the problem of small amounts of oily materials emulsified in effluent waters is beginning to receive considerable attention.

In liquid-liquid processes where one liquid stream is used to extract a desired ingredient from another liquid stream, troublesome emulsions are often encountered. We will now try to see how our emulsion concepts can be applied to breaking or preventing emulsions.

We have seen that, if two pure liquids are shaken together, one of the liquids will break into small droplets that become dispersed in the other liquid. However, when the mixture is allowed to stand, the droplets separate by gravity and coalesce, so that two separate phases result. We have shown that emulsions are stabilized by small particle size, charged particles, and surface-active agents, which form "skins" around the droplets. In order to prevent emulsions, therefore, one should avoid high-shear situations in which many small droplets are produced and should take steps to prevent the introduction of materials that will contribute a charge to the particles or concentrate at the interface to prevent coalescence.

It is often difficult to avoid situations in which shear occurs. Moreover, we are unfortunately often presented with circumstances in which the exclusion of all surface-active agents is impossible. We are thus usually forced to try to counteract the effect of the surface-active agent and then allow the mixture to stand quiescent until separation occurs.

We can summarize a few general principles as follows:

1. Wherever possible, it is better to avoid making an emulsion than to try to resolve it once formed.
2. In dealing with mixtures of liquids, situations where high shear can occur should be avoided. Examples include sharp corners or rough interiors in piping systems, partially opened valves, and coarse screens.

3. The introduction of surface-active agents or finely divided solids should be avoided if possible.

A good deal of present-day engineering fails to take into account these principles in designing sewers, drains, and other waste-collecting systems. All kinds of waste materials are indiscriminately dumped into a common collecting system, which may contain sharp angles, free drops of several feet, and transfer pumps whose output is controlled by partially closed valves. All these factors tend to intimately mix the materials, resulting in the production of extremely stable emulsions. In many instances a good deal of costly equipment would not be needed and time-consuming settling could be avoided if oily wastes were segregated and handled in a separate system.

Even with the best engineering, however, in many naturally occurring situations emulsions cannot be avoided and must be broken by separating the two phases. The breaking of emulsions consists of two steps that may occur separately or simultaneously. First, it is necessary to devise some means of bringing the dispersed droplets into contact with each other, and then a means must be found to allow the droplets to coalesce so that the second phase can be separated. One of the most common ways of bringing the droplets together is to allow the emulsion to stand undisturbed for a period of time. The dispersed phase either settles to the bottom or rises to the top of the tank, and in this concentrated "creamed" layer the droplets are brought into contact. However, when forced into contact, the droplets may not coalesce rapidly if their surfaces are highly charged or if they are covered with a film of surface-active material.

The charge on the droplets may be removed by adding a material with an appropriate opposite charge. This neutralizing material can be an inorganic acid or base, or an appropriately charged organic molecule. Considerable work has been done in studying the charges on dispersed solid or liquid particles. This is often

2. MAKING AND BREAKING EMULSIONS

referred to as the ζ (zeta) potential approach (23). Care must be taken in adding a neutralizing agent to be sure that just the right amount is used to neutralize the charge. If an excess is added, a charge of opposite sign may be produced on the particle, and an even more stable dispersion can result.

If an emulsion is stabilized by a film-forming surfactant, it is necessary somehow to drive this material away from the interface. This can be done either by chemically reacting with the material or by changing its solubility in one or more of the liquid phases. For instance, most materials are more soluble hot than cold. Therefore emulsions can be broken in some cases simply by subjecting them to the action of heat. When the emulsion is heated, the materials in the surface film become more soluble in the bulk liquid phase and therefore tend to migrate away from the interface, allowing the droplets to coalesce. At the same time the increased temperatures result in a lower viscosity, which also helps to promote the separation of the two phases.

The interfacial film can also be disrupted by adding a cosolvent to the system. For instance, certain materials like acetone or methanol are often soluble in both the oil and the water phases. If added to an emulsion, such materials may tend to dilute the interfacial layer and cause some of the surfactant to migrate into the bulk phase, thus destabilizing the emulsion.

Let us consider for a moment how we shall apply emulsion-breaking principles to the various types of emulsion. We have said that the most common emulsions are of the low-internal-phase-ratio oil-in-water type, and indeed these are the ones that create most of the problems in industrial waste recovery or effluent cleanup. When encountered in industrial situations, most of these emulsions are stabilized by ionic emulsifiers, usually of the anionic type. Most industrial detergents and cleaning formulations contain sulfonates or suflates or soaps, and all these anionic materials are quite effective in stabilizing low-internal-phase-ratio oil-in-water emulsions.

Consider for a moment a typical plant problem. Let us say that a manufacturing plant has a sewer system where all waste materials are discharged. Most of these waste streams are essentially aqueous, but they may contain small amounts of a wide variety of materials. Oily materials spilled in the course of operating the plant are usually flushed down the sewer drain, and the detergent solutions used for cleaning up are also flushed down the drain. If lubricants leak from bearings or gear boxes, they often eventually find their way to the sewer system. To further complicate the situation, massive spills of oily materials may be cleaned up by spreading dry adsorbants over the spill and then disposing of the oil-saturated solids. Ironically, these solids are often again flushed into the sewer system. On top of all this, sanitary sewage from the various buildings is also added to this stream.

Suppose the operators of the municipal sewate system now decided that it cannot accept this waste stream if it contains over a certain amount of oily wastes. The manufacturer is faced with the problem of trying to clean up his own effluent. He finds that, if the material is run to a settling pond, most of the solids will settle to the bottom and an oily layer will rise to the top. The oily layer will contain some oil-wet solids and some emulsified water, but it may be possible to burn it in an appropriately designed incinerator. The solids that settle to the bottom will be a mixture of inorganic and organic materials, and may also have entrained a substantial amount of oily materials. If these solids are drawn from the bottom of the settling system, they may contain too much oil to allow them to be used as landfill. Furthermore, many of the more toxic organic materials, such as chlorinated hydrocarbons, will be concentrated in this heavier fraction. Even if the manufacturer is able to find a place where the solids can be safely used as landfill and also has found a way to burn the oily layer acceptably in an incinerator, he often still finds that the effluent now consists of a low-internal-phase-ratio oil-in-water emulsion and that even a protracted settling time will not effect an acceptable separation.

2. MAKING AND BREAKING EMULSIONS

He now contacts experts in the area of emulsion breaking, and they recommend that he completely redesign his system. He, of course, objects. He says, "We are satisfactorily removing 90% of the material and have spent money for landfill and incinerator facilities. All we want you to do is tell us how to get out this last small amount of oil." We now find that the emulsion-breaking expert and the manufacturer are caught in an all too common trap because no one applied basic separation principles early enough in the process of trying to solve the problem.

Let us go back again to the point where the operators of the municipal sewage system required the manufacturer to institute a cleanup of his effluent and suppose that we first conduct a survey to determine where oily materials, toxic materials, and solids are being introduced into the disposal system. Having done this, we may find that the greater part of the load on the system is actually contributed by one or two operations in the plant and that, if the waste materials from these processes are segregated at the source and not allowed to enter the main waste system, the problem may never develop. The manufacturer may still have to provide landfill or incinerator facilities to dispose of the wastes, but the troublesome low-internal-phase-ratio water-in-oil emulsions may be either eliminated or greatly reduced.

Let us look briefly at another similar situation. Suppose a manufacturer is cleaning a large number of greasy objects and that the wash water from this process is delivered to a sewer system and the oil must be separated from the system before the waste water can be discharged. Surfactants have been used in the cleaning operation and, by their very nature, have resulted in the emulsification of the oily materials in the water. Let us suppose that about 20% oil is emulsified loosely in a water stream. The usual process is to conduct this stream to a pond or tank where it can remain quiescent for a period of time and then skim as much of the oily material as possible from the surface. The underflow, if it still contains more than an acceptable amount of oil, is then further treated. This procedure in itself is often contrary to basic emulsion principles. When we have 20% oil in the water emulsion, the droplets of oil are so numerous that, if we can arrange for their coalescence when they touch, the coalescing droplets will tend to sweep through the emulsion and bring all the oil into the coalesced phase. If, however, we separate the more loosely held oil first, the remaining stable droplets are, on the average, further apart, less likely to come into contact, and hence more difficult to coalesce. It follows, then, that, if an emulsion-breaking chemical is required, it is often best to add it at the beginning so that the more loosely held oil particles can help scavenge the more troublesome ones and a complete separation is achieved in one stage.

In considering the resolution of low-internal-phase-ratio oil-in-water emulsions we can see that two steps are necessary. First, we have to counteract the effect of the emulsifying agent so that the particles, when they come into contact, will coalesce. Then we have to provide a means for bringing the droplets of oil into contact with each other. Since many of the low-internal-phase-ratio oil-in-water emulsions that are encountered are stabilized by ionic surfactants, it is often possible to destabilize them by adjusting the pH of the water or by adding a material that will form an insoluble precipitate with the surfactant. Thus such materials as lime, clay, or ion-exchange resins can be effectively used to remove the surfactant from the system. Care must be taken in this process that only what is needed is used since an excess of reagents can often cause new difficulties. In situations where it is not economical to make a massive pH adjustment proprietary emulsion breakers may be effective. These materials are often very specific in their action, and few general rules can be given for their selection. It is best to rely on the recommendations of the supplier.

Once the treatment required to inactivate the surfactant has been determined, it is necessary to add the appropriate amount of reagent in such a way that local excesses are avoided and the

2. MAKING AND BREAKING EMULSIONS

reagent is brought into contact with all the emulsion particles. This may require the addition of a dilute solution of the reagent or the addition of the reagent to the stream just ahead of a transfer pump or other mixing devices. In designing such procedures one must be careful not to produce an emulsion of the opposite type.

Once we have destabilized the individual emulsion droplets, it is still necessary to bring them into contact in order to affect a separation. As already pointed out, this is easier to do as the phase ratio increases. It is also, obviously, easier to do as the particle size increases. Here the addition of heat may help in reducing the viscosity, increasing the density difference, and promoting thermal agitation. In some situations the incoming treated emulsion can be introduced into the settling chamber at a point where a substantial number of larger coalescing droplets have already formed, thus making it possible to use partially separated oil as a sweeping agent for the incoming fine droplets. This is equivalent to the formation of a "floc blanket" in solids-separation devices.

Another way to bring the droplets into contact is to pass the destabilized emulsion through a chamber containing a large number of oleophilic surfaces. This is essentially the same principle as that used in the hay tank mentioned in conjunction with the resolution of oil-field emulsions. In this process the oil droplets stick to the already oil-wet surface and coalesce to larger droplets, which then drain from the network of oily materials. The same effect can sometimes be achieved by placing in the line of flow coarse screens that are preferentially oil wet. These screens will allow the water to pass, but will tend to hold back the oil.

An exactly opposite approach is that where the destabilized emulsion is led to a T-shaped chamber. Each of the arms of the T is closed by a porous membrane, one that is preferentially water wet and the other preferentially oil wet. Here the water passes through the water-wet membrane and the droplets of oil are held back while the oil can pass through the oil-wet membrane and the water is barred. This system will operate satisfactorily if adequate means are provided for maintaining the membranes appropriately wetted and for draining off the excess oil that can accumulate on the water membrane. Usually such devices are not capable of high throughput.

In situations involving low volumes of waste emulsions or low oil content, the oil can actually be filtered out by using a water-wet fibrous bed. Experiments with this type of operation are described in a report issued by the U.S. Environmental Protection Agency (24). This project tested the feasibility of using fiberglass mats as a filter medium, and it was concluded that, provided the throughput rate was kept within appropriate limits and the absorptive capacity of the medium was not exceeded, essentailly 100% coalescence could be achieved. It was estimated that the cost of cleaning up effluent streams by this technique would range from 1 to 13 cents per 1000 gal, depending on how often the filter medium could be reclaimed and reused.

In general the same basic principles apply to the resolution of low-internal-phase-ratio emulsions whether they are oil or water external, with the exception that electrical resolution methods are limited in most cases to oil-external systems.

Turning to the medium-internal-phase-ratio emulsions, we find a number of commonly occurring problem emulsions. As already mentioned, it is common practice to allow streams containing mixtures of immiscible liquids to pass into settling basins or tanks where they can be at least partially separated by gravity. When this is done, it may be possible to draw relatively clean liquids from the bottom and the top of the tank, but an emulsion layer gradually accumulates at the interface. This emulsion layer can be extremely difficult to resolve since the emulsifying agents are concentrated and the phase ratio is usually high enough for the emulsion to be quite viscous. Wherever possible, it is preferable to avoid the formation of such interface layers rather than to try resolving them separately. This is another situation in which the

2. MAKING AND BREAKING EMULSIONS

emulsion expert is often called in and told that present methods are achieving a better than 80% cleanup and that "all they need" is a way of getting out the remaining 20%. In an even worse, but all too frequent, situation the small amount of interface layer is drawn off and stored in a separate vessel. Here it is allowed to accumulate and age until there is enough of it to create a problem. By this time resolution can be quite difficult.

If the oil in the interface layer is sufficiently valuable, the problem may be solved by distillation, but in the case of most waste oils the cost of distillation is too high. Methods have been studied whereby these thick emulsions are diluted with fuel oil and burnt in incinerators. This obviously resolves the emulsions.

In situations where low-internal-phase-ratio emulsions have been allowed to "cream" so that the emulsion droplets form a concentrated layer without coalescing we are faced with the problem of resolving a medium- to high-internal-phase-ratio emulsion. Emulsions of this type, whether occurring accidentally or deliberately, are usually stabilized by a film-forming surface-active material. It is therefore necessary to disrupt the film around the particles before resolution can be achieved. The films around the particles vary considerably in strength and rigidity. Some films occurring in oil-field emulsions have such mechanical strength that, if a layer of oil is poured over a layer of water and allowed to stand, the skin that forms at the interface can be removed intact and studied. For emulsions stabilized by such rigid films a combination of mechanical and chemical means is required. By adding a chemical that will dissolve in the interfacial film and weaken it, and then subjecting the emulsion to mechanical agitation, resolution can often be accomplished. However, a small amount of solid skins may be found dispersed in one or the other of the phases.

A number of "froths," "foams," and "scums" that occur in water-treatment plants and cause considerable trouble prove on examination to be medium- to high-internal-phase-ratio emulsions with solid particles and gas bubbles trapped in them. Since these compositions seldom contain much material of economic value and often contain too much water to be directly incinerated, they present a very serious cleanup problem. Many possible solutions to the resolution of these materials have been suggested. None of them has proved to have really universal application. This is an area where more work needs to be done. Recently attempts to resolve these compositions by freezing them and then subjecting them to a mechanical crushing operation have shown some promise.

The resolution of high-internal-phase-ratio emulsions has not received much attention for two reasons. First, the existence and natural occurrence of such emulsions was not generally acknowledged, and second, the recognition of the problem is so recent that the amount of work that has been done is small.

High-internal-phase-ratio emulsions may be pictured as foams in which the gas bubbles have been replaced by liquid droplets. It has been shown that these droplets are polyhedral and are separated from each other by thin, stable films (25). Where the emulsions are stabilized by ionizable materials such as fatty amine salts or amine soaps, the emulsions can be broken by changing the pH of the aqueous phase. Thus it has been found that high-internal-phase-ratio emulsions stabilized by an amine acetate emulsifier can be resolved by adding ammonia to the system (26). Where the stabilizing materials are essentially nonionic, a cosolvent is usually required. In situations of this type the addition of materials like methanol or acetone is often effective.

High-internal-phase-ratio emulsions, particularly those with more than 90% internal phase, may be pictured as being analogous to a conventional foam where the gas has been replaced by a liquid. Just as foams can be broken down by mechanical action, so it is found that, if high-internal-phase-ratio emulsions are subjected to extremely high shear, the individual films can be disrupted and at least partial resolution will result.

Since the external phase constitutes a relatively small portion of the total formulation, it is possible to adsorb it on a solid substrate and thereby resolve the emulsion. A patented process (27) combines shear with an adsorbing agent to secure emulsion resolution. In this process an insoluble breaker-adsorber, such as a montmorillonite type of clay or an inorganic salt that will form a hydrate, is brought in contact with the emulsion under conditions where thorough mixing is obtained. The solid material adsorbs the water and the surface-active agent, thus breaking the emulsion. Resolution can be completed either by allowing the solid material to settle or by passing the broken emulsion through a suitable filter where the water-containing solid is effectively separated from the internal phase. The use of a suitable clay with this method is particularly effective since it also removes the surface-active material from the system and tends to avoid the formation of water-in-oil emulsions. If the economics of the situation are advantageous, the surfactant can be recovered by extraction from the clay.

Certain medium- to high-internal-phase-ratio emulsions occurring as froths or scums in water-treatment systems, particularly those employing flotation techniques, can be resolved by freezing the emulsion and then subjecting the frozen slush to mechanical action, thus disrupting the external-phase films. The processed material is then allowed to melt, and the two phases are separated.

REFERENCES

1. P. Becher, *Emulsions: Theory and Practice*, 2nd ed., Reinhold, New York, 1965.
2. F. L. Moilliet, B. Collie, and W. Black, *Surface Activity*, 2nd ed., Van Nostrand, New York, 1961.
3. J. W. McCutcheon, *Detergents and Emulsifiers Annual*, J. W. McCutcheon, Inc., Morristown, N. J.
4. W. C. Griffin, *J. Soc. Cosmetic Chemists, 1*, 311 (1949).

5. Atlas Chemical Industries, Inc., *The Atlas HLB System*, 2nd ed., Wilmington, Del., 1963.
6. K. J. Lissant, *J. Chem. Doc.*, *3*, 103 (1963).
7. K. J. Lissant (to Petrolite Corp.), U.S. Pat. 3,083,232 (1963).
8. K. J. Lissant (to Petrolite Corp.), U.S. Pat. 3,539,406 (1970).
9. W. J. Wiswisser, *A Line-Formula Chemical Notation*, Crowell, New York, 1954.
10. K. J. Lissant (to Petrolite Corp.), U.S. Pat. 3,352,109 (1967).
11. P. Becher, op. cit., Chap. VII.
12. W. Clayton, *The Theory of Emulsions and Their Technical Treatment*, 3rd ed., Blakiston, Philadelphia, 1935, Chap. XI.
13. Petrolite Corp., St. Louis, Mo., mimeographed instruction sheet, undated.
14. K. J. Lissant (to Petrolite Corp.), U.S. Pat. 3,565,817 (1971).
15. Petrolite Corporation, British Pat. 1,227,346 (1971).
16. C. M. Blair, *Chem. Ind. (London)*, 538 (1960).
17. I. Pincus, K. F. Ockert, and C. R. Kinney, *Ind. Eng. Chem.*, *43*, 521 (1951).
18. L. T. Monson and R. W. Stinzel, *Colloid Chemistry*, Reinhold, New York, 1931, p. 538.
19. L. T. Monson, *Pet. World*, *43*, No. 4, 41 (1946).
20. C. M. Blair, *Oil Gas J.*, *44*, 416 (1945).
21. P. Becher, op. cit., p. 367 ff.
22. W. Clayton, op. cit., Chap. XII, p. 355 ff.
23. T. M. Riddick, *The Control of Colloid Stability through Zeta Potential*, Livingston, Wynnewood, Pa., 1968.
24. Illinois Institue of Technology, Department of Chemical Engineering, *Experimental Evaluation of Fibrous Bed Coalescers for Separating Oil-Water Emulsions*, Water Pollution Control Research Series, 12050, U.S. Environmental Protection Agency, Washington, D.C., November 1971.
25. K. J. Lissant and K. Mayhan, *J. Colloid Interface Sci.*, to be published.
26. J. C. Harris and E. H. Steinmetz, *Optimization of JP-4 Fuel Emulsions and Development of Design Concepts for Their Demulsification*, USAAVLABS Tech. Report 68-79, p. 79.
27. K. J. Lissant (to Petrolite Corp.), U.S. Pat. 3,378,418 (1968).

Chapter 3

MICROEMULSIONS

Leon M. Prince

Research Center
Lever Brothers Company
Edgewater, New Jersey

I.	INTRODUCTION.	126
II.	IDENTIFICATION.	127
	A. Definition.	127
	B. Commonplace Examples.	134
	C. Recognition	139
III.	THEORY.	145
	A. Historical Background	145
	B. The Transparent Emulsion.	148
	C. Negative Interfacial Tension.	150
	D. Graphical Characterization.	154
	E. Mechanism of Curvature.	157
	F. Recent Developments	159
IV.	PRACTICAL APPLICATIONS.	161
	A. Emulsifier Selection: The HLB Scheme	162
	B. Rheological Applications.	168
	C. Hints on Formation of Microemulsions.	169
	D. Microemulsification as an Analytical Tool	173
	REFERENCES.	175

I. INTRODUCTION

Microemulsions are an extraordinary kind of emulsion that forms spontaneously. Products consisting of these systems are prized for their stability and small particle size, which gives them special consideration in the marketplace.

It is the object of this chapter to describe the physical properties that identify these systems as microemulsions, to offer a theoretical explanation of the special conditions required for their formation, and to discuss some practical applications of this theory. In this endeavor there will be some unavoidable overlapping of the subject matter of other chapters, but it is hoped that the outlook of the present chapter will put the microemulsion phenomena discussed elsewhere in a fresh perspective.

The tools of the microemulsion formulator, as with the formulator of macroemulsions, are to be found in the disciplines of physics as well as chemistry. In the case of microemulsions, however, the interest in physics is at a dimensional level one order lower, involving significantly different phenomena, than with macroemulsions. For this reason the elementary principles of light scattering and sedimentation are presented as they apply to small spherical droplets. This knowledge is essential for the recognition of microemulsions and for the perspective needed to set feasible goals for their performance.

Finally, one cannot discuss emulsions without recalling the outstanding contributions made by the late Professor Jack H. Schulman, O.B.E., at his death Stanley-Thompson Professor of Chemical Metallurgy, Henry Krumb School of Mines, Columbia University, and formerly Director of the Ernest Oppenheimer Laboratory, Department of Colloid Science, University of Cambridge. His original application of studies on the Langmuir film balance to the theory of emulsions set the stage for a much more complete understanding of the subject. He also recognized that the transparent "soluble oil" concentrates of commerce consisted of spherical micelles having a core of water and a mixed monomolecular film of emulsifiers as the stabilizing membrane. After identifying such transparent, optically isotropic, dispersions of water and oil as emulsions, he coined the term "microemulsions" to describe them. Even at this late date implications of his experimental acumen are still being evaluated, as is illustrated in the discussion of the theory in Section III.

II. IDENTIFICATION

Microemulsions are defined in thermodynamic terms, and examples are given of such systems as they are utilized, in whole or in part, in industrial and consumer goods. Their optical properties and their behavior in gravitational fields are described to facilitate easy recognition.

A. Definition

Microemulsions are stable dispersions of one liquid in another in the form of spherical droplets, the diameters of which are less than one-quarter the wavelength of white light (0.25 λ), or approximately 0.14 μ (1400 Å). Although this size requirement means that light can pass through it, the system is not necessarily transparent. ("Translucency" is the general term and includes "transparency," which may be described as excellent translucency.) Indeed, many microemulsions (e.g., carnauba-wax or cutting-oil emulsions) exhibit pronounced Tyndall scattering. The essential feature of the microsystem is that a mixed film adsorbs to the interface between the oil and water phases, creating a transient, negative free energy and, causing the adsorbed monolayer to spontaneously achieve zero interfacial tension. This ensures that the system will remain disperse and will not, as macroemulsions do, achieve equilibrium by separating into the original, mutually insoluble liquid phases.

1. MICRO VERSUS MACRO SYSTEMS

Our definition puts microemulsions in the same category as soap solutions or colloidal dispersions. The aggregates, or micelles, of these systems are invisible to the human eye in the light microscope and assume their condition of equilibrium reversibly. Until recently these systems of oil-in-water (o/w) or water-in-oil (w/o) microdroplets were generally considered to be outside the realm of emulsions. Indeed, Becher (1) defined an emulsion as "...a heterogeneous system, consisting of at least one immiscible liquid intimately dispersed in another in the form of droplets, whose diameters, in general, exceed 0.1 μ (1000 Å). Such systems possess a minimal stability, which may be accentuated by such additives as surface active agents, finely divided solids, etc." Obviously this definition was not intended to include thermodynamically stable systems.

2. EMULSIONS VERSUS MICELLAR SYSTEMS

Historically, there were several reasons for this exclusion. Emulsions, as they were known to such workers in the field as Pickering, Donnan, Bancroft, Clowes, Harkins, or Hildebrand and even to such recent investigators as Schulman, Cockbain, and Clayton (1,2), appeared milky white and rapidly separated into their original two phases. Exceptions were systems in which both phases had the same index of refraction; although these were transparent, they were short lived. Against this background it was difficult to classify a translucent or transparent system that exhibited stability measured in years. The first reaction was to call such systems "solubilized" or to consider that the small volume of dispersed liquid was incorporated into a larger volume of detergent or soap, which in turn was dispersed in the second liquid in micellar form.

According to this concept, the dispersed liquid was layered between or interpenetrated among the heads or tails of the molecules of a swollen, lamellar micelle (3). This, at least, furnished an explanation for the existence of the transparent systems.

As time went on, however, it became increasingly clear to the emulsion formulator that the behavior of emulsifier in the form of an interfacial film, rather than in the form of a micelle, controlled the properties of these microdroplet systems. In micellar form, swelling of the array of emulsifier molecules with water or oil could only occur within certain limits (e.g., 10%). On the other hand, experiments indicated that well beyond this limit, the droplet size of the dispersed phase, as manifested by Tyndall scattering, was dependent on the amount and kind of emulsifier. To the physical chemist the fact that these systems were isotropic (i.e., the dispersed aggregates possessed the same dimensions in all directions) sharply differentiated them from lamellar or cylindrical micelles, which were anisotropic (i.e., the aggregates possessed different dimensions in each direction). Thus the spherical oleophilic hydromicelle and the hydrophobic oleomicelle (4) became the accepted models for these water-in-oil and oil-in-water emulsion systems. According to these models, the transition from macroemulsion to microemulsion merely involved an increase in the total interfacial area. The effect of increasing emulsifier content on the decrease in droplet size of the alkyd emulsions of Fig. 1 graphically illustrates this pattern (5).

It is not likely that these micrographs of emulsions are optical illusions (artifacts) since they were made without shadow casting (see Section II.C.1.d). For Example, the black, irregular-shaped objects in Figs. 1(c) and (d) are believed to be pieces of osmium metal reduced from osmium tetroxide that did not react with the alkyd. Thus these photographs, portraying the dispersed phase in microdroplet form, not only demonstrate the effect of emulsifier content on droplet size but also constitute strong evidence that these systems are indeed emulsions, and not made up of lamellar micelles.

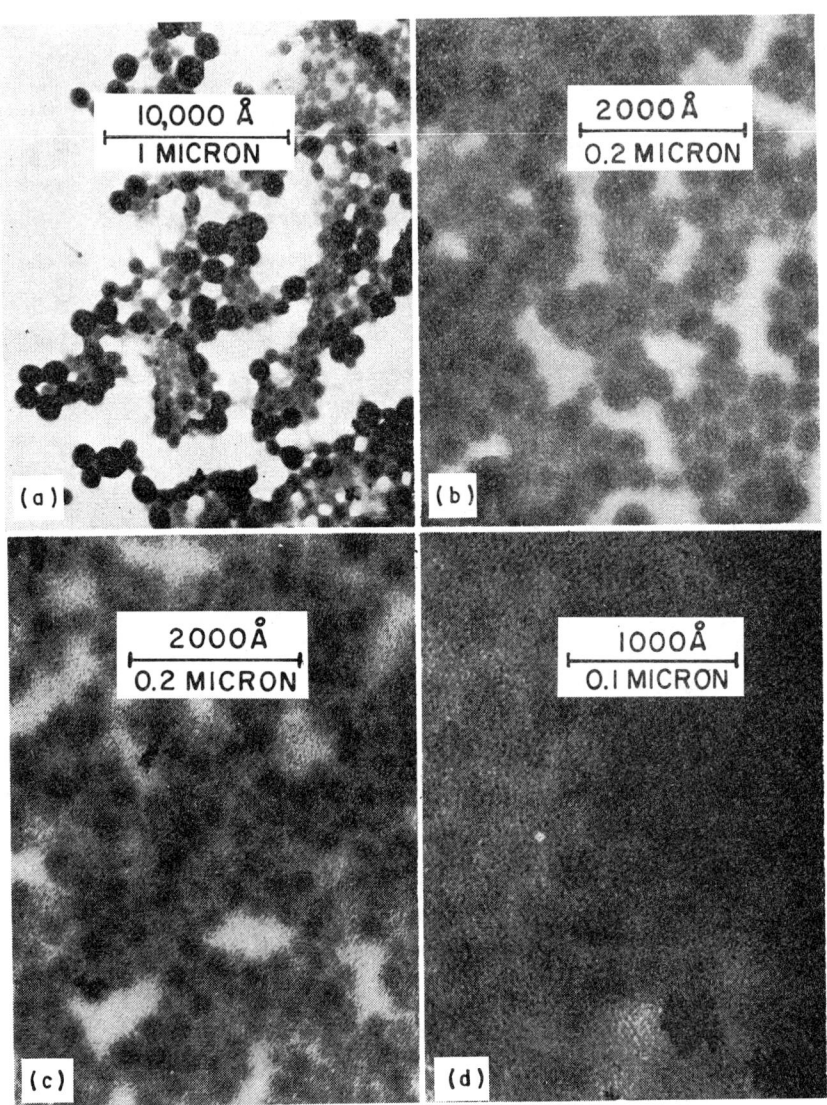

FIG. 1. Electron micrographs of alkyd-in-water microemulsions: (a) 20% emulsifier on weight of alkyd, magnification 35,000x (taken at 10,000x), av diam of droplets 1200 Å; (b) 35% emulsifier on weight of alkyd, magnification 140,000x (taken at 40,000x) av diam of droplets 450 Å; (c) 50% emulsifier on weight of alkyd, magnification 140,000x (taken at 40,000x), av diam of droplets 300 Å; (d) 50% emulsifier of weight of alkyd, magnification 280,000x (taken at 80,000x) av diam of droplets 300 Å. The droplets were stained with osmium tetroxide and photographed directly in the electron microscope without shadow casting. These micrographs were taken by Dr. Walther Stoeckenius of the Rockefeller Institute and reprinted here from Ref. (5) by courtesy of MacNair Dorland, Inc.

In spite of this visual evidence, the terms "micellar emulsion" or "micellar solution" have again appeared in the literature to describe these small-particle-size dispersions. Adamson (6) prefers the term "micellar emulsions" for his water-in-oil systems "as more indicative of the combination of properties involved." Rosoff (7) was concerned that, when the size of the emulsion droplets gets small enough, the concept of phase loses its conventional meaning and the state of the system can no longer be described in terms of phase equilibria or surface tension. Friberg et al. (8) feel that, when the interphase tension is zero, the so-called microemulsions are solutions with solubilized water or solubilized hydrocarbon. This viewpoint coincides in many respects with that of Winsor (9), who considered his clear solutions to be true solutions.

Schulman saw nothing incongruous about microemulsions. He stated (10) that the difference between a micelle and a microemulsion droplet is that in the micelle the dispersed phase is in equilibrium with a saturated solution of the molecular units comprising the micelle, whereas in the microemulsion system the internal phase is not in equilibrium with the external phase (i.e., there are no molecular species of the dispersed phase in the external phase), and therefore there can be no critical micelle concentration in this phase.

A typical emulsion phenomenon serves to illustrate the difference between solubilized systems and microemulsions. Consider the inversion of a microemulsion from a water-in-oil to an oil-in-water system. At first a small amount of water is dispersed in the form of microdroplets in the bulk oil phase. This system is isotropic, exhibiting the viscosity of the oil phase. As more water is added, the system becomes viscoelastic and exhibits streaming birefringence in crossed Nicols, indicating anisotropy. Now, instead of water droplets being surrounded by an interfacial monolayer, water is held among the polar heads and oil among the nonpolar tails of the emulsifier molecules, which are arrayed in the form of lamellar or cylindrical micelles. However, only a limited amount of water and oil can be dispersed in this manner. As more water is added, the system becomes fluid and isotropic, indicating that the emulsifier molecules are again arrayed in the form of an interfacial monolayer surrounding spheres of oil. One may consider, therefore, that a microemulsion is a special case of a solubilized system in which the micelles have assumed a spherical form. This is brought about by the liquefaction of the micelle due to the penetration of the array of soap or detergent species by alcohols (see Section III). The surface forces thus released impose the appropriate direction and degree of curvature on the monolayer adsorbed between the two liquid phases. In this form emulsifier molecules can disperse much larger volumes of oil or water than in micellar form.

3. PARTICLE SIZE VERSUS INTERFACE RELATIONSHIPS

With the full realization of the importance of the interphase, or third phase, in the development of these small-particle-size emulsions, a closer look was taken at the relationship between the area at the outer and inner sides of the interphase as well as between the volume of the interphase and the total volume of the droplet (internal phase plus interphase), as the droplet size decreased. The data of Table 1 reveal the geometric realities of this analysis. In the micron size range of droplets stabilized with soap the interphase is only 1.5% of the volume of the droplets, and droplet curvature 1.01, is negligible. On the other hand, when the droplet diameter is 100 Å and the 25-Å-thick interphase consists of a mixed film of soap and stearyl alcohol, the curvature is 4, and the volume of the interphase constitutes 87.5% of the volume of the whole microdroplet. This volume relationship points up the dependence of the physical behavior of the system as a whole on interphase properties, and the enormous curvature indicates the complexity of the molecular interactions among the interphase tenants and the molecules of the adjacent liquids.

The data of Table 1 also demonstrate how decreasing the thickness of the interphase extends the range of microemulsions almost to

TABLE 1

Geometry of the Emulsion Interphase

Droplet diameter[a]		Interphase thickness (Å)	Area at outer side[b] / area at inner side	Volume of interphase[c] / volume of droplet
μ	Å			
1.0	10,000	25	1.010	0.015
0.2	2,000	25	1.05	0.07
0.1	1,000	25	1.11	0.14
0.05	500	25	1.23	0.27
0.033	330	25	1.40	0.39
0.01	100	25	4.0	0.875
0.008	80	25	7.1	0.95
0.01	100	5	1.23	0.27
0.003	30	5	2.25	0.70
0.002	20	5	4.0	0.875

[a] Internal phase plus interphase.

[b] This ratio is defined as the curvature of the interphase, $[R/(R-T)]^2$, where R is droplet radius and T is film thickness.

[c] Defined as $[R^3 - (R-T)^3]/R^3$.

the solution (molecular dispersion) state. By using Wyandotte's Pluronic F-68 emulsifier (a block copolymer of ethylene oxide and propylene oxide) so as to achieve an interphase thickness of only 5 Å, droplet size can be reduced to 20 Å.

B. Commonplace Examples

Many products and components of products in the marketplace today are microemulsions. They are there because the thermodynamic stability of these systems makes the products uniformly acceptable with time, and the small particle size confers some special performance attribute on the dispersion, such as the ability to deposit a glossy film or exhibit a fashionable optical effect due to Tyndall scattering. Some commonplace examples with rationales for their *raison d'etre* are discussed below.

1. CARNAUBA-WAX EMULSIONS

An emulsion of carnauba wax blended with about 10% of an ammoniacal solution of shellac was marketed in 1928 in St. Louis, Missouri, by George Rodawald, doing business as the Miracul Wax Company. The trade name was Dri-Brite floor polish.[*] The carnauba-wax-emulsion portion of this product was, if not the first microemulsion, certainly the first oil-in-water microemulsion. The example shown in Table 2 is typical of the formulas made during that period.

The product was made by melting the wax, soap, and a small portion of water together until a homogeneous blend was obtained. The balance of the water was then added at 180 to 200°F. After cooling, the wax-in-wax system possessed a characteristic gray

---

[*]This information was kindly furnished by Mr. Jack T. Hohnstine, Vice-President of the Boyle-Midway Division of the American Home Products Corporation, which acquired the Dri-Brite formula in 1939.

TABLE 2

Typical Early Carnauba Wax-Emulsion Formulation

Ingredient	Quantity (g)
Carnauba wax, Prime No. 1	100
Oleic acid, red oil	12
Potassium hydroxide	4
Borax	7
Water, deionized	800

opalescence. Although the internal phase was not liquid at room temperature, these systems were rightly called emulsions because they were achieved via an emulsification process.

This product became the cornerstone of a still burgeoning water-based floor-polish industry. It owed its success to permanent stability, which enabled its tiny droplets to coalesce on the floor to a shiny film, regardless of the age of the dispersion. If the droplets were not small enough to coalesce by themselves, there would have been no spontaneous gloss in the first place; if the droplets did not remain stable in size for years, the product would have lost its gloss-imparting properties with time. A product that needed to be buffed probably would not have survived in the marketplace.

In retrospect, Rodawald's carnauba wax emulsion seems to have been the precursor of a number of other products that, although not identified as such at the time, were also microemulsions. From 1930 until 1943, when Hoar and Schulman (11) recognized "soluble oils" (neat oils) as transparent water-in-oil emulsions (oleophobic hydromicelles), a number of articles of commerce appeared that utilized the principles of the carnauba wax microemulsion.* Among

---
*The author is indebted to Dr. Alan Beerbower, Esso Research and Engineering Company, Linden, New Jersey, who graciously made available his recollections of events during this period.

the first of these was a clear emulsion of pine oil (Yarmor) stabilized with soap. At about the same time (in the early 1930s) a clear sassafras oil emulsion was put on the market. This was considered to be "solubilized." This was followed by a series of clear essential oil microemulsions. From the food field these emulsions moved into the machine tool industry, where in about 1937 stable cutting oil microemulsions made their appearance. These, like the carnauba wax emulsion, exhibited a pronounced Tyndall effect.

2. CUTTING OILS

These emulsions of lubricating oils in water evolved from the "soluble oils" made with petroleum sulfonates (mahogany oils). They are discussed elsewhere. It need only be said here that cutting-oil emulsions were made micro to ensure that a uniform mixture of lubricant and (cooling) water was at all times in contact with tool and work. Thus the higher efficiency of the product defrayed the higher cost of the microemulsion.

3. FRAGRANCE OILS

It is often desirable to emulsify water-insoluble fragrances for mouthwashes, shaving lotions, or similar products in order to prevent them from precipitating out of the alcohol-water mixtures in cold weather. Microemulsification in transparent form is essential to maintain the "polished" clarity of these products. Cost is no object.

4. CLEANING FLUIDS

Water-in-oil emulsions of cleaning fluids make it possible for dirt and stains to be exposed to both liquids during the cleaning operation. Microemulsions increase the efficiency of the mixture and prolong its life.

5. PAINT VEHICLES

Possibly the largest single use of microemulsions is in the paint industry. Many paint vehicles are oil-in-water microemulsions of alkyds or polymers. The stability of these vehicles ensures uniformity of application and coherent films. As in the case of floor polishes, their small particle size enables high-gloss finishes to be easily formulated. The use of water instead of smelly and noxious solvents has enormous consumer appeal and, in industrial use, saves money on fire-insurance premiums and reduces air pollution.

Emulsion-polymer vehicles for paint may frequently be used in other fields. Examples are acrylic latices, which also find application in paper coatings, and polystyrene latices, which are useful in floor-polish formulations. Although their dispersed phases are not liquid, these latices are understandably called emulsions because they are made by an emulsification process. Many of these emulsion-polymer systems exhibit Tyndall scattering; in some cases the monomer-water-surfactant system before, or during the early stages of, polymerization is transparent, indicating that the theory described in Section III may apply to these systems also.

6. TOXICANT AND ANTISEPTIC FORMULATIONS

Where accurate dosage or some other special property is required, microemulsions may be indicated. Otherwise their use is much too expensive. Chlordane, pentachlorophenol, pine oil, etc., are the kind of oils involved in this category.

7. COSMETICS AND TOILETRIES

Many products in this area (e.g., cold creams, shampoos) are microemulsions. Usually they are used to ensure stability or transparency, but frequently advantage is taken of the light-scattering properties (translucency or opalescence) of microdroplets in the 1000-Å range to appeal to the consumer.

8. FOODS AND FLAVORS

Microemulsions of water-insoluble raw materials are utilized in the food industry primarily to meet stringent stability requirements. Any breakdown of the emulsion during storage or shipment with consequent deterioration of appearance could be disastrous.

For this reason transparent systems with large excesses of edible emulsifiers are usually employed. Microemulsions of flavor oils in cola, cream soda, and other confections are typical examples.

9. PHARMACEUTICALS

Many drugs are insoluble in water, but it is desirable for one reason or another to administer them in an aqueous medium. Where accurate dosage is required so that uniformity of dispersion is essential, microemulsification is indicated. In the case of consumer items for oral ingestion (e.g., vitamin oils) the transparency or Tyndall scattering of a stable microemulsion has sales appeal.

10. HOMOGENIZED MILK

Homogenized milk is not a microemulsion, although for all practical purposes it appears to be one. This emulsion is stable for the life of the product, and a thin film of it on the sides of a glass exhibits Tyndall scattering indicative of droplets in the submicron size range. This behavior raises some interesting points that serve to highlight the criteria of microemulsification.

Homogenized milk is not a microemulsion, because it is not in thermodynamic (reversible) equilibrium. External work must be done on the original, coarse emulsion of fat in water to reduce droplets to the submicron size range (12, 13). This process prevents creaming of the droplets for a week or so. If, however, the emulsion is centrifuged, creaming will occur, and the creamed droplets will not spontaneously redisperse.

The milky appearance of the product is also a clue to the fact that it is not a microemulsion. The work done on the system by a colloid mill apparently does not reduce the diameter of *all* the droplets below 0.5 micron. Even a relatively few of these large droplets will impart a milky appearance to the emulsion by scattering white light and opacifying the transparent system. Normally such droplets would cream in a day or so, but, dispersed as they are with a majority of smaller droplets in violent Brownian movement, they remain in suspension until the product spoils or is consumed. How these phenomena can be interpreted to differentiate a microemulsion from a macroemulsion is described in detail in the next section.

C. Recognition

Microemulsions are recognized by the physical properties displayed by their small droplets. The optical properties of these microdroplets and their behavior in a gravitational field easily differentiate them from macrodroplets. For this reason light scattering and sedimentation rates as they apply to these systems are described in some detail. Means for measuring these properties are indispensable tools for identifying and working with these systems.

1. OPTICAL PROPERTIES

Some of the simpler and practical ways of identifying microemulsions by optical means are discussed. Emphasis is placed on the conditions for transparency and Tyndall scattering, as well as the estimation of particle size by means of the naked eye, the light microscope, and the electron microscope.

a. Transparency versus translucency. The criterion that the droplets of a microemulsion be smaller than 0.25 λ is an arbitrary one. It was chosen because droplets of this size permit white light to pass through the disperse system, making it translucent. From a practical viewpoint, transparency as distinguished from translucency is characterized by complete clarity, or lack of Tyndall scattering, to the naked eye and does not usually occur until the diameters of the spherical droplets are smaller than 100 $\overset{\circ}{A}$. Transparency is, in addition, dependent on the relative refraction indices of the oil and water. For example, in an oil-in-water emulsion in which the refraction index of the oil is as high as 1.55 (polystyrene) a clear and transparent system is usually not obtained until the droplet diameter is as small as 80 Å. It must also be kept in mind that light scattering is dependent on the concentration of the dispersed phase. Lower concentrations favor stronger light scattering. Thus an emulsion that is clear and transparent at 40% dispersed phase may very well scatter light when diluted.

In actual practice it is very difficult to obtain oil-in-water microemulsions that do not exhibit some light scattering. More than 100% emulsifier on the weight of oil is required to achieve clarity, and this is usually too expensive. Considerably less emulsifier is needed to obtain clear, transparent water-in-oil systems.

b. The Tyndall effect. Particles that are large in comparison with the wavelength of light, λ, reflect and refract light in a regular manner, whereas particles that are small in comparison with λ scatter incident light in all directions. The scattered light is plane polarized, and each particle becomes the source of a new wavefront. Thus in white light (the visible spectrum), the wavelength of which is 0.4 to 0.7 μ or 4000 to 7000 Å, large particles like escaping steam (10 to 100 μ in diameter) appear dead white, whereas small (0.01 to 0.15 μ in diameter) particles, such as fresh smoke from a cigar or chimney, appear blue to reflected light and orange red to transmitted light. This latter behavior is called the Tyndall effect and is similar to Rayleigh scattering for particles of molecular dimensions. This effect is responsible for the blue of the sky and the orange red of sunsets.

The application of these physical principles to emulsions is obvious. Macroemulsions appear milky white. Microemulsions that are not transparent but translucent exhibit strong Tyndall scattering that can clearly be seen with the naked eye.

c. Methods of measurement. Although the eye is a remarkable optical apparatus, it is an imperfect light-scattering instrument. For this reason some help is needed in estimating the size of microemulsion droplets. A few of the simpler pieces of equipment are discussed.

Microphotometers lend themselves well to measurement of particle size in the 100 to 2000 Å range. With good standards they have been used on a routine basis for quality control.

The light microscope can only resolve particles greater than 0.2 μ in size. It is therefore useless in microemulsion studies except to establish whether or not there are some large droplets in the system.

The ultramicroscope, or dark-field microscope, however, makes it possible for the human eye to detect particles in the size range from 2000 down to 100 Å as a brilliantly scintillating sea of tiny light flashes. Unlike the microphotometer, these instruments do not permit of a simple estimation of droplet size. They merely make positive identification of droplets as being in the Tyndall range. Since the naked eye can perform the same function on a qualitative basis, these instruments have limited value in this application.

In spite of the imperfections of the human eye for judging the intensity and wavelength of scattered light, there are some very useful guidelines for estimating droplet size in emulsion systems with the naked eye. These are indicated in Table 3.

TABLE 3

Guidelines for Visual Estimation of Droplet Size

Color of dilute emulsion	Tyndall effect to eye		Droplet diameter	
	Reflected	Transmitted	μ	Å
Dead white	None	None	>0.5	>5000
White to gray	Weak blue	Weak red	0.3-0.1	3000-1000
Gray to translucent	Intense blue	Intense red	0.14-0.01	1400-100
Clear transparent	None	None	<0.01	<100

If one puts an emulsion suspected of being a microemulsion in a test tube and holds it up to a strong light source (preferably sunlight), a Tyndall cone is visible if there are any droplets in the 100- to 2000-Å range. Dilution will yield further information, as already indicated. To the extent that words can describe what each individual sees in a Tyndall cone, Table 3 serves as a guide for those who would rely on their "eyes" to assign sizes to microdroplets. With practice, a fair degree of accuracy can be achieved.

   d. Electron microscopy. In certain special cases droplets of microemulsions can be seen directly in the electron microscope. One of the alkyd paint emulsions was discovered to be quite photogenic in this instrument. Because of the alkyd's double bonds, it was easily stained by osmium tetroxide, a biological staining agent. Exposing the alkyd-in-water microemulsion to this agent instantly turned the alkyd droplets into tiny cannon balls. When these were placed on the stage of the electron microscope and exposed to the electron beam under vacuum, the organic matter burned off, leaving behind an osmium-metal skeleton of the original spherical droplets. These are seen directly, without shadow casting, in Fig. 1.

   These photographs show that the droplets of these systems range in diameter from 0.12 to 0.03 μ (1200 to 300 Å). It will be noticed that, as the droplets become smaller, they are more uniform in size. In Figs. 1(b), (c), and (d) one can clearly see the hexagonal packing characteristic of the theoretical closest-packing ratio of internal to external phase of 74:26 for uniform spheres.

2. STABILITY

   The visual signals (colors) emitted by emulsion droplets too small to be seen in the light microscope should not obscure the fact that the real criterion of a microemulsion is its thermodynamic stability (i.e., that the tension of the interphase be zero). In such a system any displacement from its equilibrium condition spontaneously generates surface forces that return it to equilibrium.

In the present state of our knowledge this behavior has not yet been related to a droplet size that characterizes a microemulsion. All that can be said at this time is that the forces that account for this equilibrium also appear to be responsible for small particle size and that the transition from macroemulsion to microemulsion seems to be an abrupt one.

The techniques for determining the behavior of small particles in a gravitational field, their sedimentation rates, are for the most part quite simple and inexpensive. They provide rapid and almost foolproof identification of a system as a microemulsion. The effect of particle-size distribution, Brownian movement, and coalescence on the interpretation of these measurements is reviewed.

a. Sedimentation measurements. There are three ways to measure sedimentation rates that are readily applicable to systems suspected of being microemulsions: settling rates in a gravitational field, the centrifuge, and the ultracentrifuge. Of these, an inexpensive laboratory centrifuge is almost ideal. The use of the ultracentrifuge is a much too complicated and time-consuming operation. On the other hand, settling rates in a 1-G field are too slow.

If a sample of an emulsion exhibits no separation after being spun for a few minutes in a laboratory centrifuge at 100 G, there is a very good chance that it is a microemulsion. The exceptions to this rule are treated in the following discussion.

b. Brownian movement. Droplets with diameters smaller than 0.2 μ cannot be seen in the optical microscope. Even the largest droplets of the alkyd emulsion of Fig. 1(a) could not be resolved in a light microscope, although they were seen as a sea of tiny light flashes in a dark-field microscope. This behavior, associated with particles smaller than 0.5 μ, is called Brownian movement. Einstein and Smoluchowski proposed a molecular kinetic explanation of this phenomenon, which for our purposes simply means that emulsion droplets in this size range are small enough to absorb kinetic energy from bombardment by the molecules of the dispersion medium. It has been calculated that this causes such particles to change direction 10^{24} times per second. This keeps the dispersed droplets in a state of violent motion, preventing their settling in a gravitational field.

So long as they do not coalesce, it is Brownian movement that prevents the droplets of a microemulsion from settling or creaming. As fast as they tend to rise, due to a density lower than that of the dispersion medium, or to fall, due to a density higher than that of the dispersion medium, bombardment by the molecules of the medium counteracts this tendency and maintains them in a state of perpetual motion, just like the molecules of the medium. The effectiveness of Brownian movement in doing this is demonstrated by the following experiment: A microemulsion of droplets in the 100- to 200-Å size range was ultracentrifuged at 130,000 G. Although this effected some sedimentation of the droplets, they did not coalesce, as indicated by the fact that the emulsion soon returned to its original condition in a normal gravitational field.

c. Coalescence. The reason that the droplets do not coalesce in circumstances like those described is due to the surface free energy of the microemulsion system. Just as soon as two microdroplets coalesce to form a single droplet of larger size, the interfacial tension of the new droplet becomes negative; that is, the system has a negative free surface energy. The large droplet now spontaneously increases its curvature to effect zero interfacial tension again, and two droplets of the original equilibrium size result. This process continues as relentlessly as does the bombardment of the droplets by the molecules of the dispersion medium. It is this dynamic equilibrium that keeps a microemulsion stable.

d. Particle-size distribution. It is seen from the micrographs of Fig. 1 that the homogeneity of the droplet sizes of the microemulsions increases as the average size decreases and that quite a uniform distribution is obtained in the 300-Å size range.

Conversely, it is apparent that in the average size range of 1000 Å, in which there is a distinct economy of emulsifier, there could be some droplets, as in the case of homogenized milk, that exceed the Brownian movement limit of 5000 Å.

Such heterodispersity can cause some difficulties in interpreting the results of sedimentation measurements. Just because there are a few macrodroplets that separate out does not mean that the emulsion is not a potential microemulsion. Probably more careful mixing during formation or a slight change in formulation will eliminate the large droplets. The trick in such cases is to formulate in such a way, as in Fig. 1(a), that none of the droplets is larger than 0.2 to 0.3 μ, although there may be a very wide distribution. This keeps the cost of emulsifier at a minimum and its quantity in the system from interfering excessively with the performance of the product.

III. THEORY

As soon as it was established that the properties of microemulsions were dependent on the properties of the interfacial film, or interphase, attention was directed to the nature of the forces at play among the molecular species of this third phase and between those of this phase and of the adjacent liquid phases. Out of these studies emerged a theory of microemulsions that accounts for most of the observed phenomena.

A. Historical Background

The theory of microemulsion evolved naturally from that of macroemulsions. A number of outstanding colloid chemists concerned with the phenomena exhibited by these systems had developed a reasonable explanation for the behavior of these coarse, classical emulsions (1, 2). The contributions of these men provided the solid foundation on which it was possible for Schulman to superimpose his concepts of surface chemistry as they applied to microemulsions. It is in this perspective that the work on macroemulsions is briefly reviewed.

The idea was grasped at an early stage that the physical forces at play in a monomolecular film between two insoluble liquid phases could be responsible for emulsion formation. Donnan's work (1899-1915) showed that soap was adsorbed to such an interface in accordance with Gibbs' law (promulgated in 1878) and that the consequent lowering of the interfacial tension between the liquids was associated with the emulsifying power of the adsorbate. Bancroft (1913-1927) treated this monomolecular film as a separate phase having separate tensions at each of its sides. By this concept he explained how oil-in-water and water-in-oil emulsions could be formed. When the tensions were different, the film would bend, the side with the higher tension becoming concave. This would envelop liquid at that side and make it the internal phase. In this view film curvature was the prime mover in the process of emulsification.

The oriented-wedge theory as advanced by Harkins, Davies, and Clark (1917) and Finkle, Draper, and Hildebrand (1923), although eventually abandoned, left an important legacy. In spite of its being based on the false premise that the size and shape of monovalent and polyvalent soaps are responsible for the direction of film curvature, its conclusions that molecules occupy wedge-shaped areas in the interface was intuitively correct. The film-balance studies that were initiated by Schulman (1935) and led to the Schulman and Cockbain papers (1940) gave ample proof of this. These demonstrated that molecular forces at the two sides of the interphase replace molecular shape and size as the agencies of wedge formation.

For emulsions stabilized by molecularly dispersed species (as opposed to finely divided solids) time has borne witness to the validity of the proposal by Schulman and Cockbain (14) that reactions taking place at the oil-water interface were closely analogous to the corresponding reactions at the air-water interface, which had been investigated by the method of surface films. Schulman applied the knowledge gained from studies of monolayers of mixed films at the air-water interface to the oil-water interface with astonishing success. In the 20 years since Langmuir's original work in 1917 much had been learned about the size and shape of molecules like oleic acid, cetyl alcohol, cholesterol, and sodium cetyl sulfate as well as the physical forces acting among their nonpolar tails and polar heads when they were spread as monolayers on a trough of water.

Schulman and Cockbain used the information regarding the molecular association (complexes) between alcohols and detergent species that had been obtained in film-balance studies to formulate stable Nujol-in-water macroemulsions in the 1- to 4-μ size range. To explain the stability of these systems they summarized:

> The stability of a "nujol"-in-water emulsion depends markedly on the nature of the two components chosen to form the interfacial complex film. The stability of such films runs parallel with the stability of the same films at an air-water interface, and depends on the van der Waals forces of attraction between the non-polar parts of the two components and on the interaction between the respective head groups. Hence, alterations in the size of the non-polar residues, the stereochemical configuration of these residues, the nature of the polar heads, etc., cause alterations in the molecular packing and stability of the interfacial complex films and alterations, therefore, in the stability of the *"nujol"-in-water* emulsions, the stability of the emulsions can be deduced from a knowledge of the stability of the corresponding molecular complex at an air/water interface.

It is difficult to improve on this statement even today.

Based on the performance of these mixed films in these systems, a spherical micellar structure was proposed for the transparent, oil-continuous, soluble-oil systems of Hoar and Schulman (11). It is of interest to observe in retrospect that, had Schulman and Cockbain used kerosene or decane instead of Nujol in their experiments, they probably would have made oil-in-water microemulsions in 1940.

B. The Transparent Emulsion

The concept that transparent* oil and water systems were microemulsions developed slowly. In the 17-year period from 1943 to 1960 Schulman and his collaborators made it unmistakably clear that transparent and translucent systems exhibiting a Tyndall effect were a different kind of emulsion than the coarse, classical ones. Rather critical conditions were required for their formation. It was soon shown (15) that when the alcohol of the mixed film was of low molecular weight, the continuous phase was water, and when the alcohol was longer than six carbon atoms, the continuous phase was oil. The droplet sizes were found by low-angle X-ray measurements (16), light scattering (17), ultracentrifuge measurements (18), and electron microscopy (19, 20) to be in the 100- to 1000 Å range. As already indicated (4), it was deduced from their optically isotropic character that these systems were emulsions, and not swollen lamellar or cylindrical micelles.

During this period the interfacial film came to be treated as an interphase having all the properties of a third phase. Bowcott and Schulman (18) used the principles of the phase rule to study the composition of the continuous phases and interphases of transparent w/o emulsions. They proposed that the third phase was in

---

*Except when both phases have the same index of refraction (see Section II.A.2).

equilibrium with the oil and water phases, and that the emulsifiers could be distributed among all three phases. In 1959 Schulman, Stoeckenius, and Prince (19) suggested that the mechanism of formation of these microsystems consisted of the penetration of the highly ordered soap or detergent micelles by any molecular species capable of producing sufficient disorder in, and hence liquefaction of, the bimolecular soap leaflets to enable the micelle to swell unlimitedly. In the presence of oil and water phases surface-tension differences could then impose the appropriate direction and degree of curvature on the interphase. It was submitted that the necessary degree of disorder could be produced as a result of the penetration of a mixed film of soap and alcohol by a nonpolar hydrocarbon originally derived from the oil phase. This was the first inkling that the oil phase played a direct role in microemulsification.

There followed in 1964 experiments by Cooke and Schulman (21), fashioned after those of Bowcott and Schulman (18); these gave further indication of the importance of the oil phase in controlling emulsion properties. These authors studied water-in-oil microemulsions of hexane, decane, and hexadecane stabilized by mixed films of potassium oleate and n-hexanol. In measuring the distribution of alcohol between the oil phase and the interphase they found that the degree of partitioning of the alcohol between these two phases was directly related to the molecular weight of the hydrocarbon oil: the longer the hydrocarbon, the less alcohol remained in it. This was so despite the fact that the alcohol was completely soluble in all three of the oils. What this meant was that the chemical nature (chemical potential) of the oil phase controlled the amount of alcohol in the interphase and, equally importantly, the amount that remained in the oil phase.

C. Negative Interfacial Tension

These considerations had direct application to the development of the interfacial tension between the two bulk phases. Common to all of these microemulsion systems was a mixed film consisting of a soap or detergent and a hydrogen bonder, such as an alcohol, free fatty acid (19), or amine (14). Such an interphase was considered to be in the liquid state, and its two-dimensional pressure Π determined the interfacial tension γ_i in accordance with the equation

$$\gamma_i = \gamma_{o/w} - \Pi \tag{1}$$

in which $\gamma_{o/w}$ is the original tension between the oil and water before the addition of emulsifying agents.

In 1960 Stoeckenius, Schulman, and Prince (20) had proposed that when $\Pi > \gamma_{o/w}$, γ_i would be negative, this being the criterion of a microemulsion. In their words, "The negative interfacial tension produced by the mixing of the components will, at equilibrium, become zero and dispersion not separation will be the equilibrium condition." The high film pressures required to achieve this were considered to result from the penetration of the mixed film of soap and alcohol by hydrocarbon molecules of the oil phase.

In 1961 Schulman and Montagne (22) measured the surface pressures of duplex films between oil and water on a Langmuir trough. They found that films of soap and alcohol together gave higher pressures at a given area per molecule of soap than did films of the soap or alcohol alone. This was attributed to the interpenetration of the mixed films by oil molecules. The subsequent experiments by Cooke and Schulman indicated, however, that hydrocarbon molecules were probably ejected from these films at high pressures, although their contribution to higher areas per molecule at low pressures could be considerable.

This evidence forced some modification of the negative-interfacial-tension concept because, attractive as this concept was, it was now difficult to believe that film pressures as high as 55 dynes/cm could spontaneously be developed, even momentarily, in an uncompressed emulsion interphase. A much more reasonable suggestion made by Prince (23) in 1967 was that the initial, negative, interfacial tension is the result not so much of a high initial film pressure Π as of a large depression of the tension $\gamma_{o/w}$ between the water and oil phases due to the partitioning of the alcohol between the mixed film and the oil phase. This concept led to a new appraisal of the causes of microemulsification.

In order to visualize this, consider first a mixed monomolecular film consisting of a 1:1 complex of soap and fatty alcohol adsorbed at a *flat* oil-water interphase as in Fig. 2. The soap and alcohol species are oriented with their heads in the water and tails in the oil phase. As the number of these molecules per unit area is increased, they begin to crowd one another, thereby developing the two-dimensional, lateral pressure Π. In accordance with Eq. (1),

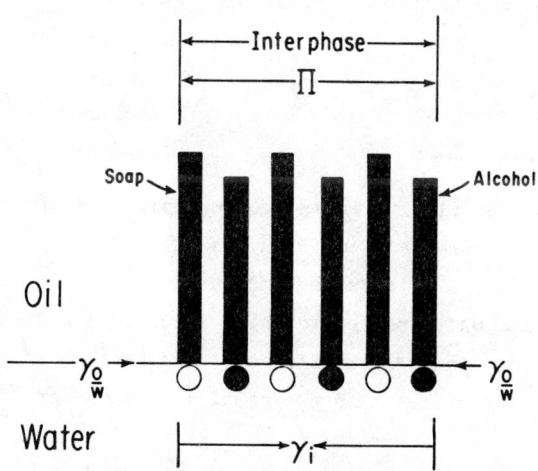

FIG. 2. Schematic of the two-dimensional pressure and tensions developed in a mixed film of soap and long-chain alcohol adsorbed at the interphase between oil and water. Increasing the pressure Π decreases γ_i in accordance with Langmuir film-balance protocol as expressed in Eq. (1). Redrawn from Ref. (24), p. 196, by courtesy of the *Journal of the Society of Cosmetic Chemists*.

this crowding and concomitant increase in pressure decreases γ_i by opposing the tension $\gamma_{o/w}$.

So long as there is no alcohol in the oil phase, γ_i will remain positive. At an oil-water interface, however, this is an unreal situation because, in order for a long-chain alcohol to reach the interface, there must be alcohol in the oil phase. This is a major difference between the behavior of mixed films on a Langmuir trough at the air-water interface and at the oil-water interface of an emulsion.

How partitioning of the alcohol between the oil phase and the interphase produces negative tensions in these systems is shown in Fig. 3. The reservoir of alcohol residing in the oil phase distributes alcohol to the interphase as well as to the direct interface between oil and water (before it disappears). This depresses the original $\gamma_{o/w}$ to $(\gamma_{o/w})_a$. At the same time the fraction of alcohol in the interphase interacts with soap to form a molecular complex that develops the pressure $\Pi_{\overline{G}}$ before curvature, the subscript denoting that the pressures at the two sides of the interphase may be different. Thus the expression for the transient interfacial tension γ_ϕ in this flat film becomes

$$\gamma_\phi = (\gamma_{o/w})_a - \Pi_{\overline{G}} \qquad (2)$$

This equation, as will be shown, comfortably accounts for negative tensions in the range -25 to -35 dynes/cm.

Values of $(\gamma_{o/w})_a$ have been measured. They are the tensions between oil and water phases to which alcohol is adsorbed. In general for long-chain fatty alcohols at a hydrocarbon-water interface $(\gamma_{o/w})_a$ approaches 15 dynes/cm at low areas per molecule of adsorbate (25-27). Lower values have been reported.

Although values of $\Pi_{\overline{G}}$ can only be estimated, there is some degree of certainty as to the conditions under which maximum pressures are developed. In these interphases there appear to be three stages of pressure development (23). In the first, with zero or

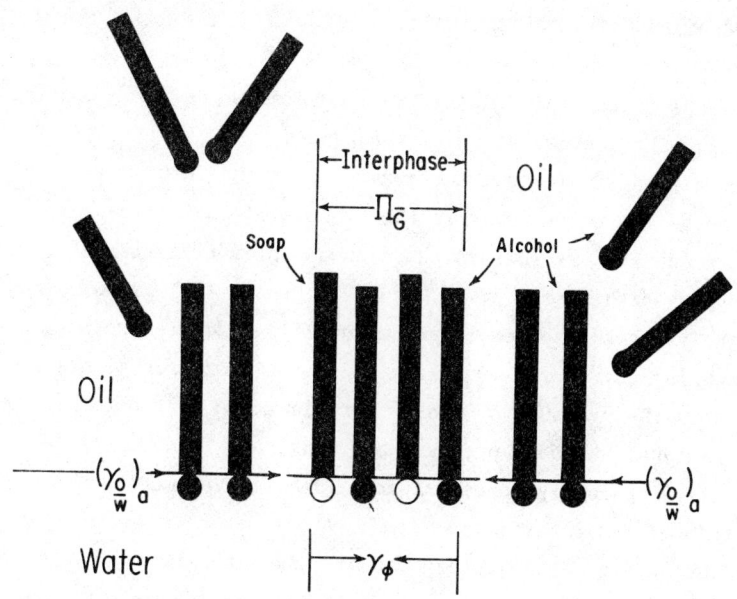

FIG. 3. Schematic of the two-dimensional pressure and tensions developed in a mixed film of soap and long-chain alcohol adsorbed at an oil-water interphase when partitioning of the alcohol occurs. The fraction of the alcohol in the oil phase depresses the original oil-water interfacial tension from $\gamma_{o/w}$ to $(\gamma_{o/w})_a$ at the same time that the fraction of alcohol in the interphase forms a mixed film with the soap there, as in Eq. 2. Increasing the pressure $\Pi_{\bar{G}}$ decreases γ_ϕ in accordance with Eq. (2). Redrawn from Ref. (24) p. 197, by courtesy of the *Journal of the Society of Cosmetic Chemists*.

only a low concentration of alcohol, the pressure is low. The second stage has been identified with the formation of a 1:1 ratio of alcohol to soap. A complex, or molecular association, is considered to form and to be responsible for the development of pressures as high as 50 dynes/cm (22). In the third stage, as the alcohol-to-soap ratio increases, pressures are depressed, and they appear to be depressed more abruptly in oil-in-water emulsions than in water-in-oil systems.

D. Graphical Characterization

In the context of Eq. (2) these considerations make possible a graphical characterization of the process of emulsification.

As already indicated, the curvature of the interphase was shown to be the prime mover in the process of emulsification. Bancroft saw the *curved* interphase as a duplex film with different tensions at either side of it. In 1969 Prince (28) proposed that only the freshly adsorbed, *flat*, mixed film is duplex in nature and that, during curvature, equalization of the pressures at its sides eliminates the pressure or tension gradient, so that both sides are finally at the same pressure and tension. To demonstrate this mechanism graphically, pressure-area (Π-A) curves were drawn for each side of the interphase.

Thus in Fig. 4 the interphase-oil and interphase-water surfaces are each characterized by their own individual Π-A curves. Curves AB and CD are the hypothetical Π-A relationships at the water and oil side, respectively, of the interphase of a microemulsion of hexadecane in water. This emulsion was stabilized by a mixed film of octadecanol and 2-amino-2-methyl-1-propanol (AMP) stearate. Curve EF is the actual Π-A relationship of this mixed film measured on a Langmuir trough by Schulman and Montagne (22). Curve EF is the sum of curves AB and CD. It is seen that the position and shape of curves AB and CD determine the character of the emulsion.

A plausible mechanism of curvature suggests itself if one considers Π'_w and Π'_o to be the film pressures of the *flat*, duplex film at the water and oil side, respectively, and Π_w and Π_o to be the pressures at the two sides of the *curved* film enveloping the emulsion droplet. The initial pressure gradient $\Pi_{\overline{G}}$ across the film then equals $\Pi'_w + \Pi'_o$. A reasonable value of the area per fatty-acid molecule corresponding to these pressures might be 50 Å. A value for Π'_w might be 30 dynes/cm; for Π'_o, 10 dynes/cm. Under the stress of these pressures, expansion at both sides of the interphase spontaneously takes place. This expansion will continue, to different

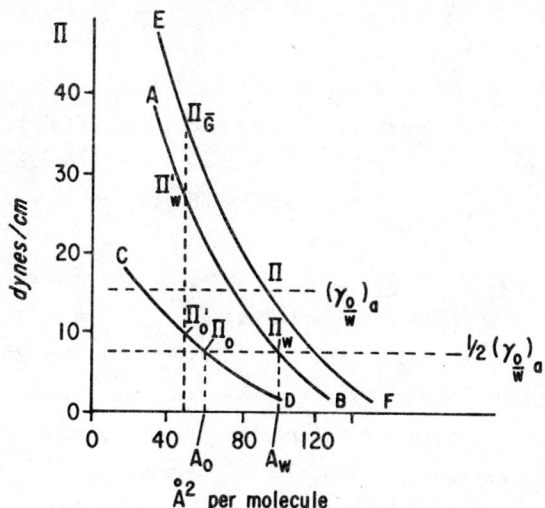

FIG. 4. Pressure-area (π-A) curves of the mixed film of an o/w microemulsion. Curve AB represents the water side and curve CD the oil side; curve EF is the sum of curves AB and CD. Because $\pi_{\overline{G}} > (\gamma_{o/w})_a$, expansion of the film occurs spontaneously from the original π'_w and π'_o at 50 Å to the final π_w and π_o at A_w and A_o. Curvature is effected as the ratio of the area per molecule at the two sides of the film changes from 1:1 to $A_w:A_o$. Redrawn from Ref. 28, p. 217, by courtesy of *Academic Press, Inc.*

degrees at each side, until these pressures become equal and the total pressure π in the film has fallen to the value of $(\gamma_{o/w})_a$. Since $\pi = \pi_o + \pi_w$, expansion will occur until $\pi_o = \pi_w = \frac{1}{2}(\gamma_{o/w})_a$ A reasonable value for $\frac{1}{2}(\gamma_{o/w})_a$ is 7.5 dynes/cm, as already indicated. Graphically what has happened is that the pressure at the water side has slid down along the curve AB to its final value, and the pressure at the oil side has slid along the curve CD to this same value. The driving force for this behavior is the pressure difference $\pi_{\overline{G}} - \pi$, or $\pi_{\overline{G}} - (\gamma_{o/w})_a = -\gamma_\phi$. This makes energy $-\gamma_\phi dA$ (where A equals surface area) available to spontaneously increase the total interfacial area.

Given these pressure values and 15 dynes/cm for $(\gamma_{o/w})_a$, the hypothetical tensions at each surface can be calculated from Eq. (2). For example, the tension γ'_w at the water side of the *flat*, duplex film ($A = 50 \; \text{Å}^2$) is equal to $\frac{1}{2}(\gamma_{o/w})_a - \Pi'_w$, or -22.5 dynes/cm. By the same token, γ'_o, at the oil side, is -2.5 dynes/cm. Adding these together makes γ_ϕ equal to -25 dynes/cm.

This graphical exposition of the mechanism of curvature also suggests the means by which particle size is determined. After pressures in the film have been equalized, the areas per molecule at the sides of the curved film are A_w and A_o, or more generally A_e and A_i at the external and internal sides of the curved interphase. The areas at each side of the interphase will thus be these terms multiplied by the number of fatty-acid species in the interphase. Now film curvature may be expressed as the ratio of the areas of the external and internal sides of the curved interphase, $[R/(R-T)]^2$, where R is the external radius of the spherical droplet (internal phase plus interphase) and T is the thickness of the interphase. Hence

$$\left(\frac{R}{R-T}\right)^2 = \frac{A_e}{A_i} \qquad (3)$$

For the oil-in-water emulsion of Fig. 4, $A_w/A_o = 100:60$. If a value of 25 Å is used for the thickness of the stearic acid film, R is 110 Å. Although hypothetical, this is a reasonable value for the droplet size of a translucent emulsion exhibiting slight Tyndall scattering.

A similar characterization, with variations, applies to water-in-oil microemulsions. In this case curve CD of Fig. 4, representing the water side, is almost vertical as for a condensed film of a long-chain alcohol. At the same time the tails of the interface species, corresponding to curve AB, would be expected to develop higher areas per molecule at low pressures. Thus in these water-in-oil systems an A_o/A_w ratio of 168:42 would not be unexpected. With a film thickness of 25 Å, this ratio corresponds to a clear, transparent emulsion. The ease with which this occurs is probably related to the fact that it is less difficult to form a transparent water-in-oil emulsion than an oil-in-water one.

E. Mechanism of Curvature

Such considerations led to the mechanical model shown in Fig. 5. This schematic illustrates how the tenants of an interphase can physically interact to effect microemulsification (24). The stress provided by different pressures at the oil and water sides of the interphase causes the adsorbed tenants to occupy different areas at each boundary, the difference in these areas being determined by the magnitude of $-\gamma_\phi$. Without the stress of these pressures there can be no curvature and hence no emulsion. Only the flat interphase is duplex in nature (i.e., has different pressures at each side); curvature makes the final pressures on both sides equal. The direction of curvature depends on the relative values of the original pressures, and whether the system is

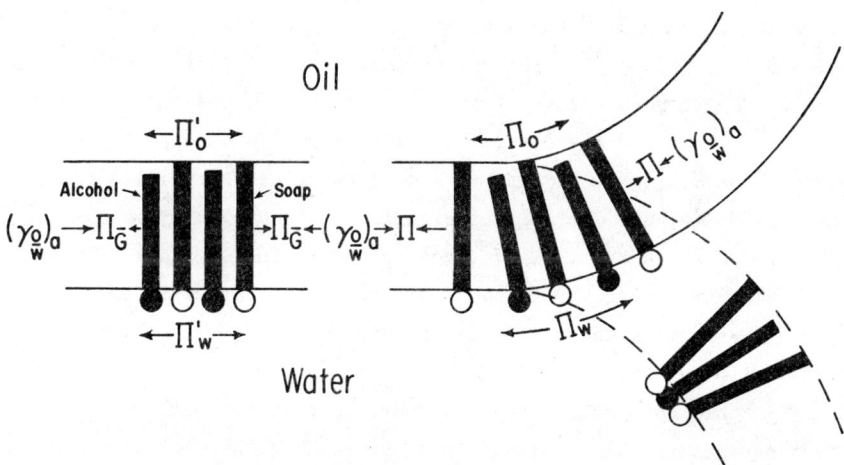

Fig. 5. Diagram illustrating the mechanism of curvature of a microemulsion film. In accordance with Fig. 4, the sum of the pressures at the sides of the flat film is $\Pi_{\overline{G}}$, and the sum of the pressures at the sides of the curved film is Π. The stress of the pressure gradient due to Π_O' and Π_W' is relieved by bending until $\Pi_O = \Pi_W$ or $\Pi = (\gamma_{O/W})a$. The direction of curvature is determined by the relative magnitudes of Π_O' and Π_W'. The degree of curvature is dependent on $\Pi_{\overline{G}} - (\gamma_{O/W})a$. Redrawn from Ref. 24, p. 198, by courtesy of the *Journal of the Society of Cosmetic Chemists*.

macro or micro is determined by the magnitude of the pressure gradient $\Pi_{\bar{G}}$ relative to $(\gamma_{o/w})_a$. In accordance with Fig. 5, if $\Pi_{\bar{G}} > (\gamma_{o/w})_a$, a microemulsion will form spontaneously; if $(\gamma_{o/w})_a > \Pi_{\bar{G}}$, work will have to be done on the system to effect macroemulsification.

In the interest of clarity, the tension $(\gamma_{o/w})_a$ has been shown in Fig. 5 to be the same before and after curvature. This is probably an oversimplication.

This model does not rule out the possibility of penetration of the mixed film by nonpolar oil species (20, 23). In accordance with Eq. (2), the film pressure Π after curvature is on the order of only 10 to 20 dynes/cm. At such low pressures oil-phase molecules could easily penetrate a film unrestricted by barriers (as in a Langmuir trough).

This picture of events at the interface is essentially a mechanical one. It differs from a rigorous thermodynamic treatment in that it requires a stepwise process rather than a single, unified one to account for interphase curvature. In this view soap in the *flat* interphase is penetrated by alcohol and possibly by hydrocarbon molecules derived from the oil phase. This increases the interphase pressure, tending to expand the *flat* film. Acting by another route, but in the same direction, the alcohol remaining in the oil phase is available for adsorption to the original oil-water interface, reducing its tension and thus its tendency to contract. The net effect is to greatly expand the interface, resulting in sharp film curvature.

In this connection it is appropriate to point out that these tensions and pressures are no less operable, and their effect no less viable, in macroemulsion systems. If there is no pressure or tension gradient across the original interphase in these systems, no emulsion will form regardless of the amount of work done on the system. Moreover, when the unrestricted contractile force $(\gamma_{o/w})_a$ in the curved film is opposed by a film pressure Π that is lower in magnitude, the total interfacial area will shrink with time, causing the two phases eventually to separate.

F. Recent Developments

In the past several years a number of papers on microemulsions have appeared, but because they were generally unrelated to the development of the mechanical model of Fig. 5, they have not been discussed up to this time. The diversity of the approaches to the problem of microemulsions embodied in these studies testifies to the wide range of interest in these systems.

Adamson (6) adroitly pointed out that water-in-oil microemulsions can exist in equilibrium with an essentially noncolloidal, aqueous second phase and that this circumstance provides a model for making realistic thermodynamic measurements in these systems. For example, measurements of the distribution of water and electrolyte can be made, and since the aqueous phase is noncolloidal, the activities of the various species can be determined. It is assumed that, in a system containing an ionizable surfactant, the interfacial tension of the water-in-oil microemulsion is positive and not extremely low (about 5 dynes/cm). At equilibrium Adamson balanced the Laplace pressure across the emulsion interphase with the osmotic pressure difference between the oil and water phases. On this basis, his model accounted for the general properties of water-in-oil microemulsions.

Using Adamson's model, Tosch, Jones, and Adamson (29) have extended the measurements of distribution equilibria to a wide range of phase compositions. This work supports the conclusion that systems of water, gasoline fractions, sulfonated hydrocarbon, and isopropyl alcohol can be in equilibrium with an aqueous external phase providing the latter is saline. In other words, electrolyte in the aqueous phase serves to pump water into the oil phase in the form of microdroplets. In fact, these authors observed that unless electrolyte was present in the external phase, the water-in-oil microemulsion would pass through a viscoelastic-gel stage and then invert to an oil-in-water macroemulsion. This knowledge would appear to have important applications in an oil refinery.

Shah (30) traced the inversion of microemulsion systems by means of electrical conductivity, birefringence, and nuclear-magnetic-resonance (NMR) measurements. He found that the conductivity measurements passed through several inflection points corresponding to changes in the kind of colloidal aggregation. The original optically isotropic, transparent water-in-oil emulsion was considered to consist of spherical droplets, as was the final transparent oil-in-water emulsion. During inversion the viscoelastic system was turbid and optically birefringent. The NMR data suggested that water existed in two distinct molecular environments in this region, first in the form of water cylinders and, as more water was added, in the form of water lamellae.

Earlier work on microemulsions using NMR measurements elucidated some of the details of the structure of the interphase of the spherical droplets. Cooke and Schulman (21) interpreted their results as indicating that the hydrocarbon chains of the potassium oleate in their water-in-oil microemulsions were free (not restricted) and that the interphase was therefore in the liquid state. Zlochower and Schulman (31) found evidence from NMR spectra to indicate that 2-amino-2-methyl-1-propanol (AMP) binds water at the interface of a water-in-benzene microemulsion. They concluded that this immobilization was primarily the result of AMP's participation at the interface in a hydrogen-bonded network with water.

Gillberg, Lehtinen, and Friberg (32) studied the NMR spectra and the phase equilibria of water-surfactant-alcohol compositions in the presence of a fourth, hydrocarbon, phase. These authors considered that, when hydrocarbon dissolves in the homogeneous alcohol phase to form a "micellar solution" (L_2 phase), the system is, in fact, a water-in-oil microemulsion. The NMR data were interpreted to indicate that (1) solubilization of water in the micelles is increased by the presence of hydrocarbon intercalated among the tails of the alcohol and surfactant species and (2) the distribution of alcohol between the intermicellar solution (oil phase) and micelles (interphase) determines the water solubilizing power of the systems. This experimental evidence may support the earlier proposal regarding the distribution of alcohol between the oil phase and the interphase in both water-in-oil and oil-in-water microemulsions, and the development of negative interfacial tensions (23).

Further insight into the effect of the nature of the interphase on the formation of microemulsions was obtained by Rosano, Peiser, and Eydt (33). In an extension of the experiments of Cooke and Schulman (21) they found that the formation of microemulsions stabilized by salts of dodecyl sulfate was insensitive to the nature of the cation, but when soap was the stabilizer, microemulsification was strongly cation dependent. They also pointed out that the addition of water to water-in-oil microemulsions does not increase the total interfacial area. Moreover, the low values of the calculated free energy of distribution of long-chain alcohol between the interphase and the oil phase led them to conclude that the alcohol acts to lower the original oil-water interfacial tension rather than to form a stoichiometric complex in a mixed film with soap or detergent. They suggested that the roles played by the osmotic pressure and the Laplace pressure account for the experimental observation that water-in-oil microemulsions are easier to form than oil-in-water ones.

IV. PRACTICAL APPLICATIONS

It has been said that nothing is so practical as a sound theory. In this context the soundness of the theory just presented can be judged by its ability to put new tools in the formulator's kit and to make suggestions for new or improved products. To the extent that this exercise yields positive results, it adds stature to the theory.

In this assessment it should be remembered that the interphase of a microemulsion is 10 to 100 times greater than that of a macroemulsion and hence has a profound effect on the system as a whole.

As shown in Table 1, a 1-µ-diameter droplet requires 1.5% emulsifier on the volume of the droplet, whereas a 0.1-µ-diameter droplet needs 14%. This geometry is responsible for the rule of thumb that a minimum of 10% emulsifier on the weight of the oil is necessary for microemulsification under ideal conditions. It is within such a framework of expanded interfacial areas that the following applications are considered.

A. Emulsifier Selection: The HLB Scheme

Perhaps no single tool of the emulsion formulator has been so widely used as the hydrophile-lipophile balance (HLB) scheme of Griffin (34). This emulsifier-selection system, originally designed for oil and water emulsions stabilized with nonionic surfactants, has been widely utilized in all areas of emulsion formulation. By this scheme, an emulsifier is matched by its calculated HLB number with the "required" HLB of the oil to be emulsified. As helpful as it is, however, this scheme does not eliminate trial-and-error experiments to establish the best chemical type of emulsifier to use for maximum stability, and it does not apply with any pertinence to emulsions other than those stabilized with nonionics.

Since the techniques for using HLB are well known, only those aspects of the scheme in which microemulsions are involved will be discussed here. These fall into two categories: (1) the application of microemulsion studies to the theoretical significance of HLB and (2) how to use the HLB scheme in formulating microemulsions. In this connection it may be observed that, although HLB is not very valuable in finding a microemulsifier for a given oil, once one has been found by trial and error, HLB applies with rigor because the microemulsion system is a stable one.

1. *MICROEMULSION THEORY VERSUS THE HLB SCHEME*

The study of microemulsions has made several contributions to the understanding of the enigmatic HLB scheme. The use of microemulsion systems as models has amplified the effect of the composition of the interphase on the properties of the emulsion as a whole and has clarified a number of heretofore unexplained aspects of HLB. Three of these are discussed here.

a. Additivity. Ordinarily the HLB of a mixture of nonionic emulsifiers is the weight average of the HLB of its components. This rule will hold providing the chemical types are the same. When they are different, however, the additivity rule does not hold, as the work of Schott (35) has demonstrated. Nevertheless, in actual practice, the formulator has little choice but to make the assumption that the behavior of his mixture corresponds to its calculated HLB, regardless of its composition. This has led to considerable confusion in the practice of the HLB system.

The root of this problem is that each component of the emulsifier mixture will be driven to partition among the oil phase, water phase, and interphase according to its own chemical potential (23, 24). This thermodynamic function is quite dependent on the nature of the tail of the nonionic. The problem is further complicated by the fact that commercial nonionics are seldom homogeneous compounds, but are complicated mixtures of long- and short-chain ethylene oxide adducts (36). The longer chained adducts will generally partition between the water phase and the interphase, leaving a minimal fraction in the water phase. The shorter chained species, which are water insoluble, will, on the other hand, partition between the oil phase and the interphase. In this case, however, a substantial fraction will be left behind in the oil phase. *This changes the composition of the oil phase, raising its required HLB.*

The subtlety that the required HLB of an oil depends on the fraction of low-HLB components in the emulsifier mixture explains two empirically determined formulating principles. One is that the required HLB of an oil increases as the amount of emulsifier used in the system is increased. The other is the widely accepted recommendation to deliberately blend high- and low-HLB nonionic emulsifiers (which are already mixtures).

In the first case this means that, as more emulsifier is used to emulsify a given oil, one should increase the calculated HLB of the emulsifier mixture. The reverse applies, of course, and has even more practical significance since it leads to the achievement of optimum stability at lower emulsifier costs.

In the second case the presence of more low-HLB emulsifier in the mixture makes it easier to emulsify the oil. This results from the depression of the original interfacial tension $\gamma_{o/w}$ to $(\gamma_{o/w})_a$ by the oil-soluble surface-active agent. This, in turn, lowers the interfacial tension γ_ϕ [Eq. (2)], reducing the amount of work $-\gamma_\phi \, d\underline{A}$ necessary to create the new interfacial area.

These considerations apply with equal force to macroemulsion and microemulsion systems. The effect on the latter, however, is much greater due to the much larger interfacial area and thus the amount of emulsifier involved.

b. Soaps versus nonionics. Although soaps are not included among the emulsifiers for which the HLB scheme is normally operable, microemulsion studies have developed data that suggest a way to make HLB applicable to soaps. This is important because soaps are among the cheapest of emulsifying agents.

Soap is an unusual member of the ionic class of emulsifiers in that it can assume a number of forms, depending on the pH of the system. For example, a dilute laurate soap exists as the cation and the laurate ion above pH 10.5. However, at pH 8.8 one-half the "acid" soap is in the form of the laurate ion and the other half in the form of free fatty acid. Finally, at pH 6.8 there are two free lauric acid molecules for each laurate ion. At intermediate pH values the ratio of free acid to ion is proportional to the pH.

The effect of this behavior on the formulation of an emulsion can be translated directly into HLB terms. At any given pH the cation and carboxylate-ion portion of the mixture will act as a soap.

This will have an HLB above 20, but probably less than the value of 40 assigned to sodium lauryl sulfate (34). At the same time the free-fatty-acid portion of the mixture will act as water-insoluble alcohol with an HLB in the range 0.5 to 1.5. Thus the ratio of free fatty acid to carboxylate ion will determine the HLB of the mixture. It is clear that a soap can be adjusted to the required HLB of almost any oil by raising or lowering the pH of the system.

A simple and effective way to lower the pH of the soap is with a weak acid like boric acid. In the formula of the oil-in-water kerosene emulsion given in Table 4, omission of boric acid resulted in the formation of an unstable macroemulsion; with boric acid, a microemulsion was formed and exhibited strong Tyndall scattering, indicative of droplets in the 0.1-µ size range.

The emulsion was made by dissolving the soap in the kerosene and adding the boric acid dissolved in a small aliquot of warm water. A clear water-in-oil microemulsion was first formed and on further addition of water passed through a clear-viscoelastic-gel stage. Further addition of water resulted in a fluid, translucent oil-in-water microemulsion.

Borax is a common source of boric acid in such systems (cf. the carnauba-wax emulsion of Section II.B.1). The effectiveness of borax is due to the fact that at elevated water temperatures it decomposes into boric acid and sodium hydroxide.

TABLE 4

Kerosene-in-Water Microemulsion[a]

Ingredient	Quantity (g)
Kerosene (deodorized base oil)	20.0
Oleic acid	4.0
2-Amino-2-methyl-1-propanol	2.5
Boric acid to bring pH to 8.7	
Water	200.0

[a]From Ref. 19.

c. Chemical type. As indicated, one of the serious shortcomings of the HLB scheme is its failure to relate emulsion stability to the chemical nature of the oil phase or emulsifier. Although the microemulsion theory advanced here does not fully explain this, it does offer some useful suggestions to the formulator for effecting stability. For example, it was demonstrated (19, 23) that a change of only a few CH_2 groups in the length of the lipophilic moiety of an emulsifier effects an abrupt change from a macroemulsion to a microemulsion. This applies to both oil-in-water and water-in-oil systems. It was also shown that, when the emulsifier is constant, a small change in the structure of the oil phase, as from mineral oil to kerosene, can effect the same abrupt, beneficial transition. This knowledge is valuable to the formulator as a way of improving his macroemulsion or microemulsion by slightly altering the composition of his emulsifier or oil, and it tells him in which direction to do so. This can then be related to the HLB of the emulsifier or the required HLB of the oil.

2. FORMULATION OF MICROEMULSIONS

The simplest and most straightforward recommendation for finding a microemulsifier for a given oil is to test a wide variety of chemical types whose HLB values approximately match the required HLB of the oil as indicated in the literature (37). For example, 1 part of emulsifier is dissolved in 1 part of oil (with heating if necessary) and then an equal volume of water is slowly poured into this mixture. If the system does not tolerate this water--or if it does, but turns white--it may be assumed that this particular emulsifier will not form a microemulsion in the normal concentration range. If, however, the system tolerates water and remains clear or opalescent, a water-in-oil microemulsion has been formed. It remains to be seen if this system will invert to an oil-in-water microemulsion.

On further addition of small aliquots of water to this water-in-oil microemulsion, an oil-in-water macroemulsion will usually form, as indicated by the appearance of whiteness. Much less frequently the system will pass through a stiff-viscoelastic-gel stage, followed by thinning to a fluid oil-in-water microemulsion. Such a system will exhibit a pronounced Tyndall effect or, with luck, will be transparent.

Once an emulsifier has been found that produces a microemulsion with the given oil, whether oil in water or water in oil, a second screening is undertaken to try to reduce the amount of emulsifier. In the course of this fine adjustment, if the viscosity of the emulsion is high, the HLB of the emulsifier should be increased. On the other hand, if the emulsion is fluid and the particle size is larger than desired as indicated by the guidelines of Table 3, the HLB of the emulsifier should be lowered.

Although this method of finding a microemulsifier is a workable one, it is obviously time consuming and its chances of success without a great many experiments are small. Experience has shown that most oils cannot be microemulsified without a major alteration of their composition. This can occasionally be accomplished by utilizing an emulsifier mixture containing an important fraction of low-HLB components or by modifying the required HLB of the oil.

Where permanent stability is indispensible, as in an essential-oil formulation or a fragrance for a shaving lotion, 200 to 500% emulsifier on the weight of the oil is sometimes successful where lower concentrations are not. In these cases it is important to be certain that the HLB of the emulsifier system is reasonably temperature resistant.

In the case of nonionic emulsifiers temperature plays a prominent role, along with HLB, in determining the partitioning of emulsifier between phases. Microemulsions are particularly sensitive to temperature. Raising the temperature of the system decreases the solubility of nonionics in water, thereby increasing their concentration in the interphase and/or oil phase. This actually decreases the HLB of the emulsifier. Lowering the temperature produces the reverse effect. Neglect of the temperature dependence of the HLB of a given emulsifier can cause a microemulsion stabilized with it at one temperature to quickly break at higher or lower temperatures not far removed from the original one.

B. Rheological Applications

The consistency of an emulsion as measured by its viscosity or yield value may frequently make an important contribution to the value of an emulsion in the marketplace. This applies to both oil-in-water and water-in-oil emulsions, be they macro or micro. For example, high viscosity may be a vital attribute of a furniture polish, or a high yield value may determine the sales appeal of a hair pomade, just as fluidity is essential to a floor-wax emulsion.

Whereas control of consistency is relatively simple in the case of macroemulsions, it presents problems in microemulsion systems. An appropriate thickening agent in the external phase of a macroemulsion can be expected to regulate its consistency without interfering with the stability of the system. In fact, it may prolong it. This is not so in microemulsion systems. These formulations are critically dependent on the amount and kind of emulsifier. The addition of thickening agents to these systems will more often than not destroy their stability.

In the absence of thickening agents the flow properties of an emulsion can be conveniently controlled by the properties of the interfacial film. Schulman and Cockbain (14) made a correlation between the rheological properties of the finished emulsion and the state of the interfacial film. This correlation uniquely applies to microemulsions in which a higher ratio of interphase to disperse phase magnifies the effect many times. In the original work it was found that a liquid expanded film yielded a fluid emulsion; a liquid condensed film resulted in a viscous emulsion or cream. Subsequent work with microemulsions has indicated that a simple way to condense a film to make a more viscous emulsion is to add a long-chain alcohol like cetyl to the existing emulsifier system. Alternative thickeners would be low-HLB nonionic emulsifiers. Very small amounts of these in the interphase have a pronounced effect on the viscosity of the emulsion.

The association of microemulsions with a liquid crystalline (mesomorphous) phase (8) has also introduced the possibility that the consistency of these may be influenced by factors other than the state of the interphase. Friberg and Solyom (38), for example, suggest that the general increase in viscosity with increasing emulsifier content may be related to the presence of a lamellar mesophase.

C. Hints on Formation of Microemulsions

The theory points out several formulation variables that are critical to the formation of microemulsions. Some of these have already been discussed; others are considered here.

1. DEPRESSION OF $(\gamma_{o/w})_a$

In view of the importance attached to the depression of $(\gamma_{o/w})_a$ in the theoretical discussion of Section III, considerable attention has been directed to finding practical applications of this concept. At present there appear to be two ways to depress $(\gamma_{o/w})_a$: one is to change the composition of the oil phase, and the other is to change the chemical type of the emulsifier. Both techniques have their place in the formulator's repertoire.

Information concerning the depression of the interfacial tension between oil and water by alcohols is available (25-27) and is indicative of the effectiveness of partitionable material in depressing $(\gamma_{o/w})_a$ in emulsion systems. The extent of the depression of $(\gamma_{o/w})_a$ for a given oil and the rate at which this occurs as the concentration of alcohol is increased are different for each class of alcohol, although they are substantially independent of alcohol-chain length in any homologous series. In general it can be estimated that a reasonable amount of alcohol will asymptotically depress $(\gamma_{o/w})_a$ to about 15 dynes/cm. In some instances, however, above a certain concentration of alcohol in the particular oil, hydrogen bonding will cause alcohol association into aggregates, limiting the depression of $(\gamma_{o/w})_a$ (39). This is similar to the critical micelle concentration (CMC) of a detergent in water. In this case surface tension becomes essentially constant beyond the CMC. Thus combinations of alcohols and oils that aggregate should be avoided when maximum depression of $(\gamma_{o/w})_a$ is desired.

There are, however, aspects of the depression of $(\gamma_{o/w})_a$ as it relates to emulsion stability and the formation of microemulsions that remain obscure. For example, the interfacial tension against water of most paraffinic hydrocarbons, like n-hexane or n-tetradecane, is depressed from 55 to 15 dynes/cm by a small amount of n-alkanol irrespective of the alcohol-chain length. In spite of this, a change in alcohol-chain length of a few CH_2 groups effects an abrupt change from a microemulsion to a macroemulsion. This was attributed to the shape of the wedge formed among the tails of the alcohol and soap: the longer the alcohol tail, the sharper the wedge and the higher the $\pi_{\bar{G}}$. A similar pattern was observed with benzene as the oil phase. The surface tension of this oil against water can easily be depressed from 35 to 15 dynes/cm by n-alkanols or by methyl cyclohexanol. Methyl cyclohexanol, however, appears to be more effective with soap than an n-alkanol in achieving water-in-oil microemulsions (18, 23). Observations like this suggest that the interaction between the oil phase and emulsifier, which is responsible for stability and the formation of microemulsions, has not been completely identified.

Nevertheless, the concept of alcohol partitioning between the oil phase and the interphase as an index of microemulsification remains a useful tool. For example, in the case of oil phases like the paraffin or microcrystalline waxes to which ordinary alcohols fail to partition in sufficient amount because of their high polarity, theory teaches that higher-molecular-weight alcohols, which are less polar, should effect microemulsification. Alcohols like ceryl and myricyl do, indeed, microemulsify these oils with soaps (5, 23). The same principle applies to the glyceride oils, which can be put in the same category, emulsion-wise, as the paraffin waxes.

It follows that oils to which ordinary long-chain alcohols, the free fatty acids of soap, low-HLB emulsifiers, or long-chain amines partition favorably can easily be microemulsified. Such oils are pine oil, turpentine and related terpenes, kerosene, mineral oil, benzene, toluene, xylene, chlordane, o-phenyl phenol, cyclohexane, indene, carnauba wax, candelilla wax, and the oxidized synthetic waxes. These oils can be microemulsified with appropriate soaps or nonionics to yield oil-in-water emulsions of commercial value. This list is relatively small.

Where it is economical and practical to do so, this list may be enlarged by modifying the composition of the oil. This may consist of diluting the oil with another oil in order to increase the solubility of the alcohol in the mixed oil phase. On the other hand, it may be necessary to decrease the amount of alcohol in the oil phase. An example of this is the dilution of ouricury wax with paraffin or microcrystalline waxes (40). This technique is applicable, with modifications, to many other situations.

In utilizing these principles it is important not to confuse the solubility of an emulsifier in the oil phase with its chemical potential or partitioning coefficient. Most alcohol-like materials are soluble in most oils. Exceptions are nonionics of high HLB or alkali-metal soaps. The significant point is that an alcohol that is soluble in the oil phase may, when water is added, almost completely leave this phase in order to achieve partitioning equilibrium in the system: oil-interphase-water. It must be remembered that emulsification depends on partitioning, not solubility.

2. THE INFLUENCE OF CATIONS

The kind of cation used with a soap or anionic detergent appears to play an important role in the kind of emulsion formed. This is particularly true of microemulsions. Early investigators of emulsion systems soon learned that polyvalent cations usually resulted in water-in-oil emulsions--and monovalent ones, in oil-in-water emulsions. From this they made certain theoretical deductions. Subsequently it has been discovered that emulsions are more complicated than they suspected. For example, a monovalent cation will form water-in-oil or oil-in-water microemulsions, depending on the volume of water in the system. Also, Na^+, K^+, Li^+, Rb^+, and Cs^+ each behaves emulsionwise in its own way (33, 41), as do ammonia and the various amines and amino alcohols commercially used as cations for fatty acids, alkyl sulfates, etc. Since at this time no full explanation for these differences in behavior is available, we present a number of empirical observations that may be helpful to the formulator.

There seems to be considerable evidence that 2-amino-2-methylpropanol (AMP, Commercial Solvents Corp.) is probably the most powerful soap cation for the microemulsification of an oil-in-water system. For example, in the kerosene-in-water emulsion formula given in Table 4, in which boric acid was used to evolve free oleic acid from the AMP oleate, no cation other than AMP was able to microemulsify this oil. Several explanations of this have been essayed (22, 31), but no completely satisfactory explanation is yet available.

Another empirical fact is that in many instances at least 2 moles of cation for each mole of soap anion is required to form an oil-in-water microemulsion. This is also shown in the kerosene emulsion example.

Finally, Sears and Schulman (41) have pointed out that electrostatic repulsion between adjacent soap molecules in a monolayer probably plays only a small role in determining the expansion of an interfacial film. Instead, they suggest that the degree of expansion is related to the size of the hydrated counterion associated with the soap and that the area per cation increases in the order Li^+, Na^+, and K^+. This may explain why a K soap is often more effective in making an oil-in-water microemulsion than is a Na soap.

3. *INVERSION TECHNIQUES*

Oil-in-water microemulsions stabilized with ionic emulsifiers can be formed by an inversion technique. The practical advantage of this is that it can save emulsifier. For example, when the kerosene emulsion of Table 4 is made by the inversion process, it requires less soap than when the kerosene is added to the aqueous soap solution. It is an empirical rule that oil-in-water emulsions made by an inversion route require the least emulsifier. On the negative side, the inversion process may often be cumbersome or time consuming. A balance must be struck.

D. Microemulsification as an Analytical Tool

Because a microemulsion is a three-phase system in equilibrium, it is possible to make certain inferences concerning the properties of, and interactions among, the molecules of the oil phase, interphase, and water phase. To this extent the theory may be used as an analytical tool. Considerable information concerning thermodynamic functions or such parameters as partition coefficients, activities, interfacial tension, and surface free energy can be gleaned from the simple fact that the several phases are in equilibrium. In addition, the equilibrium state can reveal the approximate magnitude of electrostatic and van der Waals forces. These are reflected in the behavior of the system as a whole.

In addition, a microemulsion system can be used to obtain information about the nature of the oil phase. A case in point is carnauba wax. This wax was studied by means of surface-chemistry techniques to determine why it is unique among waxes in regard to emulsifiability (5). In the course of this work it was found that when 1 part of myricyl alcohol was mixed with 3 parts of paraffin wax, a microemulsion comparable to carnauba wax was obtained with the same emulsifying system. It was also found that shorter chain alcohols like stearyl could not be substituted for myricyl. This immediately suggested that a very-long-chain-alcohol constituent of the original carnauba wax was responsible for its extraordinary emulsifiability. A study of the components of carnauba wax indicated that the components most likely to act in this manner were the ω-hydroxycerotic acid ester of myricyl alcohol and its homologs (42). The surface-chemistry analysis then placed the hydroxy ester in the interphase, oriented with its alcohol in the water phase and its ester group in the middle of the interphase. At first the presence of the alcohol in the interphase was seen as causing long-chain nonhydroxy esters to penetrate the interphase and expand the film. This was equivalent to raising $\Pi_{\overline{G}}$.

However, in view of the theory presented in Section III, it is now much easier to believe that the ester group of the very long hydroxy ester also takes its position at the water side of the interphase alongside the ω-hydroxy group. Under these circumstances it is this ester group that causes expansion of the film along with the ester group of the normal ester constituents of the wax. At the natural concentration of the hydroxy ester the interphase may be presumed to be liquid condensed. The fact that carnauba-wax microemulsions will tolerate only small amounts of a diluent like paraffin wax seems to substantiate this analysis. Apparently the film is naturally a well-balanced one for microemulsification and is extremely sensitive to any changes in composition.

The light that this study sheds on the relationship between the molecules of carnauba wax and its emulsifiability is helpful in understanding the performance of other emulsifiable materials. A similar behavior probably explains why waxes such as candelilla, ouricury, and the synthetic, oxidized, microcrystalline Fischer-Tropsch and polyethylene waxes form microemulsions. Other instances of how such an analytical technique can have practical applications easily come to mind.

REFERENCES

1. P. Becher, *Emulsions: Theory and Practice*, 2nd ed., Reinhold, New York, 1965.
2. W. C. Clayton in *The Theory of Emulsions and Their Technical Treatment*, (Sumner ed.), 5th ed., Churchill, London, 1954.
3. J. W. McBain, *Colloid Science*, D. C. Heath, Boston, 1950, p. 265.
4. J. H. Schulman and J. A. Friend, *J. Soc. Cosmetic Chemists (Great Britain)*, 1, 381 (1949).
5. L. M. Prince, *Soap Chem. Specialties*, 36, No. 9, 103 (1960); ibid., No. 10, 99 (1960).
6. A. W. Adamson, *J. Colloid Interface Sci.*, 29, 261 (1969).
7. M. Rosoff, paper presented at 25th Anniversary Meeting, Society of Cosmetic Chemists, New York, May 25-26, 1970.
8. S. Friberg, L. Mandell, and M. Larsson, *J. Colloid Interface Sci.*, 29, 155 (1969).
9. P. A. Winsor, *Trans. Faraday Soc.*, 44, 376 (1948); ibid., 46, 762 (1950).
10. D. F. Sears, Tulane Medical School, private communication, 1970.
11. T. P. Hoar and J. H. Schulman, *Nature (London)*, 152, 102 (1943).
12. P. Walstra, *J. Colloid Interface Sci.*, 27, 493 (1968).
13. P. Walstra and H. Oortwijn, *J. Colloid Interface Sci.*, 29, 424 (1969).

14. J. H. Schulman and E. G. Cockbain, *Trans. Faraday Soc.*, *36*, 551 and 661 (1940).
15. J. H. Schulman and T. S. McRoberts, *Trans. Faraday Soc.*, *42B*, 165 (1946).
16. J. H. Schulman and D. P. Riley, *J. Colloid Sci.*, *3*, 383 (1948).
17. J. H. Schulman and J. A. Friend, *J. Colloid Sci.*, *4*, 497 (1949).
18. J. E. Bowcott and J. H. Schulman, *Z. Elektrochem.*, *59*, Heft 4, (1955).
19. J. H. Schulman, W. Stoeckenius, and L. M. Prince, *J. Phys. Chem.*, *63*, 1677 (1959).
20. W. Stoeckenius, J. H. Schulman, and L. M. Prince, *Kolloid-Z.*, *169*, Heft 1-2, 170 (1960).
21. C. E. Cooke, Jr., and J. H. Schulman, *Proc. 2nd Scandinavian Symp. Surface Activity, Stockholm, November 1964*.
22. J. H. Schulman and J. B. Montagne, *Ann. N.Y. Acad. Sci.*, *92*, 336 (1961).
23. L. M. Prince, *J. Colloid Interface Sci.*, *23*, 165 (1967).
24. L. M. Prince, *J. Soc. Cosmetic Chemists*, *21*, 193 (1970).
25. E. Hutchinson, *J. Colloid Sci.*, *3*, 219 and 235; ibid., *6*, 521 and 531 (1948).
26. J. L. Jasper and B. L. Houseman, *J. Phys. Chem.*, *67*, 1548 (1963).
27. R. S. Valentine and W. J. Heideger, *J. Chem. Eng. Data*, *8*, 27 (1963).
28. L. M. Prince, *J. Colloid Interface Sci.*, *29*, 216 (1969).
29. W. C. Tosch, S. C. Jones, and A. W. Adamson, *J. Colloid Interface Sci.*, *31*, 297 (1969).
30. D. O. Shah, *Science*, *171*, 483 (1971).
31. I. A. Zlochower and J. H. Schulman, *J. Colloid Interface Sci.*, *24*, 115 (1967).
32. G. Gillberg, H. Lehtinen, and S. Friberg, *J. Colloid Interface Sci.*, *33*, 40 (1970).
33. H. L. Rosano, R. C. Peiser, and A. Eydt, *Revue Française des Corps Gras*, *16*, 249 (1969).

34. W. C. Griffin, in *Encyclopedia of Chemical Technology*, Vol. 8, 2nd ed., Wiley, New York, 1965, p. 131.
35. H. Schott, *J. Pharm. Sci.*, *58*, 1443 (1969).
36. N. Shachot and H. L. Greenwald, in *Nonionic Surfactants* (M. J. Schick, ed.), Dekker, New York, 1967, p. 31.
37. P. Becher, in *Nonionic Surfactants* (M. J. Schick, ed.), Dekker, New York, 1967, p. 612.
38. S. Friberg and P. Solyom, *Kolloid-Z., Z. Polymere*, *236*, Heft 2, 173 (1970).
39. R. Aveyard and R. W. Mitchell, *Trans. Faraday Soc.*, *65*, 2645 (1969).
40. L. M. Prince, U.S. Pat. 2,441,842.
41. D. F. Sears and J. H. Schulman, *J. Phys. Chem.*, *68*, 3529 (1964).
42. K. E. Murray and R. Schoenfeld, *Australian J. Chem.*, *8*, 437 (1955).

Chapter 4

AGRICULTURAL EMULSIONS

Paul L. Lindner

Witco Chemical Corporation
Chicago, Illinois

I. INTRODUCTION. 181
II. CALCULATION OF FORMULATIONS 183
    A. American Unitary System 183
    B. British Unitary System. 184
    C. Decimal Unitary System. 184
III. SOLVENTS FOR AGRICULTURAL FORMULATIONS. 185
    A. Characteristics of Required Solvents. 185
    B. Determination of Solvent Polarity 186
IV. SURFACTANTS FOR AGRICULTURAL FORMULATIONS 187
    A. Federal Regulations for Surface-Active Agents
       for Agriculture 187
    B. Cationics 188
    C. Nonionics 188
    D. Anionics. 190
V. EMULSION STABILITY. 190
    A. Rate of Sedimentation: Effects of Particle
       Size and Viscosity. 191

	B.	Behavior of Emulsions with Aging.	192
	C.	Effect of Charge.	192
	D.	Effect of Solvation	194
	E.	Effect of Surfactant Location	195
	F.	Effect of Multicomponent Systems.	196
	G.	Stability of Water-in-Oil Emulsions	196
	H.	Effect of Critical Micelle Concentration.	196
	I.	Spontaneous Emulsions and Quasiequilibrium.	197
	J.	Selection of Emulsifier for Optimum Emulsification.	198
	K.	Competitive Kinetics of Surface Adsorption and Surface Packing with Blended Emulsifiers.	199
VI.	BLENDING EMULSIFIERS AND ECONOMICS.		200
VII.	WATER AND AGRICULTURAL FORMULATIONS		201
	A.	Effect of Electrolytes.	201
	B.	Effect of Temperature	206
	C.	Effect of Dilution Rate	207
	D.	Effect of Mixing.	207
VIII.	TESTING		208
	A.	Testing Methods for Emulsifiable Concentrates.	208
	B.	Correlation of Laboratory Aging Tests and Field Behavior.	209
IX.	SUGGESTED OPTIMIZATION OF EMULSIFIABLE CONCENTRATES		210
X.	SELECTION OF EMULSION TYPE.		214
	A.	Water-in-Oil Formulations (Inverts)	214
	B.	Tight or Loose Emulsions.	216
	C.	Fruit-Tree-Spray Formulations	217
	D.	Aerosol Sprays.	217
	E.	Flowable Formulations	218
	F.	Solubilizations	219
	G.	Cattle Dips	219
XI.	SOME SPECIAL FORMULATIONS		219
	A.	Tank Additives.	219
	B.	Aquatic Herbicidal Oil Formulations	221

4. AGRICULTURAL EMULSIONS

      C. Mosquito-Control Formulations. 221
      D. Triple-Phase Systems and Tank Blends 222
      E. Formulations with Liquid Fertilizers 224
XII. SOIL PENETRATION, ADSORPTION, AND MOBILITY 226
XIII. APPLICATION TECHNIQUES 227
      A. Direct Surface Application 227
      B. Meristemic Control 228
      C. Spray Application. 229
REFERENCES 236

I. INTRODUCTION

Agriculture is basically a competitive low-profit industry and as such requires a constant revision of expenditure against the economy of crops. The main expenditures are for machinery, labor, seeds, feed, fuel, fertilizer, toxicants, and often water. The farmer weighs these expenditures against the income from his operation. Any industry geared to agriculture will take this into account. One of the cost-saving operations is diminution of transport costs by using concentrates that can be readily diluted in the field with water. It also allows using the same spraying equipment and the same concentrate for different field requirements. For this reason herbicides, insecticides, and fungicides used to protect plants and animals from pests are formulated mostly in the form of emulsifiable concentrates or wettable powders that can be diluted in the field in order to ensure uniformity of distribution.

The protective materials, which are more or less specific toxicants for the pest with minimum harmfulness for the protected crop, have to be brought in contact with the pest. Certain insecticides whose vapors can be introduced through the breathing tract and contact poisons (i.e., where the walking insect will come in contact with the stationary poison) can be sprayed directly. In most cases, however, it is necessary to add surfactants to the system to improve the distribution of the toxicant over the area, its spreading over the surface, and its penetration into plants, soils, and insects.

Many surfactants have sufficient water solubility to be washed off crops and undoubtedly will be removed without contamination. Establishment of tolerances requires different toxicity data, finding a specific analysis for determining the materials and products of breakdown. Since the available analytical procedures for surfactants are not specific enough and the expenses involved for the investigations that could be required by the FDA to establish tolerances would be economically prohibitive for the amount of surfactants used, there is no probability that surfactant manufacturers will try to establish tolerances. For this reason in the near future only materials exempt from tolerances are expected to be used in agriculture.

The toxicants are used in the following forms:

1. Ultra-low-volume (ULV) liquid toxicants using droplets of 50 to 150 µ uniformly sprayed where no surfactant is required

2. Dusts used mostly with a low concentration of toxicant adsorbed on a clay; cheap, but mostly uneconomical where transportation is involved

3. Wettable powders, used quite often

4. Emulsifiable concentrates used whenever possible

Wettable powders are used where there is no suitable economical solvent and emulsifier system. They are composed of the toxicant and a suitable filler-absorbent, a wetter, so that it can be incorporated into water, and a dispersant to keep it from fast flocculation. The problems with wettable powders are clogging nozzles, flocculation, caking of sediment, and mainly slow production. This is why the formulator prefers whenever possible to formulate toxicants as emulsifiable concentrates.

4. AGRICULTURAL EMULSIONS

II. CALCULATION OF FORMULATIONS

The applicator usually depends on the suggestions of agricultural research stations for the amount of a particular toxicant to counteract or prevent an endemic development of a pest that could harm a particular crop. These amounts are given in weight per area (e.g., lb/acre or kg/ha). It is inconvenient to use a scale in the field, but easy to measure volume, and for this reason formulations are labeled in pounds per gallon or imperial gallon, or percentage of weight by volume. At the same time the emulsifier is mostly specified in weight percentages, and the solvent is measured by volume. Such a system is difficult to handle even with a single toxicant system, while often a multitoxicant (1) system is required. To calculate a formulation the procedure described below is suggested.

A. American Unitary System

In formulating a multitoxicant system in pounds and U.S. gallons we begin by stating the problem as follows: To make a batch take P_1 lb/gal of toxicant 1 (technical) of a_1% activity and density d_1; write P_2, a_2, d_2 for toxicant 2, P_3, a_3, d_3 for toxicant 3, and so on; use e% emulsifiers of density d_e in a solvent of density d_s.

The density of toxicant is determined in solution to minimize technically significant differences. For emulsifier density we suggest taking d_e = 1.035, small differences from that value having a negligible effect on the technical calculation. Based on this suggestion, we calculate the solvent factor of the emulsifier

$$f_e = 1 - \frac{d_s}{1.035} \tag{1}$$

For a 100-gal batch we use T_1 = 10,000P_1/a_1 lb of toxicant 1 with a volume $V_1 = T_1/8.337 d_1$ gal. We calculate T and V for every toxicant $(T_1, V_1; T_2, V_2; T_3, V_3; \ldots)$, where the specific gravity of water $D_w = 8.337$ lb/gal. We sum up the total toxicant weight per batch, $\Sigma T = T_1 + T_2 + T_3 + \cdots$, and the total volume, $\Sigma V = V_1 + V_2 + V_3 + \cdots$. We calculate the specific gravity of the formulation in pounds per gallon using the formulation

$$D = \frac{\Sigma T + 8.337 d_s (100 - \Sigma V)}{100 - ef_e} \qquad (2)$$

The amount of emulsifier in the 100-gal batch will then be $E = eD$ lb per 100 gal. The amount of solvent per such batch will be $S = 100D - E - \Sigma T$ lb, and its volume,

$$V_s = \frac{S}{8.337 d_s} \text{ gal} \qquad (3)$$

Having thus sufficiently exact values for the weight of each ingredient in the 100-gal batch, the calculation of percentual distribution is simple.

B. British Unitary System

For a system using pounds and Imperial gallons we have only to recall that 1 Imperial gallon of water weighs 10 lb and use the same type of calculation with replacement of 8.337 by 10:

$$V = \frac{T}{10d}; \quad D = \frac{\Sigma T + 10 d_s (100 - \Sigma V)}{100 - ef_e}; \quad V_s = \frac{S}{10 d_s} \qquad (4)$$

C. Decimal Unitary System

For a system using percentage of weight per volume and liters we make a calculation for a 100-liter batch. If P_1 in this system represents the required percentage of weight per volume of the given toxicant in the formulation and all other designations remain

4. AGRICULTURAL EMULSIONS

the same, we start the calculation by finding the weight of each toxicant per 100-liter batch, $T_1 = 100P_1/a_1$ kg, and its volume, $V_1 = T_1/d_1$ liters. After summing up the total toxicant weight per 100-liter batch, ΣT, and its total volume ΣV, we calculate the density of the formulation from the equation

$$d = \frac{\Sigma T + d_s(100 - \Sigma V)}{100 - ef_e} \tag{5}$$

Accordingly, we calculate the weight of emulsifier per 100 liters as $E = ed$ kg, the weight of the solvent as $S = 100\,d - E - \Sigma T$ kg, and its volume as $V_s = S/d_s$ liters.

III. SOLVENTS FOR AGRICULTURAL FORMULATIONS

Solvents for agricultural formulations are selected as the most economical systems that have sufficient solvency for the toxicants. Important factors are phytotoxicity (2-5), cold stability of the concentrates, volatility, and behavior in the emulsion. The theory of solutions in organic solvents was developed by Hildebrand and Scott (6,7) and Scatchard (8,9), and the effects of the solubility parameters on micelle formation in nonaqueous solvents have been shown by Little and Singleterry (10,11) and also discussed by Fowkes (12) and Van Valkenberg (13).

A. Characteristics of Required Solvents

More polar toxicants require more polar solvents and will require more polar, or rather more hydrophilic, emulsifier systems. Since the toxicants are formulated as weight per volume (i.e., 4 lb/gal), it may happen that a less polar but more dense solvent will show a better solvency and at the same time lower volatility.

Basically, to form emulsions the solvent has to have no water solubility; otherwise it may be leached out, and the toxicant can form crystals that will clog the spray nozzles. Sometimes it is desirable to use a main solvent, which will keep the emulsion droplets liquid at the temperature of application, and an additive, which may be even a water-soluble, highly polar solvent, whose sole function is to improve the cold stability of the formulation. Such formulations have to be tested for the highest practical field dilution since the distribution of the additive in the water and oil phases may be important for application.

The most frequently used solvents are kerosene, diesel oils, Stoddard solvents, heavy aromatic naphthas, and xylene. The more expensive or "exotic" solvents are ketones (methyl isobutyl ketone, dibutyl ketone, methyl isoamyl ketone, cyclohexanone, isophorone, mesityl oxide), chlorinated hydrocarbons (aliphatic and aromatic), and others (14,15).

B. Determination of Solvent Polarity

Ullrich (16) suggested determining the hydrophilic character of solvents and toxicants by the water-titration method based on work by Greenwald et al. (17). One gram of the organic toxicant or solvent is dissolved in 30 ml of a mixture consisting of 96% dioxane and 4% benzene. The solution is titrated with double-distilled water until a defined cloud point at exactly 20°C is reached. Water-drop size should be less than 0.04 ml, and the speed of titration 6 ml/min. The cloud point is quite sharp. The values can be expressed in milliliters of water as "WW" (Wasserwert meaning water value).

The dioxane-benzene mixture gives a WW of 14.4; white oil, WW 3.35; xylene (technical grade), WW 11.05; DDT (technical grade), WW 11.2; Trithion, WW 11.5; DDVP, WW 23.6; and dimethoate, WW 27.1.

IV. SURFACTANTS FOR AGRICULTURAL FORMULATIONS

A. Federal Regulations for Surface-Active Agents for Agriculture

In the United States, prior to April 3, 1969, surfactants used in pesticide formulations were not regulated and were treated as inert ingredients, with toxicities that are negligible in comparison with those of the pesticides. It has been shown (18,19) that the LD_{50} toxicities of nonionic and anionic surfactants used in detergents and emulsifiers are on the order of 3 to 12 g/kg, whereas ordinary sodium chloride has an LD_{50} of 3 to 4 g/kg and sodium bicarbonate has an LD_{50} of 4 g/kg.

On December 2, 1962, the Food and Drug Administration (FDA) issued the first of the regulations (21CFR120.1001) exempting certain inert ingredients of pesticide formulations from the requirement of tolerances. On April 2, 1969, that regulation was amended to include various surfactants on which sufficient toxicity data were available at that time to exempt them from the requirement of tolerances (20).

The regulation was again amended on February 20, 1970, to include additional surfactants (21) and will undoubtedly be amended in the near future to include additional materials on which sufficient toxicity data have been or are in the process of being obtained.

Many of the surfactants for which toxicity data have not been determined, because of the cost of such procedures, may be withdrawn from use if they can be replaced by available materials and if the market for such materials will not warrant the expense.

B. Cationics

The surfactants used in the formulation of agricultural sprays have to be economical. The cationics are generally avoided because they are much too expensive; might show biological activity (e.g., complexing proteins), which could require setting of tolerances; and show preferential wetting changing with concentration and conditions. Monomolecular layers are wetting, polymolecular layers act hydrophobically. Cationics are also incompatible with cheaper anionic formulations.

C. Nonionics

Nonionic emulsifiers, mostly in the form of ethylene oxide adducts of materials having an active hydrogen capable of being ethoxylated or esterified with a polyethylene glycol ether, were widely used in agricultural formulations and are still exclusively used where the presence of an ionic surfactant may greatly contribute to the breakdown and deactivation of the toxicant. Such cases were substantiated by the investigations of the Shell Chemical Company on DDPV and Chemagro Incorporated on Meta Systox R, and these companies insist on the use of only nonionic emulsifiers for the formulations of these two toxicants.

1. THE HLB METHOD

The concept of a hydrophilic-lipophilic balance (HLB) was developed in patents by Harris and Epstein (22,23) and discussed by Clayton (24). Griffin (25-27) developed a quantitative approach to the basic character of amphiphilic materials, the balance of the hydrophilic and hydrophobic (lipophilic) part of the molecule to form a surfactant, and put them in an arbitrary scale that he called the HLB number system. He also described methods of calculation for some systems. Higher HLB numbers denote a more hydrophilic character of the molecule.

4. AGRICULTURAL EMULSIONS

Racz and Orban (28) suggested determining HLB by its linear dependability on the heat of hydration and found that in using sorbitan and ethoxylated sorbitan esters, HLB = 0.42Q + 7.5, where Q is the enthalpy of hydration in gram-calories per gram measured at 20°C. Middleton (29) suggested determining HLB numbers by a titration method with an oil phase, a surfactant of known HLB, and carbon black as indicator. In this method Middleton finds the locus of correlation between Y, the HLB required for oil-in-water emulsification of the particular oil phase, and C, the HLB number average for the carbon-black transfer to the oil phase, to be Y = 0.45X + 8.3, which suggests also a method of determining the required HLB for the emulsification of a particular oil-phase system. Becher (30) mentions other methods for determining HLB numbers and the ranges of their applicability. Griffin (25) suggests for water-in-oil (w/o) emulsifiers an HLB number of 3 to 6, for wetters an HLB of 7 to 9, for oil-in-water (o/w) emulsification an HLB of 8 to 15, and for solubilization an HLB of 15 to 18.

2. DRAWBACKS OF THE HLB SYSTEM

The HLB system is supposed to help in choosing the right emulsifier for a given system, but in fact it gives only an approximate indication of where the optimum behavior should be and leaves it to the chemists to find the right blend that will really do the job. The HLB number often shows that the approximation is far off the expected point.

Kaertkemeyer (31) investigated optimum systems for the emulsification of "agricultural phytobland oils" (nonphytotoxic, unsulfonatable oils) with blends of eight nonionic surfactants and found that good formulations could be made for one oil with an HLB ranging from 5 to 9, but in the same region poor formulations were also found. The iso-HLB zones of good emulsification of different blends showed no agreement or parallelism. Neither were other systems for classifying nonionic materials useful. Lin (32), in testing optimum emulsification of an oil and water, found the importance of placement. If the emulsifier was placed in the oil phase, an HLB number higher by 2.5 to 3.5 units was required for emulsification. Peterson and Hamill (33) found the HLB system unsuitable for nonaqueous emulsions having an HLB range of 1.8 to 16.7 for oil-in-glycerol emulsions and 4 to 11 for glycerol-in-oil emulsions.

To show how limited the system is for developing agricultural formulations, we can demonstrate that a difference of 0.2 in HLB number may be the difference between an acceptable and unacceptable formulation.

D. Anionics

The other system of surfactants most important in agricultural formulations consists of anionic surfactants, often called soaps because the first surfactants of this class to be used were fatty-acid sodium salts. To this group belong carboxylates, sulfonates, sulfates, and phosphates. They can also be systematized by giving them HLB numbers or water-titration numbers [the WW values of Ullrich (16)], but, as Ullrich has shown in one example (16), a blend of nonionics and anionics has a different WW value than would appear from the additivity principle expressed by Griffin (25,26). Greenwald et al. (17) also showed that water titration in the same family of surfactants shows a linear additive function, but different families form different lines.

V. EMULSION STABILITY

Bancroft (34) introduced the rule that the continuous phase of an emulsion will be determined by the phase in which the emulsifier is more soluble.

4. AGRICULTURAL EMULSIONS

A. Rate of Sedimentation:
   Effects of Particle Size and Viscosity

The stability of a disperse system depends on the rate of the process of reversible aggregation, where the emulsion particles settle (flocculation and sedimentation) and which will depend on Stokes' law (35). The rate of fall of a particle is

$$u = \frac{2gr^2(\rho - \rho')}{9\eta} \tag{6}$$

where g is the gravitational constant, r is the radius of the particle, ρ, is its density, ρ' is the density of the medium, and η is its viscosity. The total rate of sedimentation expressed as the rate of settling of the center of gravity of all emulsion droplets is

$$\bar{u} = \sum_i \frac{4\pi}{3V} r_i^3 n_i u_i \tag{7}$$

where V is the volume of the internal phase and the summation is over all i types of particle. Thus, using Eq. (6), we get

$$\bar{u} = \sum_i \frac{8\pi}{27\eta V} g n_i r_i^5 (\rho - \rho') \tag{8}$$

which shows the tremendous importance of the dimension of the oil droplet unless the density is brought to the density of the medium. At higher concentrations we may use Moulik's (36) "beyond Einstein viscosity region," and at high concentrations the viscosity will depend on the geometry of the emulsion (37,38). With the charges on the oil droplets the aggregation will come to a critical equilibrium distance in a close-packed system (creaming).

B. Behavior of Emulsions with Aging

In emulsions in which the specific gravities of the dispersed and continuous phases have a difference $\Delta\rho = 0.1$ and the diameter of emulsion droplets is 2 μ, the rate of sedimentation will be 0.7 mm/hr (39), but the following phenomena will appear in dilute emulsions:

1. Aggregation. Emulsion droplets come closer without coalescence, and these aggregates have a much higher rate of sedimentation.

2. Sedimentation-Creaming. The droplets settle into a closely packed system (37,38) without coalescence, which is a reversible process.

3. Coalescence. Sometimes the aggregates coalesce into bigger droplets before sedimentation if their protection is not sufficient. In concentrated emulsions the droplets may be already in the sedimented state. In the presence of volatile solvents of high mobility the exchange of materials from the emulsion droplets will be enhanced and the coalescence rate will be increased.

C. Effect of Charge

Van den Tempel (40), Greenwald (35), and Sonntag et al. (41, 42) investigated the stability of oil-in-water emulsions on the basis of the Debye-Hückel model of the electrical double layer. The Stern layer, which contains part of the counterions, is assumed to have a uniform potential ψ. This area contains strongly bound water molecules polarized with the positively charged hydrogen near, and the negatively charged oxygen in the dipole away from, the negatively charged anionic head of the surfactant. The layer is assumed to be only a few to 10 Å thick.

4. AGRICULTURAL EMULSIONS

In oil-in-water emulsions the droplets are stabilized by charges and solvation. Van der Waals attractive forces pull the droplets together, whereas Coulombic forces cause repulsion. The charge by itself is insufficient to prevent coalescence (43). The droplets may be stabilized by the Plateau-Maringani-Gibbs effect, often called the Gibbs elasticity (44-47),

$$E = 4RT \sum_{i=2} \frac{\Gamma_i}{C_i} \frac{1 - (d \ln /d \ln C_i)}{h + 2(d\gamma_i/dC_i)} \quad (9)$$

where Γ_i is the excess surface concentration of species i, C_i is its concentration, γ_i is its activity, and h is the thickness of the stabilizing film. Changes in the rate of film thinning between droplets arise from interfacial flow as a result of a gradient in interfacial tension produced by unequal distribution of the diffusing surfactant over the drop surface. This occurs because there is no adsorption equilibrium at the moment of emulsion formation, and the kinetics of the adsorption requires a certain time to come to equilibrium.

The interaction of Coulombic repulsion and van der Waals attraction will tend to bring the oil droplets to a critical point before the rupture of the continuous-phase film between them will cause coalescence (48-50). Since such distances for nonstabilized emulsions are 0.1 to 5 μ, these emulsions coalesce immediately. In surfactant-stabilized emulsions the droplets may come to a quasiequilibrium position at 50 to 200 Å without coalescence, giving a stable emulsion with a film-rupture probability of 10^{-7} sec^{-1}. Such an equilibrium point will then exceed by 10 times the thickness of the Debye-Hückel distance.

Tartar (51) assumes that the distance of closest approach is proportional to the thickness of the ionic atmosphere (TIA), where

$$TIA = \frac{DT}{\Sigma c_i r_i^2} \frac{1000k}{4\pi eN}^{\frac{1}{2}} \quad cm \quad (10)$$

D being the dielectric constant, c the concentration of ionic species i with charge r, e the electron charge, N Avogadro's number and k Boltzmann's constant. Thus TIA is proportional to $\sqrt{DT}$, and since the temperature coefficient of D is negative, the effect of temperature could be small.

The second process that can occur is irreversible coalescence with increase in surface. Surfactants may in certain cases where solvation is low and the specific gravity is high have little effect on the rate of flocculation, but may retard coalescence, and this will depend on the electrostatic energy barrier. The addition of electrolytes will have an effect on both flocculation and coalescence.

D. Effect of Solvation

In respect to nonionic surface-active agents containing a polyethylene oxide chain, Schick et al. (52) showed that surfactants with a short ethylene oxide chain in the mixture behave like alcohols in the system of microemulsions developed by Schulman et al. (53), since they are less water soluble. Lengthening the ethylene oxide chains makes the molecules more water dispersible and at the same time enlarges their spatial requirements in the water phase with the hydrogen bonding of water molecules. This produces the same effect as increasing the ionic Coulombic repulsion among the charged heads of soap-type molecules. Investigating the change in the interfacial tension of a solution of nonionic surfactants in distilled water in contact with octane Becher (54) found a minimum of interfacial tension with lengthening of the ethylene oxide chain. This would show that in surfactants with a short ethylene oxide chain part of the hydrophilic chain is submerged in the oil phase, and the minimum is the point at which the submersion disappears.

The theory of micelle formation of nonionic surfactants is much more difficult to represent than that of ionic ones. Poland and Scheraga (55-57) tried to treat it from the standpoint of

4. AGRICULTURAL EMULSIONS

hydrophobic bonds, but certain data still do not fit. Boehmke and Heusch (58) assumed that each ether linkage is hydrated with four to four and a half water molecules, a value that varies with different nonionics. In Becher's view (59,60) the ethylene oxide part extended into the water phase is in zigzag form, sheathed with water molecules and dipole-dipole bonded, with a small micellar charge extending for many angstroms into the water phase. Only a weak charge can be found, and there is no Stern layer, only a weak Gouy layer. The hydrogen-bonded water molecules are crowded and increase the spatial requirements of the molecules. Roesch (61) concluded that hydration changes in the zigzag form into a meander form (helix?) that shortens the extension of the molecule into the water phase and more than doubles the dipole moment, and also increases greatly the spatial requirements.

In surfactants with long ethylene oxide chains the barrier to coalescence may be sterical and entropical. Where the stabilized droplets approach each other, there will be some interlocking, which stops the free movement and lowers the entropy. Such interlocking requires activation energy. This energy barrier will protect the emulsion droplets (62-64).

E. Effect of Surfactant Location

From the work of Lin (32) and Middleton (29) it would appear that the placement of the surface-active agent is of great importance. When placed in the oil phase, a part of the ethylene oxide chain will be almost always submerged, and not hydrated, and for this reason surfactants show a higher HLB behavior when placed in water than when placed in the oil phase. Lin and Lambrechts (65,66) also found that the addition of a high-HLB surfactant together with a low-HLB one may prevent the high-HLB surfactant from migrating into the water phase.

F. Effect of Multicomponent Systems

Using one surfactant in the water phase and one in the oil phase, Schulman and Cockbain (67) formed very stable microemulsions and found that to obtain a stable water-in-oil emulsion the emulsifying agent must be brought to the interface and form a rigid film. In the case of a nonionic surfactant the film will be uncharged. A different surfactant type will tend to fit between the positions of the other surfactant, giving the effect of packing the surface and improving the stability of the emulsion droplets, thus preventing coalescence.

G. Stability of Water-in-Oil Emulsions

In water-in-oil emulsions electrostatic stabilization is not efficient, only solvation. The low dielectric constant will make the electrical potential more diffused, and there will be no significant electrostatic potential barrier to prevent the water droplets from coming nearer each other. However, if the water has to break through an interlocking long chain of hydrophobic molecules (steric effect), it may require activation energy. This barrier and the increase in system viscosity may stabilize the emulsion. The stability is caused by the wetting resistance of segments of adsorbed film (68) and will depend on temperature, viscosity of the continuous phase, particle size, and the nature of the adsorbed film, which has to be displaced from the surface to cause coalescence.

H. Effect of Critical Micelle Concentration

Cockbain and McRoberts (68) found that, with the emulsifier placed in the water phase, the optimum stability was at concentrations slightly higher than the critical micelle concentration.

4. AGRICULTURAL EMULSIONS

Since the submersion of surfactant in the oil phase (54) changes the character of the surfactant, there is a complicated situation with the surfactant placed in the oil phase. This becomes even more complicated with emulsifier systems composed of a multiplicity of surfactants.

I. Spontaneous Emulsions and Quasiequilibrium

Prince (69-71) developed the theory that microemulsions form spontaneously when the high initial surface pressure Π is much greater than the interfacial tension $\sigma_{o/w}$, giving a negative total interfacial tension $\sigma_L = \sigma_{o/w} - \Pi$ that causes the emulsion to form. Since the difference in film pressure on both sides of the interface causes the surface to bend, the emulsion droplets are formed with increased surface curvature, until the system comes to a stable equilibrium with the film pressure equal on both sides of the interface, and the interfacial tension positive, but small. The fact that the interfacial tension between miscible liquids is zero would cause us to think that a negative interfacial tension is an improbability for an emulsion. Prince's concept of negative interfacial tension should be regarded as a mathematical approach corresponding to a microscopic reversibility of a kinetic process.

Prince's approach, based on work by Stackelberg et al. (72), explains the spontaneity of emulsification as being caused by negative surface tension. Van't Hoff already suggested that emulsions may be stabilized by negative surface tension. There is no known system with a negative surface tension, as it would have to proceed to a molecular dispersion. Volmer (73) proved that a round emulsion droplet could not be explained without a positive surface tension. Stackelberg et al. thought it possible for the surface tension to drop to zero with a subsequent raise to a positive value since the two-dimensional pressures are dependent on the curvature of the droplet.

Winsor (74) suggested that formulations emulsify

spontaneously if they are not in balance by means of chemical reaction or distribution; that is, with a fatty acid in the oil phase and an amine in the water phase, a spontaneous emulsification can be expected. It will also happen when the system develops a lot of energy by hydration--and dissociation of surfactant, a rapid drop in interfacial tension, and leaching of water-soluble solvents. All these energy changes are the function of the water phase, and this is why spontaneous enulsions can only be found for oil-in-water emulsions, no spontaneous water-in-oil emulsions being known. Spontaneous emulsification will give rise to an emulsion that may be more or less stable. It will nevertheless separate by itself or by centrifugation, and after separation it will no longer be spontaneous, which shows that though the emulsion may be kinetically stable, this is a quasiequilibrium, as the real equilibrium will be reached after separation.

J. Selection of Emulsifier for Optimum Emulsification

On the basis of work by Ross et al. (75), Becher (76) suggested finding the required HLB number of the nonionic emulsifier to get the best stable emulsions. He used the equation $S_1 = \sigma_B - \sigma_A - \sigma_{AB}$ to calculate the spreading coefficient of the system and found that the optimum oil-in-water emulsification was obtained when the spreading coefficient of liquid A over liquid B was zero to slightly negative and at the same time consistent with a low interfacial tension.

For water-in-oil systems the spreading coefficient of water over oil should have the largest negative value. In investigating the stabilization of emulsions with solid powders Schulman and Leja (77) found that the most stable oil-in-water emulsions are formed when the wetting angle is near 90° but slightly smaller for water. This is equivalent to Becher's rule.

4. AGRICULTURAL EMULSIONS

Arai, Shinoda, and Saito (78-80) suggested the phase-inversion temperature (PIT) method for choosing an emulsifier that would give the most stable oil-in-water emulsion for a given system. The method requires shaking the oil phase with water at 2 to 4°C below the PIT for the surfactant, to get the finest dispersion, and then rapidly cooling it down to the storage temperature, preserving the small-particle-size droplets of the emulsion. A PIT change from 27 to 94°C corresponds to an HLB change from 11.1 to 14.7. In agricultural formulations this system can be applied where there are difficulties with determining the HLB numbers, especially for blends of surfactants from different classes, since the HLB number additivity is very poor (16) and the spread of the HLB numbers suggested for good performance is too broad (81) for practical applications.

K. Competitive Kinetics
   of Adsorption and Surface Packing
   with Blended Emulsifiers

The spatial requirements of water molecules hydrogen bonded to the ether oxygens of the nonionic ethoxylated surfactants (55-57) will prevent them from packing closely on the oil-water interface. On the other hand, the ionic surfactants' charged heads will be affected by Coulombic repulsion and thus prevented from packing closely on the interface. Blending the two systems in the emulsifier will to an extent overcome this handicap, improving surface packing for a better stabilization of emulsion with less emulsifier. The anionic surfactants in the blends add charge to the emulsion droplet, preventing coagulation by Coulombic repulsion, but give a narrow Stern layer sphere (4 to 10 Å), with little solvation of the ions with water molecules. The added nonionic ethoxylated surfactants protrude much further into the water phase (30 to 50 Å) and form a big solvation sphere. The plane of shear that determines the Stern layer is in this way extended and stabilized by the hydrated ethylene oxide chain, resulting in improved emulsion stability.

The kinetics of formation (82,83) of emulsions is still not fully known since it depends on the rates of surfactant diffusion to the interface, which will be very complicated for a blend of surfactants of different molecular weights and charges. The ionic surfactant will move faster, depending on the charge on the interface and the electrical potential. The lower-molecular-weight nonionic material will diffuse fast to the interface to be later displaced competitively by the slower diffusing higher ethoxylated surfactants, which are more hydrophilic. Such multicomponent emulsifiers will provide a better packing of the interface and give a better, more stable emulsion (84-100).

VI. BLENDED EMULSIFIERS AND ECONOMICS

Formulations with blended emulsifier pairs or triplets give compromise emulsifier systems consisting of a more hydrophilic component and a less hydrophilic (i.e., lipophilic) one. Such a pair, often composed of multiple surfactants taken in different proportions, can cover many formulations and diminish inventories. A third emulsifier can be added for formulations not covered by the main pair. Since that approach is naturally a compromise, some formulations will not show a perfect performance and other will require an increased amount of emulsifier. Large-scale formulators and toxicant manufacturers are able to use more sophisticated systems of specific emulsifiers for specific problems.

The problem of blending emulsifiers is quite simple with a two-component system. If the blending is done in 5% increments, we get 21 possible combinations, and it is easy to correct a blend that is too hydrophilic or too lipophilic. A blend of three components at the same interval gives rise to 231 combinations, and a

4. AGRICULTURAL EMULSIONS

correction of the blend may be very difficult. In a combination of four ingredients with the same increment there would be 1771 combinations, and only a specialist could make a correction. For this reason emulsifier blends should be limited to two components. Higher range blends have to be avoided without a skilled specialist in production control.

Generally the number of possible blends S_n of n surfactants in emulsifiers taking an increment of a% gives rise to

$$S_n = \frac{\prod_{i=1}^{i=n-1} [100 + (n-1)a]}{(n-1)! \, a^n} \text{ combinations} \quad (11)$$

Thus for a = 1% increment and n = 4 surfactants the number of possible combinations is

$$S_4 = \frac{103 \times 102 \times 101}{3 \times 2} = 176,859$$

Since some emulsifiers have up to 12 ingredients, it is no wonder that the formulation of an emulsifier is still an art.

VII. WATER AND AGRICULTURAL FORMULATIONS

Water is an important factor in agricultural emulsion formulations. Whereas cosmetic emulsions are prepared with pretreated water, the farmer has to use the water as he finds it in nature. A geological survey of the public water supplies in the United States ([101]) established the pH and hardness of waters.

A. Effect of Electrolytes

The effect of electrolytes on the behavior of surfactants and emulsifiable concentrates is very important in the design of an emulsification system. Ions formed by the introduction of salts, acids, or bases into water interfere with the regular structure of water, having a stronger bond with water than water-water bonds, and in this way break down the water structure. They immobilize the nearest water molecules by hydrogen bonding and polarization. So the effect is often proportional to the polarizing power of effective charge divided by its radius. Ions also form an electric field that shortens the half-life of water clusters (102,103). Higher temperatures break down the water structure by diminishing hydrogen bonding in water clusters (104). Electrolytes also increase the surface tension and change the surface potential of water (105), with the exception that at low salt concentrations there is a minimum of surface tension (106) caused by charge stabilization. It appears that anions have a greater effect than cations. Some electrolytes will change the orientation of the water molecule on the surface, giving a positive surface potential.

With an increase in the concentration of salts, nonionic surface-active agents will lose part of the water of hydration (107-109). This is a salting-out effect. The decrease in critical micelle concentration (CMC) is proportional to the lyotropic number of the salts. Water-structure-breaking ions have a stronger salting-out effect. The lyotropic number depends on the size of hydrated ions, relative to their position in the Hofmeister series:

$$F > SO_4 > Cl > Br > NO_3 > I > CNS; \quad NH_4 > K > Na > Li > Mg$$

Here are some of the known lyotropic numbers: $K = 75$; $Na = 100$; $Li = 115$; $F = 4.8$; $Cl = 10$; $Br = 11.3$; $NO_3 = 11.6$; $I = 12.6$; $CNS = 13.3$.

The salting out of neutral molecules depends on the concentration and the ion radius of the added electrolyte. The smaller the hydrated ion (low lyophilic number), the bigger the salting-out effect, because of a stronger polarization effect. Addition of electrolytes decreases hydration of the ether linkages in nonionic

4. AGRICULTURAL EMULSIONS

surfactants and shifts the hydrophilic-lipophilic balance to the lipophilic side. Small amounts of acids also decrease the CMC, whereas bigger amounts increase it due to hydronium-ion formation with the ether linkages. Bases decrease the CMC. Urea and dimethylformamide give a mixed cluster with water and increase the CMC.

The salt effect on anionic surface-active materials depends very much on the cation charge. A dibasic cation will suppress the CMC much more than a monovalent cation. The thickness of the double layer is (110)

$$\tau = \left(\frac{DkT}{8\pi e^2 \Sigma n \nu^2}\right)^{\frac{1}{2}} \tag{12}$$

where D is the dielectric constant, k is Boltzmann's constant, e is the electronic charge, and n is the concentration of ions of valency ν. So the thickness is inversely proportional to the valency.

The addition of small amounts of salts (111-113) will cause a denser packing of the ionic surfactant on the interface and give a better emulsion. For optimal stability, which is related to the formation of a rigid film on the surface (67) or the so-called black film, there is an optimal salt concentration that varies with the ion. An increase in salt concentration will diminish the solvation and salt out the emulsion. At high salt concentrations in a water-in-oil emulsion stabilized with sodium oleate, the formation of sodium oleate-water micelles in the oil phase will lead to total coalescence.

1. HARDNESS OF NATURAL WATERS IN THE UNITED STATES

The United States contains several different regions of water hardness (101). The Mississippi Delta water, represented by water from the Greenville Agricultural station (the Stoneville well), has a hardness of 8 to 13 ppm calculated as $CaCO_3$. However, it contains natural softeners and sequestrants (114) that behave, for emulsifiable concentrates, as if it were of negative 500-ppm hardness as $CaCO_3$. It is difficult to design a synthetic water with such properties, but an equivalent is supposed to be water of the following composition per liter (115)

$$0.0094 \text{ g } CaCl_2 \cdot 2H_2O$$
$$0.0103 \text{ g } MgCl_2 \cdot 6H_2O$$
$$0.8000 \text{ g } NaHCO_3$$

The Pacific Northwest (Washington, Oregon, Idaho, and northern California) and the north Atlantic Coast have very soft waters with a hardness of 20 to 40 ppm as $CaCO_3$. In the East, New Jersey, the Carolinas, Virginia, West Virginia, Georgia, Alabama, eastern Mississippi, Tennessee, and northern Florida have waters with a hardness of 50 to 180 ppm as $CaCO_3$, with southern Florida having 120 to 130 ppm. The Corn-Belt and Wheat-Belt states have waters with a hardness of 120 ppm in the east and up to 800 ppm in the west.

The hardest waters are in west and southwest Texas, Arizona, New Mexico, southern California, and the Dakotas (600 to 1200 ppm), with waters in the Pecos River Valley wells and the Badlands of the Dakotas having a hardness of 1200 ppm and up to 3000 ppm in some instances.

2. SYNTHETIC WATER FOR TESTING FORMULATIONS

Since it is impossible to get samples of natural waters from all over the country, it is customary to use synthetic water based on calcium chloride, magnesium chloride, and magnesium sulfate. The hardness of water is normally expressed as parts per million of $CaCO_3$ equivalent, and as the molecular weight of $CaCO_3$ is 100.09, it is suggested to use approximately 50 g $CaCO_3$ equivalent or 0.9991 mole/l as a 50,000-ppm solution and dilute it proportionally to make up solutions containing the required proportion of hardness

4. AGRICULTURAL EMULSIONS

from calcium and magnesium ions. Stock solutions equivalent to 50,000 ppm as $CaCO_3$ diluted to 2 liters each have the following composition (see also Table 1):

$$146.89 \text{ g } CaCl_2 \cdot 2H_2O$$
$$246.34 \text{ g } MgSO_4 \cdot 7H_2O$$
$$203.12 \text{ g } MgCl_2 \cdot 6H_2O$$

TABLE 1

Various Specifications for Synthetic Water

Test water	Hardness	Salt	Concentration (ppm)	Stock required for 1-liter dilution (ml)
Federal Specification, 12/2/1947	292	$CaCl_2$ $MgCl_2$	160 132	3.2 2.64
Atlas, 1000 ppm	1000	$CaCl_2$ $MgSO_4$	600 400	12.0 8.0
$CaCl_2$ 1000 ppm	1000	$CaCl_2$	1000	20.0
Navy, Army, WHO hard	342.2	$CaCl_2$ $MgCl_2$	274 68	5.48 1.36
WHO soft	57	$CaCl_2$ $MgCl_2$	45.7 11.3	0.914 0.226
WHO, 1000 ppm	1000	$CaCl_2$ $MgCl_2$	800 200	16.00 4.00
Half Army (Navy, WHO)	171	$CaCl_2$ $MgCl_2$	137 34	2.74 0.68
Navy Department hard [89]	500	$CaCl_2$ $MgCl_2$	214 286	4.28 5.72
Naturally softened [89]	20	$CaCl_2$ $MgCl_2$	8.5 11.5	0.17[a] 0.23[a]

[a] With 0.8 g $NaHCO_3$ added.

All synthetic waters are neutral and are made with $CaCl_2$, $MgCl_2$, and $MgSO_4$ as sources of $CaCO_3$ equivalent hardness. However, the agricultural formulator must bear in mind that the different ions will behave differently in emulsions. At the same ionic strength, which is equivalent to the same $CaCO_3$ equivalent hardness, the salting-out effect is $Ca^{2+} > Mg^{2+}$ and $Cl^- > SO_4^{2-}$. Thus a $CaCl_2$ solution of 1000-ppm $CaCO_3$ equivalent hardness behaves much harder than an equivalent $MgSO_4$ solution (100). Natural hard waters of predominantly carbonate hardness behave much softer than synthetic waters; a natural water of 1700-ppm hardness from west Texas was behaving softer than 600-ppm synthetic water.

B. Effect of Temperature

The effect of temperature on nonionic surfactants is well known. At higher temperatures the hydrogen bonding of water molecules drops rapidly, and with hydration drop the solubility and the hydrophilic-lipophilic balance of the nonionic surfactants. At sufficiently high temperatures the balance makes them water insoluble, this being their cloud, or phase-inversion, point (78-80,116).

The cloud point is a function of dehydration and in the same series is a function of the number of ethylene oxide units giving a means for classification and standardization. The hydration is different in different products (117). Using octyl phenol with 7.3 moles of ethylene oxide, Doren and Goldfarb (118) found that added NaCl decreases the cloud point, whereas NaSCN, in concentrations of 0.062 to 0.1245 molal, increases the cloud point over that of distilled water. The cloud point is not a dehydration (119) but a phase separation in which the "oil" phase still has water solubilized in it; for example, a 1% solution of octyl hexaglycol ether has a cloud point of 78°C, and the separated phase still contains 7.5 molecules of water per each accounted ethylene oxide group. With a longer ethylene oxide chain the cloud point goes up to some

4. AGRICULTURAL EMULSIONS

maximum (117,120) where the chain bends and winds on itself, and the hydration per each ether linkage drops. Ionic surface-active agents are also affected by the breakdown in water structure that occurs with increase in temperature.

However, in agricultural emulsions with liquid fertilizers and anionic surfactants the effect of higher temperatures is improved emulsification. This can be explained by an increase in the solubility of both the anionic surfactants and the fertilizer salts, and the formation at high temperature of more "available water" (higher water activity).

C. Effect of Dilution Rate

More concentrated emulsions (lower dilution rate) must be formed at conditions nearer to the flocculation-sedimentation stage. There is also more interaction among the emulsion droplets, and both the solvation (hydration) and the CMC of anionic surface-active agents are suppressed. The effect is equivalent to increased salt concentration and requires a more hydrophilic blend than does a dilute emulsion.

D. Effect of Mixing

The way the emulsifiable concentrate is added to the water is important for the formation of the emulsion and for its stability. For an oil-in-water emulsion the oil phase has to be added to the water phase so that during the initial stage there is a large proportion of the water phase to the oil phase, which promotes an oil-in-water emulsion. On the other hand, for a water-in-oil emulsion, the water phase has to be added to the oil phase. The addition of the dispersed phase in a fast thin stream to fast-swirling water requires a different emulsifier blend than does rapid addition to an immobile liquid. It is a question (82,83) of the dynamics of surface tension on the one side and the leaching of the water-soluble parts of the emulsifier on the other side.

VIII. TESTING

A. Testing Methods for Emulsifiable Concentrates

The testing of emulsifiable concentrates is often based on the individual approach of the formulator. International organizations [e.g., the World Health Organization (121)], governments, public utilities, and mosquito-control units will have different specifications for preparation and testing as well as different criteria for the quality of approved emulsions.

Behrens and Griffin (122-124) suggested a procedure and apparatus for uniform mixing and measuring of emulsion performance. The procedure and apparatus are seldom used because most formulators find the mixing not representative of field use and the tubes too narrow to eliminate the effect of glass walls. Selz (125) and Pearce (126) recommended the use of 100-ml glass-stoppered cylinders and addition of the exact required amount with a Mohr pipette. The test dilution rates should mirror the required use in the application and will vary from 1:400 to very often 1:99, 1:19, and 1:9. In specific cases and for ultra-low-volume applications the dilution may go up to 20:80 and even 50:50.

The emulsions are judged on the ease of emulsification (spontaneity); appearance, which depends on particle size; rate of creaming at a specified temperature; and sedimentation and coalescence. Since the applicator may interrupt application for the night, the settled emulsions are often tested for the ease of reemulsification (reconstitution) and the creaming rate after reemulsification.

4. AGRICULTURAL EMULSIONS

Other test methods (115,127) recommend preparation of emulsions by adding the concentrate to water swirling at 1000 rpm, mixing for 1 min, and observing the emulsion after transferring it to a glass-stoppered 100-ml graduated cylinder. An alternative procedure (e.g., for cattle dip) is rapid addition through a funnel into a 4-ft tube with a graduated centrifuge tube attached for measuring the creaming rate.

Certain test methods use 1:99 dilution, estimation of particle size on a spatula; and sedimentation-rate determination in graduated oil tubes or in Nessler tubes. The emulsions are tested in different ranges of water hardness according to specific regional problems or by the arbitrary decision of the formulator, federal agency, or international organization. Test temperatures vary from room temperature to 30°C and occasionally from 4 to 32°C.

B. Correlation of Laboratory Aging Tests and Field Behavior

Pesticides as a rule are more or less labile materials that normally break down in prolonged storage and quite rapidly after spraying in nature under the influence of light, water, chemicals, catalysts, enzymes, and time (128). The emulsifiable concentrate should, nevertheless, have sufficient shelf-life stability between the time of production and the time of application. In most cases the emulsifiers are very stable and will often not change in years, but the toxicants are labile and in breaking down will almost always produce much more polar materials, rending the emulsifier not sufficiently hydrophilic for the new system and giving poor, fast-creaming, and oiling emulsions.

Emulsifiable concentrates intended for long storage have to be tested for shelf life. Since it is not practical to test them at room temperature, accelerated aging tests have been designed. According to a test devised by the U.S. Department of Agriculture

(129), incubating for 120 days at 43°C is supposed to be equivalent to 1¼ years at ambient temperature. Decomposition at different temperatures depends on the activation energy of the decay reaction (130,131) and is therefore different for different materials. Thus the generally used accelerated aging tests cannot be directly translated into shelf-life stabilities. Based on the Arrhenius equation and assuming a frequency factor independent from temperature, the logarithm of the ratio of rates of decay for different temperatures is

$$\log \frac{K_2}{K_1} = \frac{E_a}{R} \; 0.4343 \; \frac{\Delta T}{T_1 T_2} \qquad (13)$$

where E_a is the activation energy.

IX. SUGGESTED OPTIMIZATION OF EMULSIFIABLE CONCENTRATES

From the behavior of surfactants with changes in salt concentration and temperature, and the behavior of toxicants with time, it is apparent that, for oil-in-water systems, emulsifiers that are not sufficiently hydrophilic for the required conditions will not sufficiently stabilize the emulsion droplets. The emulsion will go through the stage of collapsing into bigger droplets, a rapid sedimentation with coalescence (oiling) and, with the increased oil-to-water ratio during the sedimentation step, a tendency to water-in-oil emulsification, causing sticking to the walls and possible clogging of spray nozzles. Emulsifiers that are too hydrophilic for the specific conditions will stabilize the emulsion droplets, but they will come close enough to cause interlocking of the hydration spheres. This will cause aggregate formation and accordingly a slight increase in the rate of sedimentation (creaming) over the optimum blend.

Since high water hardness, high temperatures, higher proportions of oil phase to water phase, and decay products will have

4. AGRICULTURAL EMULSIONS

a salting-out effect (requirement for a more hydrophilic emulsifier), the practical rule is to optimize oil-in-water formulations for most practical salting-out conditions and compromise on others. Naturally, if the formulation is designed for unnecessarily hard conditions to meet only sporadic requirements, behavior in soft water and at high dilution will be endangered and increased levels of emulsifier may be required.

Increasing the amount of an insufficiently hydrophilic emulsifier generally does little to prevent the salting-out effect, but it does increase stabilization against aggregation, improving soft-water performance.

Figures 1, 2, and 3 show the temperature and water-hardness dependence of optimum performance of three insecticide formulations.

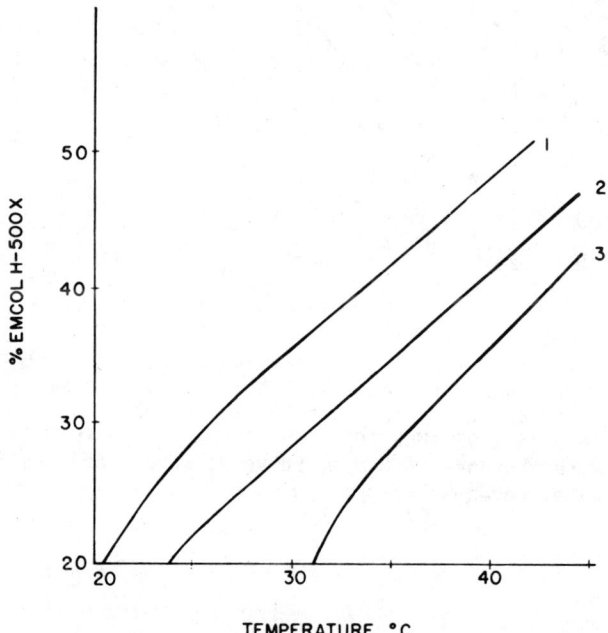

FIG. 1. Optimum performance as a function of temperature, water hardness, and hydrophilicity of emulsifier formulation of 50% malathion in xylene with 4% emulsifier blend of Emcol H-140 (HLB 11.1) and Emcol H-500X (HLB 13.5). Curve 1, water hardness 600 ppm; curve 2, water hardness 342 ppm; curve 3, water hardness 34 ppm.

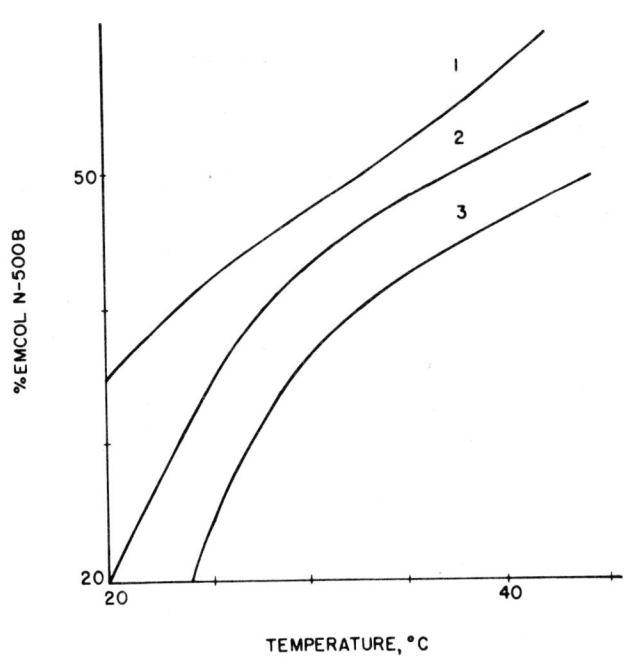

FIG. 2. Optimum emulsion performance as a function of temperature, water hardness, and hydrophilicity of emulsifier for a formulation of 40% toxaphene, 20% DDT in xylene with 5% emulsifier blend of Emcol N-139B (HLB 10.0) and Emcol N-500B (HLB 13.5). Curve 1, water hardness 600 ppm; curve 2, water hardness 342 ppm; curve 3, water hardness 57 ppm.

4. AGRICULTURAL EMULSIONS

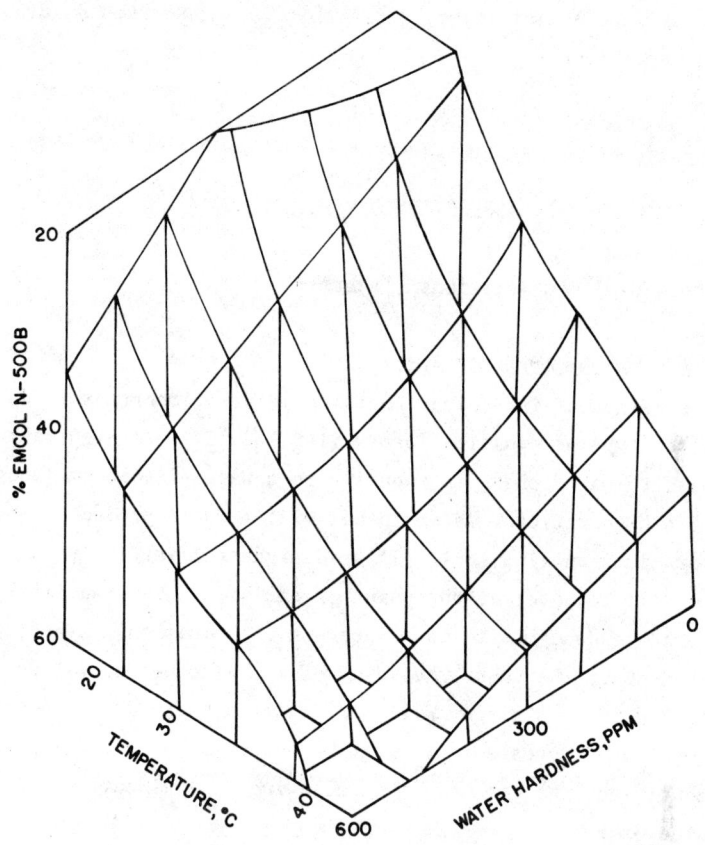

FIG. 3. Optimum emulsion performance as a function of temperature, water hardness, and hydrophilicity of emulsifier for a formulation of 40% Strobane, 20% DDT in xylene with 5% blend of Emcol N-139B (HLB ~ 10.0) and Emcol N-500B (HLB ~ 13.5) as emulsifier.

Laboratory testing methods are arbitrary, and there is usually no way to relate them to the requirements of field performance unless special difficulties develop requiring changes in test methods. Most often the laboratory requirements are much stricter than those warranted in actual use. Some formulations have to be diluted with oils for application, and such formulations may require different emulsifiers (132) that can stand such dilution without separation.

X. SELECTION OF EMULSION TYPE

A. Water-in-Oil Formulations (Inverts)

1. EFFECT OF THE ORDER OF ADDITION

The method of ingredient addition is very important in determining the type of emulsion formed (109,133,134). A high ratio of oil phase to water phase is inducive to a water-in-oil emulsion, whereas a high ratio of water to oil will tend to produce an oil-in-water emulsion (135,136). If we designate ϕ_i as the volume ratio of oil to water at the point of emulsion inversion, with $\phi > \phi_i$ the tendency is to form water-in-oil emulsions. The lower the HLB number, the lower the value of ϕ_i, although at values that are too low the emulsifier will be totally submerged in the oil phase and some increase in ϕ_i will be observed, depending on the equipment used. We can make the following generalizations about the value of ϕ_i:

1. The higher the viscosity of oil, the lower the value of ϕ_i.

2. The more oil-wetted hydrophobic surfaces, the smaller the ϕ_i.

3. The rougher the surfaces, the lower the ϕ_i.

4. The higher the rate of shear, the lower the ϕ_i.

Preferential wetting with the oil phase improves the formation of a water-in-oil emulsion. Sherman (137-141) investigated the stability of water-in-oil emulsions. The drop in interfacial tension of the oil phase with low-hydrophilicity emulsifiers is

4. AGRICULTURAL EMULSIONS   215

small. As already mentioned, anionic emulsifiers are inefficient
in stabilizing water-in-oil emulsions. This increases the amount
of surface-active agents necessary to stabilize water-in-oil emulsions. A low interfacial tension is desirable (142) because it
improves the ease and speed of emulsion formation and increases
the speed of adsorption on the interface, but it is not essential
if a strong interfacial film is formed. This film strength can
be assessed by measuring the surface viscosity (143).

2. *FORMATION OF WATER-IN-OIL EMULSIONS*

Since the more hydrophobic surface-active agents required for
water-in-oil emulsions do not develop enough energy during adsorption on the interface, a great deal of shear is necessary and care
should be taken to increase the sticking of the oil phase to the
walls during incorporation of the water phase. The best way to
form a stable high-viscosity water-in-oil emulsion is to add the
water phase slowly to the oil phase containing the right emulsifier,
with high-shear mixing, and then to use a homogenizer. During the
aging of a water-in-oil emulsion the water globules aggregate and
link to form a loose grouping with the continuous phase like a net
in the interstices. With some shear the nets are torn into irregular globs if the amount of water phase is high.

3. *DRIFT PROBLEMS IN AGRICULTURE AND INVERTS*

Emulsions of toxicant formulations are generally sprayed at
volumes of 40 to 50 gal/acre. Such sprays are called high-volume
sprays. In airplane spraying, to save on loading time and to
increase the acreage sprayed with one tankful, low-volume sprays
are used (1 to 10 gal/acre). Ultra-low-volume sprays can be used
advantageously in some cases, with 10-oz to 2-qt spray volumes per
acre (144).

If the droplets in an oil-in-water emulsion are too small,
they will be subject to rapid evaporation (13). At a wind strength
of 1 mph, 50% relative humidity, and 25°C, a 150-μ droplet of water

will evaporate after falling 13 ft. At 40% relative humidity at 20°C, a 50-μ drop will evaporate in 4 sec, and a 100-μ drop in 16 sec. The optimal droplet size seems to be 500 μ. Smaller drops evaporate too fast, and the toxicant will drift from the place it is applied. Big drops, on the other hand, will cause poor distribution of toxicant. With 500-μ droplets, 5 gal/acre will give 50 drops per square inch. This will be sufficient for systemic toxicants and emulsifiable concentrates with a good spreading coefficient, but insufficient for some wettable powders, especially fungicides. Increasing the viscosity by adding thickening colloids will keep the drops bigger, prevent breakup of droplets, and diminish evaporation, but the colloids will slow down the release of the toxicant (145). Additives that prevent water evaporation are ineffective in the presence of surfactants and interfere with them.

Herbicides, which have to prevent the competitive, undesirable plant species from growth without harming the crop plant, have to be very selective. With adjacent fields of different crops and different susceptibility to herbicides, the drifting small droplets may contaminate the other crop and destroy it. The use of a water-in-oil emulsion is especially attractive for diminishing spray drift. The use of a low-volatility oil as the outer phase in an invert emulsion (146) will prevent a decrease in the size of the droplets. The high viscosity of the system will produce big droplets spreading on the plant surface (147-153). Such application is especially desirable for spraying narrow rights-of-way.

B. Tight or Loose Emulsions

Most formulators are interested in slow-creaming formulations, which require a tight, small-particle-size emulsion. However, for better deposition with diminished runoff it is better to have a

4. AGRICULTURAL EMULSIONS

coarse emulsion. The danger is that, with insufficient suspensibility, the active ingredient may be unevenly distributed during application unless the formulation is subjected to constant mixing.

C. Fruit-Tree-Spray Formulations

For fruit-tree sprays with phytobland oils a coarse spray emulsion is mandatory for a good function. There are a great many papers dealing with the types of oil used and the emulsion requirements (154-164). The oils must be of high viscosity (60 to 100 SUS) and must have a high unsulfonatable residue (minimum 92%). The emulsions should break quickly.

The general dilution rate used for fruit-tree spraying is 1:50 or higher because the total leaf area to be covered is large. Low emulsifier levels ensure a coarse emulsion.

Another type of preparation used is a mayonnaise-type emulsion (165). This is an oil-in-water emulsion with a high ratio of oil phase to water phase (82.5:17.5), very viscous and very stable. It is prepared by slowly adding the emulsifier-containing oil phase to the water phase with very high shear (Waring Blendor, Hobart mixer, gear pumps). A very viscous, high-internal-phase, stable oil-in-water emulsion is formed. Sprayed with disk sprayers on leaf surfaces, the water evaporates and the oil spreads. It can also be introduced into a tank of water, dispersed into a coarse emulsion with minimum shear, and sprayed in diluted form with centrifugal sprayers.

D. Aerosol Sprays

Aerosol applications are too expensive and cumbersome for agricultural use. They do find use, however, in pest-control operations, in gardens and greenhouses, and in animal sprays (166).

In order to have a spray in the form of droplets, not foam, the propellant has to be intimately dissolved, which requires a stable water-in-oil emulsion with the hydrophobic outer phase (166-171). The propellants used are fluorinated hydrocarbons or butanes (172). In accordance with the rules of formation of water-in-oil emulsions, the water phase has to be slowly added to the oil phase with good mixing.

E. Flowable Formulations

A flowable formulation is a stabilized concentrated emulsion or suspension in water or oil which will be diluted with water during application. It has to be pumpable, and in the event of phase separation it should be easily reconstituted without caking. The purpose is to achieve a high-concentration pumpable formulation for situations where no convenient and cheap solvent system is available for liquid formulations and where viscosity and high toxicity require that the material be handled as an emulsion. The flowables may contain gum-type materials or clays to modify viscosity and prevent recrystallization, as well as emulsifiers or wetters (173-177). Some flowables may be formulated as triple-phase systems of a solid suspension in oil emulsified in a water phase.

The flowable emulsion can be prepared by emulsification and addition of stabilizers or by the phase-inversion-temperature method, often using homogenizers, gear pumps, or bypass pressure pumps. Flowable suspensions are often prepared in ball or colloidal mills, with surfactants and stabilizers added. During milling addition of antifoams will be necessary, and often a different antifoam is required at the end of the operation to release the incorporated air and improve packing characteristics. Flowables can also be formed by reversed encapsulation (178).

F. Solubilizations

In cases where the dispersion cannot be allowed to separate the formulation has to form a transparent microemulsion or solubilization. The toxicant is solubilized into micelles of surfactant, or the emulsion droplet has to be smaller than the wavelength of light or smaller than 0.4 µ. Such formulations require large amounts of highly hydrophilic emulsifiers and are seldom used except for pest control, mildew proofing, and animal-drinking-water additives (179).

G. Cattle Dips

Specially stable emulsions have to be made for treating cattle on the grazing area. The cattle are herded through a dip vat containing an emulsion of toxicants that should kill the parasitic flies, grubs, ticks, etc. At the same time the toxicant should not be deposited in appreciable amounts on the animal's skin, which may destroy the animal. The emulsion must be extremely stable and have a small particle-size distribution; it must be capable of properly wetting the hairs and skin, and must not break down for a long time in contact with all the materials introduced by the cattle (115, 180-183), A good mechanical spraying system may replace the cattle dips (184).

XI. SOME SPECIAL FORMULATIONS

A. Tank Additives

1. POSITIVE EFFECTS

Surfactants in certain concentrations may show a phytotoxic

effect on young seedling plants and young leaves (2-5). Surfactants also enhance the activity of herbicides in addition to wetting and solubilizing poorly soluble materials. They increase spray retention and herbicide penetration (185-201). The surfactants may be, to a certain degree, specific, and some are better with specific herbicides than with others. Surfactants also improve the uptake and retention of iron chelate through foliage (202), in nonionic series the optimum uptake being at HLB 13 to 15; with other types of surfactants there is no such straight relationship, and the optimum uptake is a function of the structure and chemical character of the surfactant. On the suggestion of the agricultural station many applicators will add the recommended surfactants or blends to the water phase of the herbicide spray tanks in order to exploit the enhanced effect and use less toxicant.

Surfactants also show insecticidal activity, and some surfactants that are perfectly innocuous for warm-blooded animals are more effective against southern house mosquitoes than reference insecticides (203-206). They also improve uptake of insecticides by insects.

2. NEGATIVE EFFECTS

By using tank-additive surfactants care has to be taken that the surfactants are compatible with the other formulations. The amount of surfactant, 0.1 to 1.0% of the water phase, is so big in comparison with the amount incorporated into the toxicant formulation as wetters or emulsifiers that the interference may cause in some cases a fast flocculation and sedimentation, which has to be avoided.

3. WETTERS, SPREADERS, AND SORPTION PROMOTERS

The solubility of some herbicides in water does not mean that their efficacy cannot be increased by the addition of surfactants

4. AGRICULTURAL EMULSIONS

to promote wetting and absorption (207). In many cases the herbicidal function is directly correlated to the wetting, which can be measured by the Draves test (208) or the Shapiro test (209,210) at the field dilution rate. The surfactants are added directly to the concentrated formulation or in the form of tank additives to the diluted concentrate. The wetting test should be made at the use-level dilution.

The toxicant (211-214) has to be adsorbed on the surface, penetrate into the plant or the insect, be transported through membranes by diffusion or convection into the phloem, and reach the point of action (215). These functions are enhanced by sorption promoters: solvents and surfactants acting as spreaders, wetters, or solubilizers.

As the toxicants undergo detoxification in the organism, the rate of accumulation to the toxic level is determined by two factors: (1) the dynamics of the rate of toxicant penetration to the point of action after the induction period and (2) the rate of detoxification and excretion of the toxicant and toxic products (216). This toxic level has to be reached if the toxicant is to take effect. Synergists will improve this ratio.

B. Aquatic Herbicidal Oil Formulations

Aromatic oils, especially xylene, are phytotoxic (2,3) and are used to control submerged aquatic weeds in irrigation ditches (217). With the right emulsification system the aquatic weed will be uprooted and easy to remove from the canal.

C. Mosquito-Control Formulations

Mosquito control requires treating stagnant-water breeding areas with toxicant formulations. Many oil companies sell for this purpose petroleum-oil cuts with a surfactant added (218). These

are much more effective than the regularly used diesel fuel No. 2.
For the best effect as a pupicide and larvicide the oil should be
a blend of certain cuts of paraffinic and aromatic oils (219,220)
and have good spreading characteristics, which, according to the
Antonoff rule, are obtained when $\sigma_{LV} = \sigma_{WA} - \sigma_{LW}$, where σ_{LV} is the
surface tension of oil, σ_{WA} is the surface tension of water, and
σ_{LW} is the interfacial tension.

Since the shrinking of surface coverage into lenses or
globules could be correlated with diminished activity, the kinetics
of spreading (221) is of utmost importance.

D. Triple-Phase Systems and Tank Blends

Oils have been shown to improve the function of herbicides and
insecticides (222-224). Especially interesting was the work
initiated in 1963 by G. Jones of the University of Guelph, Ontario,
on the effectiveness of diminished amounts of herbicide atrazine
80W, a wettable powder, with an emulsifiable phytobland oil (225).
This triple-phase system of water, an emulsifiable oil (often called
"corn oil"), and the wettable powder atrazine 80W was approved by
the Ontario Department of Agriculture for use in corn protection
(226). The system was rapidly picked up by many investigators and
recommended for use in the United States.

This brings us to the problem of mixed tank formulations. The
emulsifier requirement for a blend of toxicants is not an additive
function, which could allow one to use an emulsifier system that is
intermediate between the requirements of the individual toxicants.
The toxicant's requirement for a more hydrophilic emulsifier will
prevail, and the emulsifier will have to be chosen accordingly.
Using in one spray tank two different emulsifiable concentrates
optimized for the conditions will give a fairly good emulsion, but
time and strong mixing may change it to a poor emulsion.

The triple-phase system will also require changes in the emulsification system. Since atrazine 80W has a fixed wetting system,

4. AGRICULTURAL EMULSIONS

the emulsifier for the oil phase has to be modified accordingly from the regular oil emulsifier to get the right emulsion for the triple-phase system. If this is not taken care of, the effect will be that under good mixing the wettable powder will become oil wetted and the system will become inverted, clogging spray screens and nozzles. This effect is called "buttering" or "cheesing" and was known to formulators recommending tank application of spray oils with wettable fungicides. It occurs because the hydrophobic wettable powder approaching an oil droplet not sufficiently protected by hydration is absorbed on the surface and, being more submerged in the oil phase than in the water phase, will according to the Bancroft rule cause inversion. In his investigation ^{14}C-labeled atrazine Triplett (227) found that such inverted sytems induce a slower uptake of atrazine into the plant than that of plain atrazine 80W in water. It was stated in many papers (191, 192,196) that the mode of surfactant action in enhancing herbicidal activity is not well understood. It cannot be correlated with solubilization, wetting, or conductivity index of interaction (13). In regard to oil, Crafts (222) assumed that it saturates the cuticle and allows the toxicant to move in (223). Triplett (227) found that the presence of emulsifiable oil induces a more rapid penetration and translocation into the plant. Since the herbicide shows a strong hydrogen-bonding ability, it is rapidly adsorbed to the walls of the plant's capillary conduits, which slows down its penetration. Preferential adsorption of the surfactant and oil enables the herbicide to move rapidly into the water phase, an effect not dissimilar to lubrication. Since phytotoxicity is only the function of the herbicide, and not of the phytobland oils, there is no synergistic effect in the sense determined by Colby (228) as an increase in the effect of two active materials over the expected function of the individual toxicants.

E. Formulations with Liquid Fertilizers

1. *BEHAVIOR AND CLASSIFICATION OF LIQUID FERTILIZERS*

Agricultural formulations with liquid fertilizers have been reviewed by the author (229). For economic reasons liquid fertilizers are prepared with a high concentration of materials of low cost per unit of available nutrients for plant growth. The most common materials for liquid fertilizers are anhydrous ammonia, aqueous ammonia, urea, ammonium nitrate, ammonium sulfate, ammonium phosphate, ammonium polyphosphate, and potassium chloride. With the exception of the first three, all these materials are strongly ionized, and the ions are strongly solvated. The ions also exert a strong polarizing effect on the nearest water molecules, decreasing the water activity and breaking the water structure, which depends on the much weaker dipole-dipole interaction forces; this diminishes the available water, produces a strong salting-out effect, and increases the surface tension (104-107). The effect can be measured by the diminished vapor pressure of the water.

Anhydrous ammonia has a surface tension $\sigma = 20d\ cm^{-1}$ at 20°C (230); for 29.4% aqueous ammonia $\sigma = 69.5$ dynes/cm at 18°C. Anhydrous is used only infrequently as a carrier for sulfur (231), and never for other formulations. Aqueous ammonia often behaves the same way as hard water since the salting-out effect and the modification of water structure are very moderate and the available water (water activity) is high. Ammonia fits into the water structure and decreases the salting-out effect. Urea forms strong dipole-dipole bonds with water, fits well into the water cluster structure, and strengthens it. Therefore it increases water availability and salt solubility, and diminishes the salting-out effect (232,233).

For this reason we subdivide liquid fertilizers into the following groups, which may be combined with other materials in the form of emulsions:

1. Aqueous ammonia. Emulsification can be made with common-type emulsifiers.

4. AGRICULTURAL EMULSIONS 225

2. Ammonium nitrate plus urea solutions (Uran type). Urea moderates the effect of ammonium nitrate, and salting-out is partially suppressed. Emulsification is achieved with modified nonionic-anionic blends.

3. High-pressure nitrogen solutions containing ammonia, ammonium nitrate, water, and sometimes urea (Nitrana and Urana type). Since ammonia diminishes the surface tension and increases water activity, emulsification is achieved with modified nonionic-anionic blends.

4. Ammonium nitrate solutions (Feran type) and N-P and N-P-K liquid fertilizers. These produce a strong salting-out effect, increased surface tension, low water activity, and broken water structure. They form a separate class, and the emulsions have to be prepared with different emulsifiers.

2. EMULSIFIERS FOR LIQUID FERTILIZERS

Nonionic surfactants of the ethylene oxide adduct type form weak dipole-dipole type bonds with water which are stable only in moderate salt solutions. Despite the lingering opinion (234) from the time when the only anionic surfactants were carboxylic soaps that nonionic surfactants are effective at high salt concentrations, they cannot compete with ions for the available water in low-water-activity salt solutions, nor counteract polarization in the medium by the ions.

The emulsifiers for such systems have to be based on strongly hydrophilic anionic surfactants (235-248).

With low amounts of available water it is difficult to have enough hydration, and the emulsions are not as stable as regular formulations in water. The increased usage of liquid fertilizers has raised the demand for formulations compatible with both systems.

XII. SOIL PENETRATION, ADSORPTION, AND MOBILITY

Soil penetration will depend on the characteristics of the soil. The leaching of substituted-urea herbicides (249) will depend on the surfactant used with the toxicants. Surfactants that increase the toxicant's mobility will increase leaching, and some cationics that complex with the toxicant and the soil will decrease leaching. The behavior will depend on the behavior of the particular toxicant and the soil (250,251). The adsorption and phytotoxicity of diuron and simazin can be correlated to the type of soil, but not those of CIPC (252,253). The activity of CIPC, monuron, and diuron can be correlated with organic matter in the soil, but not that of atrazine and simazin. To determine the adsorption Freundlich's adsorption isotherm can be used (254,255):

$$\frac{\gamma}{M_0} = K C_W^{1/m} \tag{14}$$

where γ is the amount absorbed on the M_0 amount of the sorbent from a C_W concentration of the toxicant in water, and K and n are constants depending on the toxicant, soil type, and organic matter.

The toxicants can move through the soil by diffusion or by the diffusion of their vapors, or in the form of water solutions, suspensions, or emulsions. The soil can be considered as a set of capillaries of diameter a, with tortureous-path factor k. Adsorption and effects of streaming potential interacting with the ζ potential of the emulsion droplets and suspensions hamper soil penetration. Neglecting gravitational effects, the speed and uniformity of movement through this capillary system (256) can be expressed by the combined Laplace-Poiseuille equation (257,258) in the simplified form,

$$\frac{l^2}{t} = \frac{a}{k^2} \frac{1}{2\eta} \sigma_{AW} \cos\theta \tag{15}$$

4. AGRICUTURAL EMULSIONS

where l is the distance penetrated in time t through a capillary of diameter a with a tortuous-path factor k of solution with a viscosity η, surface tension σ_{AW}, and wetting angle θ; a/k^2 is the packing factor. The viscosity of the water solution being approximately unity, the rate of movement by capillary forces l^2/t in a given soil is thus proportional to the surface tension of the water solution times the cosine of the wetting angle. The addition of surfactants diminishes surface tension and the contact angle, increasing $\cos \theta$. For hydrophobic soils $\cos \theta$ is nearly zero, and the addition of surfactants, though diminishing the surface tension σ_{AW}, will still sizably increase the soil-penetration factor. For a hydrophilic sandy soil the wetting angle is already small and $\cos \theta$ is close to unity, and hence the addition of surfactants diminishing the surface tension σ_{AW} will slow down penetration.

XIII. APPLICATION TECHNIQUES

Pesticide formulations are mostly diluted with water to form emulsions and have to be distributed over the treated area or the animals in order to bring the spray into contact with the particular pests that are destroying the crops and diminishing yields. Emulsions can be applied and distributed directly or in the form of sprays. The size of the droplets and their behavior have the utmost effect on the biological performance, the economics of the use of the pesticides, and the danger to the ecology and adjacent vegetation.

A. Direct Surface Application

Toxicants that are introduced with liquid fertilizer, and as preemergence, lay by, or incorporated can be injected into the soil or sprayed with regular or flooding nozzles (259). It has also been suggested that herbicides be applied in foams (260). Such application increases the coverage to give improved weed control and also allows the applicator to see where the herbicide was applied. Special foam nozzles have to be used. The foam is formed by air blown through a thin film of the emulsion or suspension containing a foam stabilizer in the water. With foaming, the volume applied increases up to 350 times. In this way 60 gal of spray applied per acre can be transformed into 21,000 gal of foam, giving a coverage 0.8 in. thick. This helps the herbicide to reach all the weeds.

B. Meristemic Control

The use of emulsions to control plant meristemic growth is another type of direct application. The purpose of the treatment is to selectively inhibit the growth of axillary buds in a variety of plants (261-266). It is used in chemical tobacco suckering and pruning of ornamental flowers. Highly active for this purpose are the methyl esters of fatty acids and alcohols of a selective chain length. Since the purpose of such treatment is to prevent the growth of a part of the plant without damaging the leaves and other parts of the same plant, the system has to be worked out with utmost care. The emulsifiers used have to show no phytotoxicity. The emulsions applied directly to the plants have to be stable, without much runoff, and should drain along the stem to contact the axillary buds. The ratio of emulsifier to the active alcohols and methyl esters, and also the rate of dilution with water, can be very critical both for efficacy and possible injuries. With different plants the requirement may be drastically specific. Careful greenhouse and field tests should be undertaken before marketing such formulations, and the suggested rules of application must be followed exactly.

4. AGRICULTURAL EMULSIONS

C. Spray Application

When a highly dilute emulsion is sprayed with a boom with poor atomization, the emulsion can also behave as it does in direct application. Most of the agricultural emulsifiable formulations are sprayed in the form of droplets. The size and distribution of the droplets in the spray, the output, the speed, and the pressure applied will have a great effect on the biological function and the economics of the spraying of the pesticides.

1. DROPLET AND DISTRIBUTION MEASUREMENTS

Many methods are used to measure the distribution of the spray and of the droplet sizes. They can be assessed by the "flying-spot particle-resolver" technique (267) or collected on kaolin-covered slides (268), treated aluminum or paper sheets (269,270), or glass plates (145,271,272). Velocity can be determined with high-speed twin-flash photography (273). To make the droplets visible for measuring the size and distribution, one can add to the spray a water-soluble dye like Nigrosine (274-276) or a fluorescent water-insoluble pigment of small particle size (277-280).

As to particle size, we distinguish several categories of sprays, listed in Table 2.

TABLE 2

Classification of Sprays by Particle Size[a]

Category	Particle size (μ)
Fumes (smoke, true aerosols, haze)	0.4-1
Dry fog	2-5
Wet fog	10-40
Misty rain	50-100
Light rain	100-400
Moderate rain	400-1000
Heavy rain	1000-5000

[a] From Ref. 281.

What we customarily call an aerosol spray is not a true aerosol but a spray dispersed by using a compressed low-boiling-point propellant and is generally of the wet-fog category.

2. DROP SIZE, COVERAGE, AND DRIFT

With smaller droplets we get a better coverage of the sprayed area, the optimum diameter of an efficient insecticidal spray being 20 to 30 μ (277-279). Large droplets do not reach and impinge on the target insects with the contact toxicant and only increase the residual deposit on plants and soil. This secondary source of contact toxicity is highly inefficient. On the other hand, with small-particle-size sprays drift becomes a big problem. At a crosswind of 5 mph, a 20-μ droplet will drift 1.4 miles. In airplane and helicopter spraying (269-271) the drift problem is so grave that only droplets of medium size (150 μ) are effective with sufficient recovery on the target. The size of the droplet depends on the pressure applied, the size and form of the nozzle orifice, and the viscosity of the sprayed fluid (276). The data in Table 3 show the effect of viscosity on droplet size, percentage of recovery, and coverage.

TABLE 3

Effect of Spray Viscosity on Droplet Size, Recovery, and Coverage[a]

Parameter	Viscosity (cP)		
	1	28	45
Droplet diameter (μ)	120	170	200
Recovery (%)	53	64	70
Coverage (drops/in.^{2})	95	60	40

[a]Ultra-low-volume application (0.5 gal/acre) under the following conditions: airplane flying 3.5 ft above the field at 85 mph; 50-psi pump pressure, flat fan nozzles; liquid surface tension 32 dynes/cm.

4. AGRICULTURAL EMULSIONS

For ultra-low-volume applications the formulations are sprayed in the form of an oil phase or concentrated emulsion (281,282), and hence the factor of evaporation is not so important. If the drift is not dangerous for the surroundings, one may try to spray in small droplets and often use Micronaire rotary atomizers.

Potts (284) gives thorough information on spraying equipment and the applications of concentrated sprays.

Where the rate of dilution with water is high, the effect of evaporation on the droplet size (13,146) is very considerable. This is described in Section IX.A as the purpose for the development of water-in-oil emulsion systems for agriculture. In practice the total water sprayed locally increases the relative humidity in the spray, and the oil phase present diminishes evaporation to a degree. Hence the losses in the diameter of the droplets are much lower than expected theoretically, considering only the size of water droplets. More experimental data are necessary. Nevertheless the effect of evaporation has to be taken into account, and for highly diluted sprays it is necessary to select a nozzle that will produce suitably large droplets.

3. ATOMIZATION OF SPRAYS

In the process of atomization the fluid is broken down into droplets by the high pressure in regular spray nozzles, by interaction with high-speed air in twin-fluid atomizers, or by centrifugal force in spinning-disk atomizers (285). The fluid comes out in the form of a flat sheet from fan-jet nozzles or an empty cone sheet from cone nozzles (286). At a certain distance from the orifice, the sheet breaks up into filaments and droplets. The droplets will be further disintegrated by the impact of the air and finally on the surface sprayed.

The lowering of the surface tension of water by adsorbed surfactants enables the sheet to spread further and become thinner, with the effect that smaller droplets are formed. Since the force of sheet contraction is proportional to the square root of the surface tension, reducing the surface tension of water from 73 to 50 dynes/cm will reduce the mean droplet size by 20%.

The reduction of surface tension is produced by the adsorption of the surface-active agents on the air-liquid interface of the freshly formed surface and is not instantaneous. It is a kinetic process depending on the distribution of surface-active agents between the oil and water phases, as well as the competitive adsorption of the different surfactants in the water phase (287); the latter will depend on the diffusion coefficient, the chemical character, and the concentration of each species. As the effective life of the water sheet is on the order of 0.004 sec (286), not much of the surfactants can generally be adsorbed on the freshly formed surface before it breaks up. For twin-fluid atomizers the life of the sheet is even shorter. After the sheet breaks up into filaments and droplets, the adsorption equilibrium of the surface-active agents can be achieved. The droplets may disintegrate in air when a critical relative velocity to the air is reached; another factor leading to disintegration is the formation of instability ripples on the surface of the droplets--an effect of decreased surface tension.

4. *BEHAVIOR OF THE SPRAY ON IMPACT*
   (BOUNCING, SPREADING)

At the moment of its impact on the surface of the leaf, stem, or other solid surface the droplet has an excess kinetic energy that can be absorbed as compressive-strain energy. The droplet will also flatten with increase in surface or splash to break up into smaller droplets and so increase the surface energy. The flattened drop will elastically contract again and may bounce and leave the surface (13). The bouncing will occur when the remaining kinetic energy of the droplet is still greater than the decrease of the free energy of the contact with the solid surface (wetting). The viscous drag of the air flowing past the solid is one of the factors determining the bouncing. If the remaining kinetic energy and the surface energy are not too high, the droplet will oscillate, flattening and

4. AGRICULTURAL EMULSIONS

contracting until the excess energy is dissipated and equilibrium is established according to Young's equation:

$$\sigma_{sa} = \sigma_{sl} + \sigma_{la} \cos \theta \tag{16}$$

where σ_{sa}, σ_{sl}, and σ_{la} are the surface tensions of solid-air, solid-liquid, and liquid-air, respectively, and θ is the equilibrium contact angle. The surface-active agents in the droplet will cause the droplet to spread, with improvement of the surface coverage.

The ideal system would be if the emulsion were stable for uniformity in spraying, the oil globules in the atomized spray would stick on impact to the target surface, allowing the aqueous phase to run off, and the oil phase would spread on the surface to improve the coverage.

5. EFFECT OF VISCOSITY AND REYNOLDS NUMBER

In viscous Newtonian and non-Newtonian emulsions the diameter of spray droplets is dependent on the viscosity of the sprayed fluid. In analyzing droplet formation Ford and Furmidge (274,275) found that the fluid leaves the orifice of the nozzle mostly in the form of sheets and partially in the form of ligaments, which break up into droplets in a different way. In the sheet a sinusoidal wave is formed at right angles to the direction of sheet motion. The wave proceeding away from the orifice increases in amplitude. The sheet breaks up at the nodes of the wave to form ribbons parallel to the leading edge of the sheet (273), which then breaks up into droplets. Ford and Furmidge plotted in a log/log plot the instability function

$$\pi_1 = \frac{\sigma}{\gamma \theta \rho_1 V_e^2} \tag{17}$$

(where σ is the surface tension of the liquid; γ is the length of the sheet to the breakup point; θ, in radians, is the angle of the sheet at the orifice; ρ_1 is the density of the liquid; and V_e is the velocity of the liquid at the orifice) against the reciprocal of the Reynolds number (R_e^{-1}) characterizing the flow from the nozzle and as such a function of the viscosity of the liquid. They found four distinctive ranges of behavior. In range I the instability factor is independent of the Reynolds number, which shows that in this region of low viscosities the drop size is independent of viscosity change and the sole factor in breaking up the sheet is the rippling instability of turbulent flow. In range II the instability factor decreases to a minimum, which suggests that this is the critical region where turbulent flow changes to laminar flow and the drop size diminishes with increase in viscosity. In range III the flow is laminar and the instability factor increases with the decrease in Reynolds number and the drop size increases with viscosity. Finally, in range IV, with Reynolds numbers below 20, the sheet is not formed any more, only ligaments are, and drops are formed by the breaking up of ligaments.

6. CHOICE OF SPRAYING SYSTEM

The different types of atomizer are selected for their spray characteristics: average droplet size, distribution of droplet sizes in the spray, drift considerations, amount delivered per minute (which will depend on the type of application, ground or air), speed of movement, and number of nozzles on the boom (285, 288, 289). Fan nozzles are less susceptible to changes in viscosity than cone nozzles, but the latter give a more uniform distribution (290). Rotary atomizers give a narrower range of drop sizes than compressed-air nozzles (285) and can produce a finer spray, but the amount delivered is smaller and would not be used in, for example, orchard spraying, where big volumes of spray have to be used to cover all the foliage.

For different conditions nozzles as well as the method of application may have to be changed. For ground-crop application with a tractor boom, twin-fluid atomizers with a fan will give a very fine spray and a low-pressure pump will be sufficient (290).

4. AGRICULTURAL EMULSIONS

For airplane spraying, twin-fluid atomizers can be used for coarse sprays and rotary atomizers for very fine sprays. For the ground spraying of shrubs, compressed-air nozzles may be necessary. For spraying large trees, the single-beam spray has to be directed into the foliage from a slow tractor using compressed air. For animal spraying the spray has to be directed in such a way (184) as to reach the skin without being too much deflected by the hair.

In general, where drift is no problem, where evaporation is slow because of a high proportion of oil phase to water phase in oil-in-water emulsions, with invert emulsions, and in close-to-target spraying small-droplet-size sprays are preferred and give better area coverage and better utilization of the toxicant (277-279). On the other hand, with a high proportion of water phase or danger of drift, a bigger droplet size is imperative and more economical. For stationary insects, a bigger volume spray may be necessary to reach them all (280). Special care should be taken to clean spraying equipment to prevent rusting and clogging of nozzles (291); it is necessary to calibrate the nozzles often and to change their locations accordingly to ensure uniform coverage with the right overlap and droplet-size distribution.

To diminish the problem of drift the use of water-in-oil emulsions is especially recommended. This type of emulsion, which is desirable at high viscosities (50,000 cP), can only be pumped with difficulty through narrow booms and will not give a uniform spray with regular equipment requiring very high pressures. The bifluid spray system (146,147) was specially designed to take care of this problem. The emulsifier-containing oil phase and the water phase are brought by two different pipes to the bifluid mixing chamber, which is situated immediately before the spraying nozzle; this cuts the pumping requirement drastically. Since the liquids stay in the chamber for a very short moment (0.1 sec) the choice of emulsifier is especially important in order to get a good emulsion.

If two or more different formulations are to be applied from the same spray tank simultaneously, the greatest care should be taken to ensure that they are compatible (291). Two good emulsifiable formulations may interaxt to give a poor emulsion. The same effect could be caused by the various tank additives, which should be carefully tested for compatibility with the formulation before they are recommended. Antifoams can be added to the spray tanks if foaming is excessive. Often the foam-breaking characteristic disappears on prolonged mixing, and it may be desirable to add more to the formulated emulsion in a limited amount. Thus 1 pint of kerosene may be sufficient for 24 ft^2 of foam surface, but the function disappears if the kerosene is allowed to emulsify in the spray tank by strong mixing.

REFERENCES

1. P. Lindner, Farm Chem., April 1966.
2. R. Gast and J. Early, Agr. Chem., 4, 42 (1956).
3. R. T. Gast and J. D. Early, North Carolina State College, Raleigh, Inf. Note 103, 1956.
4. C. G. L. Furmidge, J. Sci. Food Agr., 10, 274 (1959).
5. C. G. L. Furmidge, J. Sci. Food Agr., 10, 419 (1959).
6. J. H. Hildebrand and R. L. Scott, Solubility of Nonelectrolytes, 3rd ed., Reinhold, New York, 1950.
7. J. H. Hildebrand and R. L. Scott, Regular Solutions, Prentice-Hall, Englewood Cliffs, N. J., 1962.
8. G. Scatchard, Chem. Rev., 8, 321 (1931).
9. G. Scatchard, Trans. Faraday Soc., 33, 160 (1937).
10. R. C. Little and C. R. Singleterry, J. Phys. Chem., 68, 3453 (1964).
11. R. C. Little, J. Colloid Interface Sci., 21, 266 (1966).
12. F. M. Fowkes, in Solvent Properties of Surfactant Solutions (K. Shinoda, ed.), Dekker, New York, 1967, Chapter 3.
13. W. Van Valkenberg, in Solvent Properties of Surfactant Solutions (K. Shinoda, ed.), Dekker, New York, 1967, Chapter 6.

14. H. S. Butler and C. C. Harvey, Jr., U.S. Pat. 2,768,111 (1956).
15. L. Q. Boyd, U.S. Pat. 2,543,955 (1951).
16. K. Ullrich, *Arch. Pflanzensch.*, 5, 25 (1968).
17. H. L. Greenwald, G. L. Brown, and M. N. Fineman, *Anal. Chem.*, 28, 1693 (1956).
18. K. J. Olson, R. W. Dupree, E. T. Plomer, and V. K. Rowe, *J. Soc. Cosm. Chem.*, 13, 469 (1962).
19. J. Thiernagand and W. E. Adams, *Fette, Seifen, Anstrichm.*, 71, 767 (1969).
20. Anon., *Fed. Reg.*, 34, 6041 (1969).
21. Anon., *Fed. Reg.*, 35, 3233 (1970).
22. A. K. Epstein and B. R. Harris, U.S. Pat. 1,917,253 (1933).
23. B. R. Harris, U.S. Pat. 1,917,256 (1933).
24. W. Clayton, *The Theory of Emulsions and Their Technical Treatment*, 4th ed., Blakistons, Philadelphia, 1943, p. 127.
25. W. C. Griffin, *J. Soc. Cosm. Chem.*, 1, 311 (1949).
26. W. C. Griffin, *J. Soc. Cosm. Chem.*, 5 249 (1954).
27. W. C. Griffin, *Amer. Perf. Ess. Oil Rev.*, 65, 26 (1955).
28. I. Racz and E. Orban, *J. Colloid Sci.*, 20, 99 (1965).
29. J. J. Middleton, *J. Soc. Cosm. Chem.*, 19, 129 (1968).
30. P. Becher, in *Nonionic Surfactants* (M. J. Schick, ed.), Dekker, New York, 1967, Chapter 18.
31. L. Kaertkemeyer and J. Armand, SCI Monograph No. 21, 1966, p. 28.
32. T. J. Lin, paper presented at the 16th Annual Seminar of the Society of Cosmetic Chemists, St. Louis, Mo., 1969.
33. R. V. Peterson and R. D. Hamill, *J. Soc. Cosm. Chem.*, 19, 627 (1968).
34. W. D. Bancroft, *J. Phys. Chem.*, 17, 501 (1913).
35. H. L. Greenwald, *J. Soc. Cosm. Chem.*, 6, 164 (1955).
36. S. P. Moulik, *J. Phys. Chem.*, 72, 4682 (1968).
37. K. J. Lissant, *J. Colloid Interface Sci.* 22, 462 (1966).
38. K. J. Lissant, *J. Soc. Cosm. Chem.* 21, 141 (1970).

39. H. Lange, J. Soc. Cosm. Chem., 16, 697 (1965).
40. M. Van den Tempel, Rec. Trav. Chim., 72, 419, 433, 442 (1953).
41. H. Sonntag, J. Netzel, and H. Klare, Kolloid-Z., 211, 121 (1966).
42. H. Sonntag, Tenside, 5, 188 (1968).
43. W. Albers and J. T. G. Overbeek, J. Colloid Sci., 14, 501 510 (1959).
44. L. E. Scriven and C. V. Sternling, Nature (London), 187, 186 (1960).
45. G. D. M. MacKay and S. G. Mason, J. Colloid Sci., 18, 674 (1963).
46. B. V. Derjaguin and A. S. Titiyevskaya, in Proc. 2nd Int. Congress Surface Activity, Butterworths, London, Vol. 1, 1957, p. 211.
47. M. Van den Tempel, J. Lucassen, and E. H. Lucassen-Reynders, J. Phys. Chem., 69, 1798 (1965).
48. M. Van den Tempel, in Proc. 3rd Int. Congress Surface Activity, Cologne, 1960, p. B573.
49. B. V. Derjaguin and M. M. Kusakov, Zhur. Fiz. Khim., 26, 1536 (1952).
50. M. Van den Tempel, J. Colloid Sci., 13, 125 (1958).
51. H. V. Tartar, J. Colloid Sci., 17, 243 (1962).
52. M. J. Schick, S. M. Atlas, and F. R. Eirich, J. Phys. Chem., 66, 1326 (1962).
53. J. H. Schulman, W. Stoeckenius, and L. M. Prince, J. Phys. Chem., 63, 1677 (1959).
54. P. Becher, J. Colloid Sci., 18, 665 (1963).
55. D. C. Poland and H. A. Scheraga, J. Colloid Interface Sci., 21, 273 (1966).
56. D. C. Poland and H. A Scheraga, J. Phys. Chem., 69, 2431 (1965).
57. D. C. Poland and H. A Scheraga, J. Phys. Chem., 69, 4425 (1965).
58. G. Boehmke and R. Heusch, Fette und Seifen, 62, 87 (1960).
59. P. Becher, J Colloid Sci., 18, 196 (1963).

60. P. Becher, J. Colloid Sci., 17, 325 (1962).
61. M. Roesch, Kolloid-Z., 150, 153 (1957).
62. E. L. Mackor and J. H. Van der Waals, J Colloid Sci., 7, 535 (1952).
63. E. L. Mackor, J Colloid Sci., 6, 492 (1951)
64. M. Van der Waarden, J. Colloid Sci., 5, 317 (1950).
65. T. J. Lin and J. C. Lambrechts, J. Soc. Cosm. Chem., 20, 626 (1969).
66. T. J. Lin and J. C. Lambrechts, J. Soc. Cosm. Chem., 20, 185 (1969).
67. J. H. Schulman and E. G. Cockbain, Trans. Faraday Soc., 36, 651 (1940).
68. E. G. Cockbain and T. S. McRoberts, J. Colloid Sci., 8, 440 (1953).
69. L. M. Prince, J. Colloid Interface Sci., 23, 165 (1967).
70. L. M. Prince, J. Colloid Interface Sci., 29, 216 (1969).
71. L. M. Prince, J. Soc. Cosm. Chem., 21, 193 (1970).
72. M. Stackelberg, E. Klockner, and P. Mohrhauer, Kolloid-A., 115, 53 (1949).
73. M. Volmer, Z. Phys. Chem., 125, 151 (1921)
74. P. A. Winsor, Trans. Faraday Soc., 44, 376 (1948).
75. S. Ross, E. S. Chen, P. Becher, and H. J. Ranauto, J. Phys. Chem., 63, 1681 (1959).
76. P. Becher, J. Soc. Cosm. Chem., 11, 325 (1960).
77. J. H. Schulman and J. Leja, Trans. Faraday Soc., 50, 598 (1954).
78. H. Arai, J. Colloid Interface Sci., 23, 348 (1967).
79. K. Shinoda and H. Arai, J. Phys. Chem., 68, 3485 (1964).
80. K. Shinoda and H. Saito, J. Colloid Interface Sci., 30, 258 (1969).
81. T. Mitsui and Y. Machida, J. Soc. Cosm. Chem., 20, 199 (1969).
82. W. Drost-Hansen, in Chemistry and Physics of Interfaces (S. Ross, ed.), American Chemical Society, Washington, D. C., 1965, p. 22.

83. D. A. Netzel, G. Hoch, and T. I. Marx, *J. Colloid Sci.*, *19*, 774 (1964).
84. R. L. Mayhew and R. C. Hyatt, *J. Amer. Oil Chem. Soc.*, *29*, 357 (1952).
85. R. A. Kaberg and J. S. Harris, U.S. Pat. 2,447,475 (1948).
86. R. A. Kaberg and J. S. Harris, U.S. Pat. 2,509,233 (1950).
87. H. L. Sanders, E. A. Knagg, and M. L. Nussbaum, U.S. Pat. 2,696,453 (1954).
88. R. W. Behrens, U.S. Pat. 2,786,013 (1957).
89. V. J. Keenan, U.S. Pat. 2,862,848 (1958).
90. H. L. Sanders, E. A. Knagg, and M. L. Nussbaum, U.S. Pat. 2,872,368 (1959).
91. P. L. Lindner, U.S. Pat. 2,898,267 (1959).
92. M. Quaedvlig, G. Boehmke, and H. Hempel, German Pat. 1,094,523 (1960).
93. R. W. Behrens, Swiss Pat. 353,016 (1961).
94. S. Altsher and T. F. Groll, Jr., U.S. Pat. 3,071,550 (1963).
95. S. Altsher and T. F. Groll, Jr., U.S. Pat. 3,172,750 (1965).
96. A. Stefcik and F. E. Woodward, U.S. Pat. 3,240,585 (1966).
97. O. L. Scherr, U.S. Pat. 3,256,322 (1966).
98. L. A. Joo and W. E. Kramer, U.S. Pat. 3,277,014 (1966).
99. O. L. Scherr, U.S. Pat. 3,298,912 (1967).
100. K. L. Johnson, U.S. Pat. 3,442,818 (1969).
101. E. W. Lohr and S. K. Love, *The Industrial Utility of Public Water Supplies in the U.S.*, Geological Survey Paper 1299, U.S. Government Printing Office, Washington, D.C., 1954.
102. E. Forslind, *Acta Polytechnica*, *115*, 9 (1952).
103. E. Forslind, in *Proc. 2nd Int. Congress Rheology*, Butterworths, London, 1953, p. 50.
104. O. D. Bonner and G. B. Woosley, *J. Phys. Chem.*, *72*, 899 (1968).
105. N. L. Jarvis and M. A. Scheiman, *J. Phys. Chem.*, *72*, 74 (1968).
106. G. Jones and W. A. Ray, *J. Amer. Chem. Soc.*, *59*, 187 (1937).
107. W. U. Malik and S. M. Saleem, *J. Amer. Oil Chem. Soc.*, *45*, 670 (1968).

108. J. M. Corkill and J. F. Goodman, *Adv. Colloid Interface Sci.*, 2, 298 (1969).
109. D. F. Cheesman and A. King, *Trans. Faraday Soc.*, 35, 241 (1940).
110. P. Sennet and J. O. Olivier, in *Chemistry and Physics of Interfaces* (S. Ross, ed.), American Chemical Society, Washington, D.C., 1965, p. 75.
111. E. Matijević and B. A. Pethica, *Trans. Faraday Soc.*, 54, 1382, 1390 (1958).
112. R. D. Vold and R. C. Groot, *J. Colloid Sci.*, 19, 384 (1964).
113. H. Sonntag and H. Klare, Jr. *Tenside*, 2, 33 (1965).
114. G. L. Brown and G. C. Riley, *Agr. Chem.*, 10(8), 34 (1955).
115. B. I. Sparr and C. V. Bowen, *Agr. Food Chem.*, 2, 871 (1954).
116. G. E. Mapstone, *J. Soc. Cosm. Chem.*, 12, 239 (1961).
117. H. Schott, *J Colloid Interface Sci.*, 24, 193 (1967).
118. A. Doren and J. Goldfarb, *J. Colloid Interface Sci.*, 32, 67 (1970).
119. H. Lange, *Fette, Seifen, Anstrichsm.*, 70, 748 (1968).
120. B. I. Dashevskaya, M. Ch. Gluzman, and R. G. Zaslavskaya, *Ukr. Khim. Zh.*, 35, 388 (1969).
121. *Specification for Pesticides*, 2nd ed., World Health Organization, Geneva, 1961.
122. W. C. Griffin and R. W. Behrens, *Agr. Chem.*, 3 (1952).
123. W. C. Griffin and R. W. Behrens, *Anal. Chem.*, 24, 1076 (1952).
124. R. W. Behrens, *Agr. Food Chem.*, 6, 20 (1958).
125. E. Selz, *Agr. Food Chem.*, 1, 381 (1953).
126. D. A. Pearce, *Anal. Chem.*, 27, 163 (1955).
127. B. I. Sparr and C. V. Bowen, USDA Agr. Res. Admin., Bur. Entymol. Plant Quar., E866, 1954.
128. L. Kennon, *J. Soc. Cosm. Chem.*, 17, 135, 313 (1966).
129. U.S. Department of Agriculture, Agricultural Research Service, USDA TSC-0164, 1964.
130. N. G. Lordi and M. W. Scott, *J. Pharm. Sci.*, 54, 531 (1965).
131. C. J. Clark and H. E. Hudson, *Manuf. Chem.*, 1, 25 (1968).

132. R. P. Upchurch, J. A. Keaton, and H. D. Coble, *Weeds*, , 505 (1969).
133. R. M. Dvoretskaya, *Kolloid Zh.*, *11*, 311 (1949).
134. R. M. Dvoretskaya, *Kolloid Zh.*, *13*, 432 (1951).
135. J. T. Davies in *Proc. 3rd Intern. Congress Surface Activity, Cologne*, Vol. 2, 1960, p. B585.
136. J. T. Davies, *J. Soc. Cosm. Chem.*, *12*, 193 (1961).
137. P. Sherman, *J. Soc. Cosm. Chem.*, *16*, 591 (1965).
138. P. Sherman, *J. Colloid Interface Sci.*, *24*, 67 (1967).
139. P. Sherman, *J. Colloid Interface Sci.*, *24*, 107 (1967).
140. P. Sherman, *J. Phys. Chem.*, *67*, 2531 (1963).
141. P. Sherman in *Proc. 3rd Int. Congress Surface Activity, Cologne*, Vol. 2, 1960, p. B597.
142. R. E. Ford and C. G. L. Furmidge, *J. Colloid Interface Sci.*, *22*, 331 (1966).
143. S. Biswas and D. A. Haydon, *Kolloid-Z.*, *185*, 31 (1962).
144. W. Duyfjes, *Philips Tech Rev.*, *28*, 112 (1967).
145. G. W. Ware, B. J. Estesen, W. P. Cahill, P. D. Gerhardt, and K. R. Frost, *J. Econ. Ent.*, *63*, 1314 (1970).
146. J. P. Colthurst, R. E. Ford, C. G. L. Furmidge, and A. J. A. Pearson, SCI Monograph No. 21, 1966, p. 47.
147. E. B. Stull and J. Morrow, U.S. Pat. 3,197,299 (1965); British Pat. 944,229 (1966).
148. J. A. Kelly and G. W. Scoles, U.S. Pat. 3,189,430 (1965).
149. R. L. Voda, U.S. Pat. 3,125,517 (1964).
150. J. T. Foley and R. H. Rogers, U.S. Pat. 3,244,638 (1966).
151. H. F. Wiese, U.S. Pat. 3,269,946 (1966).
152. W. Bonin, U.S. Pat. 3,442,842 (1969); German Pat. F43029 (1964).
153. F. Lachampt and A. Viout, U.S. Pat. 3,489,690 (1969); Luxembourg Pat. 47,604 (1964).
154. P. J. Chapman and G. W. Pearce, *Agr. Chem.*, March/April 1947.
155. P. J. Chapman and G. W. Pearce, *J. Econ. Ent.*, *34*, 207 (1941).
156. P. J. Chapman, S. E. Lienk, A. W. Avens, and R. W. White, *J. Econ. Ent.*, *55*, 737 (1962).

157. L. A. Riehl, *J. Agr. Food Chem.*, *15*, 878 (1967).
158. H. A. Dean and C. E. Hoelscher, *J. Econ. Ent.*, *60*, 1668 (1967).
159. L. W. Hall, Jr. and E. L. Ratledge, paper presented at National Petroleum Association meeting, New York, 1967.
160. W. H. Volck, U.S. Pat. 1,707,465 (1926).
161. G. Thorne and A. F. Millican, U.S. Pat. 3,426,126 (1969).
162. L. A. Riehl, *Proc. 1st Int. Citr. Symp.*, *2*, 897 (1969).
163. L. A. Riehl and G. E. Carman, *J. Econ. Ent.*, *46*, 1007 (1953).
164. E. R. DeOng, H. Knight, and J. C. Chamberlin, *Hilgardia*, *2*, 351 (1927).
165. W. B. Parker, U.S. Pat. 2,144,808 (1939).
166. T. H. Cheng, E. H. Frear, and H. F. Enos, *J. Econ. Ent.*, *55*, 39 (1962).
167. P. A. Sanders, *J. Soc. Cosm. Chem.*, *9*, 274 (1958).
168. P. A. Sanders, *J. Soc. Cosm. Chem.*, *17*, 801 (1966).
169. P. A. Sanders, *J. Soc. Cosm Chem.*, *20*, 577 (1969).
170. P. A. Sanders, *J. Soc. Cosm. Chem.*, *21*, 377 (1970).
171. F. A. Mina, U.S. Pat. 2,702,957 (1955).
172. K. Dixon and M. Davies, SCI Monograph No. 21, 1966, p. 14.
173. L. H. Flett, U.S. Pat. 2,205,950 (1940).
174. W. A. Schulze and J. C. Hillyer, U.S. Pat. 2,490,481 (1949).
175. H. C. Wohler and T. C. Davis, U.S. Pat. 2,521,318 (1950).
176. C. V. Coash, U.S. Pat. 2,614,061 (1952).
177. I. F. Walker, U.S. Pat. 2,519,088 (1950).
178. L. H. Princen, J. A. Stolp, and R. Zgol, *J. Colloid Interface Sci.*, *28*, 466 (1968).
179. K. Shinoda and T. Ogawa, *J. Colloid Interface Sci.*, *24*, 56 (1967).
180. C. V. Bowen, USDA Agr. Res. Admin., Bur. Entymol. Plant Quar., ET-285, 1950.
181. B. I. Sparr, J. C. Clark, E. E. Vallier, and A. H. Baumhover, USDA Agr. Res. Admin., Bur. Entymol. Plant Quar., E-849, 1952.
182. L. L. Wade, *J. Econ. Ent.*, *61*, 908 (1968).

183. A. F. Machin, SCI Monograph No. 21, 1966, p. 81.
184. J. S. Skaptason, *Agr. Chem.*, 9, 39 (1959).
185. J. W. Mitchell and P. J. Linder, *Science*, 112, 54 (1950).
186. V. H. Freed and M. Montgomery, *Weeds*, 6, 386 (1958).
187. L. L. Jansen, W. A. Gerntner, and W. C. Shaw, *Weeds*, 9, (1961).
188. C. L. Foy, *Weeds*, 10, 97 (1962).
189. C. G. McWhorter, *Weeds*, 11, 83 (1963).
190. L. S. Jordan, B. E. Day, and R. T. Hendrixson, *Weeds*, 11, 198 (1963).
191. C. G. McWhorter, *Weeds*, 11, 265 (1963).
192. R. E. Temple and H. W. Hilton, *Weeds*, 11, 297 (1963).
193. L. L. Jensen, *J. Agr. Food Chem.*, 12, 223 (1964).
194. L. L. Jensen, *Weeds*, 12, 251 (1964).
195. C. L. Foy and L. W. Smith, *Weeds*, 13, 15 (1965).
196. G. D. Hill, I. J. Belasco, and H. L. Ploeg, *Weeds*, 13, 103 (1965).
197. L. L. Jensen, *Weeds*, 13, 117 (1965).
198. L. L. Jensen, *Weeds*, 13, 123 (1965).
199. R. A. Evans and R. E. Eckert, *Weeds*, 13, 150 (1965).
200. D. E. Bayer and H. R. Drever, *Weeds*, 13, 222 (1965).
201. A. G. Dexter, O. C. Burnside, and T. L. Lavy, *Weeds*, 14, 222 (1966).
202. P. V. Nelson and H. H. Garlich, *J. Agr. Food Chem.*, 17, 148 (1969).
203. E. N. Cory and G. S. Langford, *J. Econ. Ent.*, 28, 257 (1935).
204. H. L. Dozier, *J. Econ. Ent.*, 30, 968 (1937).
205. K. E. Maxwell and W. D. Piper, *J. Econ. Ent.*, 61, 1633 (1968).
206. D. A. Wolfenbarger, M. J. Lukefahr, and W. L. Lowry, *J. Econ. Ent.*, 60, 902 (1967).
207. C. G. McWhorter, *Weeds*, 14, 191 (1966).
208. C. Z. Draves and R. G. Clarkson, *Amer. Dyestuff Reporter*, 23, 1938 (1934).
209. L. Shapiro, *Amer. Dyestuff Reporter*, 39, 38 (1950).

210. R. D. Ashworth and G. A. Lloyd, *J. Sci. Agr.*, *12*, 234 (1961).
211. W. A. Ritchel, *Angew. Chem. Int. Ed.*, *8*, 699 (1969).
212. H. B. Curier and C. D. Dybing, *Weeds*, *7*, 195 (1959).
213. S. Yamaguchi and A. S. Crafts, *Hilgardia*, *28*, 161 (1958).
214. L. M. Wax and R. Behrens, *Weeds*, *13*, 107 (1965).
215. A. S. Crafts and W. W. Robbins, *Weed Control*, 3rd ed., McGraw-Hill, New York, 1962.
216. Y. P. Sun, *J. Econ. Ent.*, *61*, 949 (1968).
217. P. A. Frank, *Weeds*, *16*, 489 (1968).
218. D. W. Micks, G. Chambers, J. Jennings, and A. Rehmet, *J. Econ. Ent.*, *60*, 426 (1967).
219. F. C. Nelson and G. W. Fiero, U.S. Pat. 2,898,263 (1959).
220. D. W. Hagstrum and M. S. Mulla, *J. Econ. Ent.*, *61*, 220 (1968).
221. T. P. Yin, *J. Phys. Chem.*, *73*, 2413 (1969).
222. A. S. Crafts, *J. Agr. Food Chem.*, *1*, 51 (1953).
223. J. L. Barrentin and G. F. Warren, *Weed Sci.*, *18*, 365 (1970).
224. H. F. Madsen and K. Williams, *J. Econ. Ent.*, *60*, 121 (1967).
225. G. W. Anderson, Res. Report Natl. Weed Committee, Eastern Section, Canada, 1963.
226. *Chemical Guide to Weed Control*, Ontario Department of Agriculture, 1965.
227. G. B. Triplett, Jr., *Weed Sci. Soc. Amer. Abstract*, 18 (1968).
228. S. R. Colby, *Weeds*, *15*, 20 (1967).
229. P. L. Lindner, in *Pesticide Formulation: The Physical Chemical Principles* (W. Van Valkenberg, ed.), Dekker, New York, 1973.
230. N. A. Lange, *Handbook of Chemistry*, 10th ed., McGraw-Hill, New York, 1961.
231. J. J. Mortvedt, *Agr. Chem.*, *20(6)*, 39 (1965).
232. L. H. Adams and B. E. Gibson, *J. Amer. Chem. Soc.*, *54*, 4520 (1932).
233. V. E. Bower and R. A. Robinson, *J. Phys. Chem.*, *67*, 1524 (1963).
234. R. W. Behrens, *Weeds*, *12*, 255 (1964).
235. P. L. Lindner, U.S. Pat. 2,976,208 (1956).

236. P. L. Lindner, U.S. Pat. 2,976,209 (1956).
237. P. L. Lindner, U.S. Pat. 2,976,211 (1958).
238. P. L. Lindner, U.S. Pat. 3,080,280 (1961).
239. P. L. Lindner, U.S. Pat. 3,236,626 (1961).
240. P. L. Lindner, U.S. Pat. 3,236,627 (1961).
241. P. L. Lindner, U.S. Pat. 3,284, 187 (1962).
242. P. L. Lindner, Canadian Pat. 818,423 (1969).
243. A. Stefcik and F. E. Woodward, U.S. Pat. 3,317,305 (1967).
244. K. L. Lynch, U.S. Pat. 2,438,092 (1948).
245. A. O. Jaeger, U.S. Pat. 2,028,091 (1936).
246. A. F. Steinhauer, U.S. Pat. 2,854,477 (1958).
247. S. E. Ainsworth, Natl. Fert. Sol. Assn., Liq. Fert. Round-up Proc., July 1967.
248. P. L. Lindner, U.S. Pat. 3,408,174 (1962).
249. D. E. Bayer, Weeds, 15, 249 (1967).
250. E. P. Lichtenstein, T. W. Fuhrmann, K. R. Schulz, and R. F. Skrentny, J. Econ. Ent., 60, 1714 (1967).
251. R. H. Beal and L. C. Fairie, J. Econ. Ent., 61, 380 (1968).
252. C. I. Harris and T. J. Sheets, Weeds, 13, 215 (1965).
253. C. I. Harris, Weeds, 14, 6 (1966).
254. S. M. Lambert, P. E. Porter, and R. H. Schieferstein, Weeds, 13, 185 (1965).
255. R. C. Rhodes, I. J. Belasco, and H. L. Pease, J. Agr. Food Chem., 18, 524 (1970).
256. R. E. Pelishek, J. Osborn, and J. Letey, Soil Sci. Amer. Proc., 26, 595 (1962).
257. J. J. Bikerman, Surface Chemistry for Industrial Research, Academic Press, New York, 1947.
258. Wetting, SCI Monograph No. 25, 1967.
259. F. P. Achorn and H. L. Baley, Agr. Chem., 24 May, 1970.
260. C. G. McWhorter and W. L. Barrentine, Weed Sci., 18, 500 (1970).
261. T. C. Tso, Nature (London), 202, 511 (1964).
262. T. C. Tso, L. G. Burk, and G. L. Steffens, Tobacco Sci., 10, 77 (1966).

263. H. M. Cathey, G. L. Steffens, N. W. Stuart, and R. H. Zimmerman, *Science*, *152*, 1382 (1966).
264. G. L. Steffens, T. C. Tso, and D. W. Spaulding, *J. Agr. Food Chem.*, *15*, 972 (1967).
265. H. M. Cathey and G. L. Steffens, SCI Monograph No. 31, 1968, p. 224.
266. G. L. Steffens and H. M. Cathey, *J. Agr. Food Chem.*, *17*, 312 (1969).
267. C. G. L. Furmidge, *Brit. J. Appl. Phys.*, *12*, 268 (1961).
268. J. R. Lake, *Chem. Ind. (London)*, 233 (1970).
269. J. M. Davies, W. E. Waters, D. A. Isler, R. Martineau, and J. W. Marsh, *J. Econ. Ent.*, *49*, 338 (1956).
270. T. R. Plumb, L. R. Green, and V. E. White, *Weeds*, *14*, 114 (1966).
271. G. W. Ware, B. J. Estesen, W. P. Cahill, P. D. Gerhardt, and K. R. Frost, *J. Econ. Ent.*, *62*, 840 (1969).
272. G. W. Ware, E. J. Apple, W. P. Cahill, P. D. Gerhardt, and K. R. Frost, *J. Econ. Ent.*, *62*, 844 (1969).
273. L. O. Roth and J. G. Porterfield, *Weeds*, *13*, 326 (1965)
274. R. E. Ford and C. G. L. Furmidge, *Brit. J. Appl. Phys.*, *18*, 335 (1967).
275. R. E. Ford and C. G. L. Furmidge, *Brit. J. Appl. Phys.*, *18*, 491 (1967).
276. W. Maas and W. deLange, *Philips Tech. Rev.*, *30*, 21 (1969).
277. C. M. Himel, *J. Econ. Ent.*, *62*, 912 (1969).
278. C. M. Himel and A. D. Moore, *J. Econ. Ent.*, *62*, 916 (1969).
279. C. M. Himel, *J. Econ Ent*, *62* 919 (1969).
280. E. M. Stafford, J. B. Byass, and N. B. Akesson, *J. Econ. Ent.*, *63*, 769 (1970).
281. B. Carrol, *Farm Chemicals*, 44, July, 1970.
282. J. D. Gilpatrick and J. Terril, *J. Econ Ent*, *63* 15 (1970).
283. A. H. Higgins, *J. Econ. Ent.*, *60*, 280 (1967).
284. S. F. Potts, *Concentrated Spray Equipment, Mixtures and Application Methods*, Dorland Caldwell, N. J., 1958.
285. W. H. Walton and W. C. Prewett, *Proc. Phys. Soc.*, *62*, 23 (1949).

286. C. G. L. Furmidge, *J. Sci. Food Agr.*, *10*, 267 (1959).
287. J. F. Baret and A. G. Bois, *Can. J. Chem.*, *46*, 3211 (1968).
288. J. B. Byass, SCI Monograph No. 2, 1958, p. 89.
289. P. Hebblethwaite and P. Richardson, SCI Monograph No. 21, 1966, p. 67.
290. L. P. Ditman, C. R. Rosenberger, and F. P. Harrison, *J. Econ. Ent.*, *47*, 600 (1954).
291. R. Howes, SCI Monograph No. 21, 1966, p. 61.

Chapter 5

FOOD EMULSIONS

Matthew J. Lynch and William C. Griffin

ICI America Inc.
Specialty Chemicals Division
Wilmington, Delaware

I. INTRODUCTION. 250
   A. Definitions 250
   B. Emulsions in Nature 251
   C. Commercial Food Emulsions 252
II. EMULSION PROPERTIES 253
   A. Appearance. 254
   B. Dispersibility. 254
   C. Viscosity 254
   D. Particle Size 255
   E. Particle Charge 256
   F. Conductivity. 256
   G. The pH. 257
   H. Stability 257
   I. Preservation. 257
III. EMULSIFICATION. 258
   A. General Remarks 258
   B. Emulsifier Selection. 260

IV.	PRODUCT DEVELOPMENT	268
V.	EQUIPMENT	274
VI.	FOOD EMULSIONS	279
VII.	OTHER USES FOR SURFACTANTS	281
	A. Foams	281
	B. Suspensions	282
	C. Demulsification and Antifoaming	282
	D. Complexing	283
	E. Crystallization Control	284
	F. Wetting	286
	G. Lubricating	286
VIII.	MULTIPLE EMULSIFIER EFFECTS	286
	REFERENCES	289

I. INTRODUCTION

Before discussing food emulsions in particular, it is necessary first to understand certain basic facts about emulsions and emulsion technology.

A. Definitions

The classical definition describes an emulsion as a two-phase system of immiscible liquids. The immiscible materials are usually water, oil, or fat. One liquid is dispersed as fine globules in the other and is referred to as the dispersed, discontinuous, or internal phase. The surrounding liquid is known as the continuous, or external phase. The food technologist broadens this definition to include colloidal dispersions or solubilizations as well as dispersions of a gas in a liquid (foams) and combinations of liquids, solids, and gases in emulsions (or batters). The typical emulsion

5. FOOD EMULSIONS

systems can be divided into the following principal classes:

1. Those consisting of droplets of oil dispersed throughout an aqueous medium, usually referred to as oil-in-water (o/w) emulsions

2. Those in which droplets of water are dispersed throughout an oil or fat medium, termed water-in-oil (w/o) emulsions

Emulsion of either type may be as fluid as water or as viscous as solid fat. Science has developed the materials and techniques to produce these and many other properties in biphase and multiphase food systems.

B. Emulsions in Nature

Food emulsions are a mechanism to make a particular food more acceptable or wholesome or both. Emulsions are usually regarded as synthetic products prepared by the food processor; however, the outstanding classical emulsions are natural systems containing many of the essential nutrients, such as fat, carbohydrate, protein, vitamins, trace minerals, and water. Milk is a typical natural thin liquid emulsion. Processed meat, such as frankfurters, represents a high-viscosity type mentioned under general emulsion definitions. As food science developed, the emulsion systems found in nature were studied and investigated. Man has learned that emulsification not only makes certain foods more palatable but that the emulsion mechanism is used biologically as a transport system to utilize fat. The bile salts in bile fluid produced by the liver play a major role in emulsifying the ingested fat, making it available for absorption and utilization. In summary, like many other advances, nature has provided man with certain excellent emulsion models to study.

C. Commercial Food Emulsions

As civilization has developed, the spectrum of prepared and processed food products has broadened. As study revealed the composition of natural products, man learned to use this knowledge to prepare new foods and improve those already available. Some examples of man-made emulsions considered major achievements at the time they first appeared were the first butter substitute, oleomargarine; salad dressings and mayonnaise; and meat emulsions (sausage). In many food emulsions the emulsifiers serve more than one function; for example, both cake mixes or cake batters and ice cream benefit from the emulsification or demulsification properties of the food emulsifiers added and also require the aeration characteristics. However, although food emulsifiers are of tremendous commercial importance, selling in the millions of pounds annually, most of them are used in nonemulsion applications: as starch antistaling agents, crystallization modifiers, wetting agents, stickiness reducers, texture improvers, and antifoaming agents. Moreover, in nature they serve to carry lipophilic materials in aqueous systems. The more recent food emulsions allow unique combinations of physical properties, especially viscosity and tase.

Considering food emulsions in particular, it is of interest to examine them with respect to the classification system outlined in Table 1. First, it is noted that the vast majority of food emulsions are of the oil-in-water, rather than the water-in-oil, type. Second, it is apparent that, of the oil-in-water emulsions, most are in the low-internal-phase-ratio category (see Chapter 1).

TABLE 1

Classification of Food Emulsions

Type	Internal-Phase-Ratio Category		
	LIPR[a]	MIPR[b]	HIPR[c]
Oil in water	Milk	Heavy cream	Mayonnaise
	Table cream	Fluid shortening	Salad dressing
	Ice cream[d]	emulsions	
	Cake (in batter form)[e]	Meat emulsions (frankfurters0	
	Coffee whitener (liquid)		
	Calf milk replacement		
	Cheese spread		
	Onion (etc.) dip[e]		
Water in oil	Butter		
	Margarine		

[a]Low (<30%) internal-phase ratio.
[b]Medium (30 to 70%0 internal-phase ratio.
[c]High (>70%) internal-phase ratio.
[d]Contains air.
[e]Contains "solids."

II. EMULSION PEOPERTIES

Before discussing emulsion theory and preparation, it may be well to discuss the properties of emulsions and how these properties can be altered by changes in formulation. The emulsion properties discussed individually below may be chemical or physical in nature, or both. Usually it is not possible or desirable to categorize every facet of an emulsion. Generally speaking, it can be said that the properties of an emulsion depend on (1) the properites of the continuous phase and (2) the proportion of the continuous phase to the internal phase.

A. Appearance

The appearance of an emulsion depends on the ingredients used, their color and the difference in refractive index, and the particle size of the dispersed phase. Opacity is best achieved (with stability) by having a particle size of about 0.5 to 5 μ with a significant difference in refractive index. The color of the continuous phase is usually the controlling color for the product. Clarity of an emulsion may be achieved by reducing the particle size to a few nanometers (i.e., less than the wavelength of visible light) or by adjusting the two phases to an identical refractive index.

B. Dispersibility

The dispersibility of an emulsion depends on the emulsion type. If the external phase is water, the emulsion can be dispersed in, and diluted with, water or aqueous solvents. If the external phase is oil, it can be dispersed in, or diluted with, oily materials. Dispersibility is useful in determining emulsion type. The most common type of emulsion encountered in food applications is the oil-in-water type.

C. Viscosity

Viscosity is in general dependent on the viscosity of the external phase, the ratio of the external phase to the internal phase, and, to a lesser extent, the particle size of the dispersed droplets. Thus it also depends on the type and concentration of emulsifier.

The viscosity of an emulsion is usually similar to that of the external phase when that phase constitutes the major part of the product (low internal-phase ratio). When the internal phase is increased in concentration, the viscosity of the product is increased, and when the volume of internal phase is greater than that of the external phase, the apparent viscosity increases markedly. This phenomenon is caused by particle crowding in the emulsion. Theoretically only 74% of the total emulsion volume can be occupied by the internal phase if the globules are of uniform spherical shape. When the internal-phase ratio is extremely high, the particles of that phase become quite distorted, and particle size and particle charges have a greater effect on emulsion viscosity.

Thickeners or bodying agents are commonly added to *high-external-phase-ratio* emulsions to increase their apparent viscosity. Reduction of particle size usually increases viscosity slightly.

D. Particle Size

Particle size is usually expressed as the diameter of the globules in the internal phase. If the size is not uniform, the size range occurring most frequently is generally used to denote the particle size of the emulsion. The values of the smallest and largest particles are said to be the range of particle sizes present. Emulsions containing particles that are small in diameter are said to be *fine* emulsions; those containing large globules are *coarse* emulsions.

The particle size depends on the type and quantity of emulsifier, the amount of work done to prepare the emulsion, and the order of addition of ingredients. Most commercially available emulsions have a particle size of 0.5 to 2.5 µ. Fine, uniform particle size is usually an indication of good stability. The particle size of an emulsion can be estimated from its appearance (see Table 2).

TABLE 2

Estimation of Particle Size by Appearance[a]

Particle size (µ)	Appearance
> 1	Milky white emulsion
0.1-1	Blue-white emulsion
0.05-0.1	Gray semitransparent emulsion
0.05 and less	Transparent emulsion

[a] Assuming the two phases have different refractive indices.

E. Particle Charge

The dispersed phase of practically all emulsions exhibits a particle charge that can be determined by electrophoresis. This charge may be caused by the dissociation of one of the ingredients (e.g., soap) or, in the case of nonionic emulsions, it might be caused by frictional electricity. The charge is much greater in ionic systems than in nonionic ones.

In small-particle size emulsions particle charge is of extreme importance in maintaining stability. In high-viscosity emulsions particle charge is of less consequence in maintaining stability than in fluid emulsions.

F. Conductivity

The conductivity of an emulsion is dependent on the conductivity of the external phase. Hence an oil-in-water emulsion conducts electricity well; a water-in-oil emulsion is a feeble conductor. For this reason a conductivity test is usually an excellent means of identifying emulsion type.

G. The pH

In the past few years more recognition has been given to the importance of the pH of emulsions. Good emulsions may be adjusted to a particular pH to achieve some desired effect. They may or may not be buffered. Nonionic emulsified products may be used in a pH range from 3 to 10, depending on the nature of the emulsifier.

H. Stability

The stability of an emulsion is evidenced by retention of the original appearance, viscosity, and odor under shipping and shelf-life conditions. Poor stability, except for lack of color and odor stability, is generally caused by the coalescence of the dispersed particles. The rate of coalescence depends on the type and concentration of emulsifier; the viscosity of the emulsion and its component phases, the size of the dispersed globules, the charge on the particles, and the storage conditions to which the emulsion is subjected.

Emulsions obey Stokes' law of sedimentation ([2]). One of the factors involved in sedimentation, though not the most important, is the viscosity of the continuous phase of the emulsion. High viscosity reduces the tendency of the emulsion to separate.

Referring again to Stokes' law, the finer the particle size, the slower the sedimentation rate and the more stable the emulsion.

I. Preservation

Emulsions are subject to contamination by microorganisms during preparation and use. Excessive growth of these microorganisms does not usually occur in commercial products because they contain bacteriostatic agents. However, it is essential that all emulsions be protected with a preservative system. Some preservatives that have been used in food products are sorbic acid, benzoic acid, sodium and calcium propionate, and propylene glycol. Many companies subject all such products to a standard spoilage test to determine the effectiveness of their preservative system (2).

III. EMULSIFICATION

A. General Remarks

When oil and water are mixed together without an emulsifier, the initial dispersion of oil is quite unstable (the oil rapidly coalesces until complete separation into a layer of oil and water results). This might be anticipated inasmuch as the increase in surface area produced by the dispersion of the oil greatly increases the free surface energy of the sytem and at such a high energy level the system is unstable. By the process of coalescence this free surface energy again attains its minimum value. Free surface energy is dependent on both the surface area (capacity factor) and the interfacial tension (intensity factor) (4). Obviously the former cannot do anything other than increase when the surface area is increased during dispersion. Therefore, if a stable emulsion is to be produced, it is necessary to add some third material (i.e., the emulsifying agent that, by its presence at the interface, prevents coalescence of the oil globules. The mechanism that prevents coalescence is explained by numerous theories advanced in the extensive emulsion literature (1,2,4).

As mentioned, emulsions are stabilized by adding a third component, the emulsifying agent. These materials may be surface-active compounds (monoglycerides and diglycerides), gums (gum acacia, methylcellulose), or finely divided clays (cocoa, gums, etc.). Although all these materials function in a different way in emulsion systems, under ideal conditions, each type is capable of stabilizing specific emulsions (5).

In forming a stable emulsion three separate mechanisms are

5. FOOD EMULSIONS

involved, any or all of which may be operative in a particular case:

1. Reduction of interfacial tension
2. Formation of a rigid interfacial film
3. Electrical charges

Because the lowering of surface tension was so obvious, early investigators were certain that this effect explained the whole matter of emulsification.

In practice one learns that, under the best conditions, a surface-active agent will lower the surface or the interfacial energy by a factor of only 20 or 25, which is rather small when one considers that the interfacial energy of an emulsified system is a 10^6 times higher than that of a broken emulsion. To explain further, Donnan and Potts (4) measured the drop numbers of a hydrocarbon oil against aqueous solutions of the sodium salts of saturated fatty acids, from acetic to lauric, with the aid of a pipette. It was found that all these soaps lowered the interfacial tension, the lowering effect increasing with increasing molecular weight. Although interfacial-tension reduction is a contributor to emulsion *stability*, it can hardly be considered to be a major one.

On the other hand, interfacial-tension reduction is all important in the ease of preparation. In fact, if the interfacial tension approaches zero, a spontaneous emulsion will result. However, a spontaneous emulsion is not necessarily stable.

The second contributing factor of emulsion stability is the formation of a rigid interfacial film. It has been accepted for many years that a surface-active agent is present in such a form that a rigid film is produced between the two immiscible phases, and the film may act as a mechanical bar both to flocculation and coalescence of the emulsion droplets.

It can be concluded that the formation of a tightly packed film contributes significantly to the stability of the emulsion. The work of onnan and Potts (4) supports the fact that in forming stable emulsions mixed emulsifiers are often more effective than single emulsifiers. Certain emulsifier blends pack tightly and contribute strength to the film, thus forming a stable emulsion. It may be that gums stabilize an emulsion system by forming a fairly dense gel structure at the interface. This, no doubt, also holds true for certain nonionic surface-active agents.

The third theoretical concept related to emulsification is that of the effects of the electrical charge contributed by certain surface-active agents. Molecules that tend to concentrate at an interface, such as that between oil and water, must be dipoles (i.e., they are oppositely charged at two points or poles). In such molecules the geometrical and electrical centers are not coincident and the molecules will be electrically asymmetrical, thus acting like small magnets.

To elaborate, a nonpolar substance, such as hydrocarbon wax, would not be expected to add stability to an emulsion. However, if the hydrocarbon structure of the wax is altered and a strong polar group like COONa is added in place of a terminal CH_3 group, the resultant molecules (called soaps) are dipoles and will develop electrical charges and will assist in stabilizing an emulsion.

The magnitude of the charge on emulsion particles, the ζ potential, is measured by means of a cataphoresis cell, or ζ meter. Although many emulsifiers show a particle charge as a result of their polarity, it is of value to observe both the charge on the emulsifier and on the emulsion particles.

B. Emulsifier Selection

Risking oversimplification, foods are generally composed of fat, carbohydrate, and/or protein. These three basic ingredients may be present in a wide range of blends involving any one, two, or all three. In addition, and of considerable importance for influencing texture, are various amounts of cellulose, water, and/or air. This concept is difficult to portray because there are

5. FOOD EMULSIONS

five or more variables. If we consider only the primary components --fat, carbohydrate, and protein--and allow air-water blends to be the fourth component, all foods should be included in the graphical representation in Fig. 1. For ease of handling the air-water blend content can be eliminated as in Fig. 2.

Foods then have needs based on the properties to be achieved, the basic ingredients and their proportions, and the aid supplied by the natural emulsifiers present. There is no overall completely systematic method of selecting emulsifiers for food purposes.

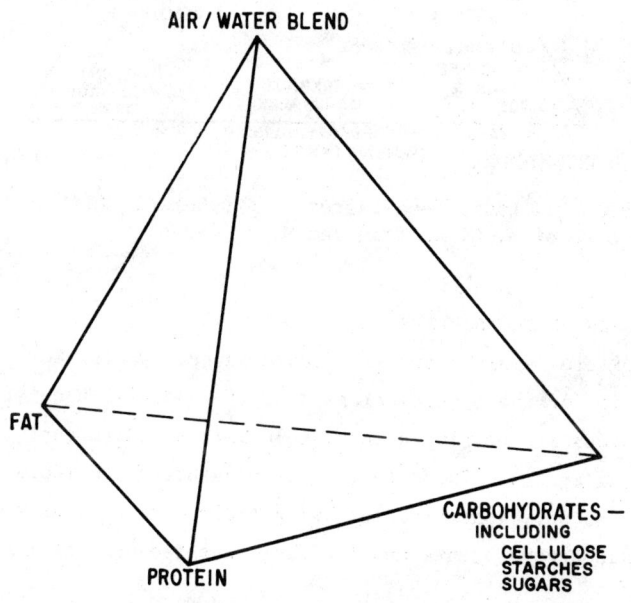

FIG. 1. Food-ingredients schematic. (From data of W. C. Griffin and M. J. Lynch.)

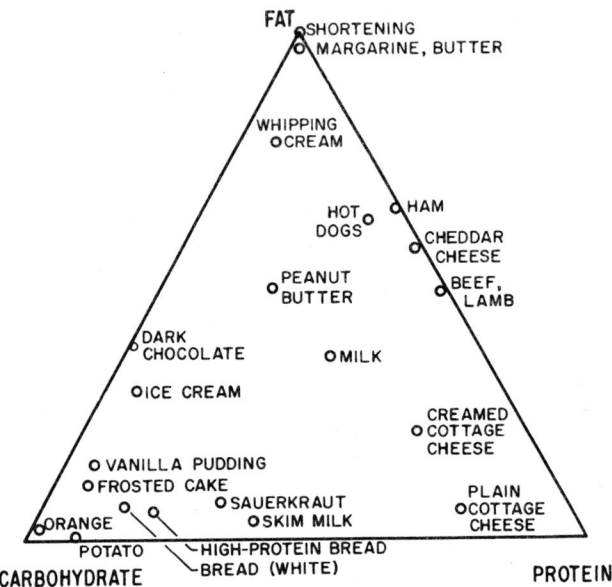

FIG. 2. Basic-food-ingredients schematic (solids basis). (From data of W. C. Griffin and M. J. Lynch.)

1. CLASSES OF EMULSIFIERS

The term "emulsifier" is often misused. Emulsifiers are a subdivision of the general class of surface-active agents. Other subdivisions are wetting agents, solubilizers, detergents, and suspending agents. These terms are frequently used indiscriminately, the only justification being their common classification as surface-active agents and the fact that the uses of many do overlap.

Emulsifiers may be divided or classified according to their ionic behavior: anionic, cationic, nonionic, and amphoteric. The ionic type of emulsifier is composed of an organic lipophilic broup and a hydrophilic group. The ionic types may be further divided into anionic and cationic ones, depending on the nature of the ion-active group. The lipophilic portion of the molecule is usually considered to be the surface-active portion. Thus in soap the surface-active fatty-acid portion of the molecule represents the anion in the molecule, and therefore soap is classed as an anionic emulsifier. As would be expected, anionic and cationic surface-active agents are not mutually compatible. Owing to opposing ionic charges, they tend to neutralize each other, and their surface-active effect is nullified.

Nonionic emulsifiers show no apparent tendency to ionize. Therefore they may be combined with other nonionic surface-active agents and with either anionic or cationic agents as well. The nonionic emulsifiers are likewise less affected by the action of electrolytes than are the anionic surface-active agents.

Amphoteric emulsifiers have the faculty of behaving either as anionic or cationic, depending on the acidity of their environment.

In addition, mention should be made of gums as well as similar emulsion and suspension stabilizers that act as protective colloids. Gum acacia, gum tragacanth, alginates, methylcellulose, sodium carboxymethylcellulose, and hydroxyethylcellulose are representative of these materials.

In the United States the most notable list of commercial emulsifiers available is that of McCutcheon (6), which includes more than 3000 items from more than 300 suppliers. This list is useful in determining the acceptability of the emulsifier by classification, molecular identity, and recommended uses. Table 3 lists the major classes of food emulsifiers.

TABLE 3

Classes of Major Synthetic Food Emulsifiers

Fatty-acid soaps	Ionic
Simple fatty-acid ester	Nonionic
Modified fatty-acid ester	Nonionic
Fatty-acid ether ester (polyoxyethylene or polyglycerol fatty-acid ester)	Nonionic
Fatty-alcohol ether (polyoxyethylene fatty alcohols)	Nonionic

The selection of emulsifiers for food emulsions must embrace at least two major general areas of consideration:

1. The products selected must be approved for the intended use by the U.S. Food and Drug Administration (FDA).

2. The emulsifier or emulsifier blend must be functional in producing the effects desired in the final product.

In addition, the emulsifier must be (a) chemically stable, (b) inert, and (c) suitably low in odor, taste, and color.

2. FUNCTIONALITY

Because emulsifiers can and do function in many ways, their selection must be related to all the properties of the intended formula. For example, improving the dispersion in a product can change its color. These various functions will be treated in greater depth in Section VII.

3. THE HLB METHOD OF SELECTION

The hydrophile-lipophile balance (HLB) is an expression of the relative simultaneous attraction of an emulsifier for water and for oil (or for the two phases of the emulsion system being considered). It would appear to be determined by the chemical composition and extent of ionization of a given emulsifier. For example, propylene glycol monostearate (pure), $CH_3CHOHCH_2OOC(CH_2)_{16}CH_3$, would be strongly lipophilic; a polyoxyethylene monostearate, $H(OC_2H_4)_nOOC-(CH_2)_{16}CH_3$, with a long polyoxyethylene chain would be slightly hydrophilic because of the greater preponderance of hydrophilic character in the molecule; and sodium stearate, $CH_3(CH_2)_{16}COONa$, would be strongly hydrophilic since it ionizes and thus provides an even stronger hydrophilic tendency.

The HLB of an emulsifier is partially related to solubility. However, two emulsifiers of similar HLB may exhibit different specific solubility characteristics.

5. FOOD EMULSIONS

The HLB of an emulsifier determines the type of an emulsion that tends to be formed. It is an indication of the behavior characteristics, and not an indication of emulsifier efficiency. Thus all emulsifiers that are strongly lipophilic will tend to make water-in-oil emulsions. For any specific problem both the best HLB and the best chemical class of emulsifiers must be found, this latter being the specificity factor mentioned above.

The HLB, even in an approximate form, may be utilized to reduce the amount of trial and error involved in emulsion formulation. The approximate required HLB is determined for the given emulsion components and system or type of product in question. With the approximate required HLB known, emulsifiers having such an HLB may be selected from several different chemical types, allowing a test for specificity. These will represent approximate optimum efficiencies in their respective chemical types, and emulsions prepared by using them will permit selection of the most efficient chemical type.

The HLB of an emulsifier of the surface-active-agent type is an expression of the size and strength of the hydrophilic and lipophilic groups that compose the emulsifier molecule. A strongly lipophilic emulsifier has a low HLB number, usually less than 10; a highly hydrophilic emulsifier has a high HLB number, over 10. The numbers for some typical food emulsifiers so far assigned (see Table 4) range from 2.8 to 40. These numbers have been determined experimentally or by calculation.

If two surfactants are used as a blend, the HLB of the combination may be found by multiplying the weight proportion of each surfactant by its HLB value to find its contribution to the total HLB and then adding the two values (see Table 5).

The HLB method is a useful tool in the preparation of most emulsions. By using HLB values, surfactant performance can often be predicted, reducing the number of emulsification tests. Table 6 shows an approximate relationship between the HLB and the end use of a surfactant.

TABLE 4

Calculated or Determined HLB Numbers of Some Surfactants Used in Foods

Chemical name	HLB number
Glycerol monooleate	2.8
Propylene glycol monostearate	3.4
Glycerol monostearate	3.8
Lecithin	4.2
Sorbitan monostearate	4.7
Glycerol monostearate, self-emulsifying	5.5
Polyoxyethylene (20) sorbitan tristearate	10.5
Polyoxyethylene glycol 400 monooleate	11.4
Polyoxyethylene glycol 400 monostearate	11.6
Polyoxyethylene (20) sorbitan monostearate	14.9
Polyoxyethylene (20) sorbitan monooleate	15.0
Polyoxyethylene (40) stearate	16.9
Sodium lauryl sulfate	~40

TABLE 5

Example of Emulsifier Blending for an Initial Test HLB Series

Test	Emulsifier Blend		Calculated HLB number
	Span 60	Tween 60	
1	100	0	4.7
2	75	25	7.3
3	50	50	9.7
4	25	75	12.3
5	0	100	14.9

TABLE 6

Relationship Between HLB and Surfactant End Use

HLB number	Application
4-6	Emulsifiers for w/o systems
7-9	Wetting agents
8-18	Emulsifiers for o/w systems
13-15	Detergents
15-18	Solubilizers

5. FOOD EMULSIONS

a. *Solubility*. The solubility of surfactants in water can usually be used as a guide in approximating their HLB and their usefulness. Although it is by no means an infallible guide, Table 7 can be used in estimating the HLB numbers of many surfactants.

b. *Cloud point*. Nonionic surfactants in aqueous solutions often are less soluble hot than cold. During the temperature increase a point of maximum turbidity occurs. This is known as the cloud point of the surfactant. The HLB numbers of surfactants, particularly those of the highly hydrophilic nonionic type, can often be estimated by determining their cloud points. Table 8 demonstrates the relationship of HLB to cloud point.

c. *Calculation of HLB numbers from composition*. The HLB of a nonionic emulsifier is most accurately expressed by the formula

$$HLB = \frac{H}{5}$$

TABLE 7

Estimation of HLB by Water Solubility

Action in water	HLB range
Not dispersible	1-4
Poor dispersibility	3-6
Milky dispersion after vigorous agitation	6-8
Stable milky dispersion	8-10
Translucent to clear dispersion	10-13
Clear	13+

TABLE 8

Relationship of HLB to Cloud Point

Cloud point (°C)	HLB number
<40	<13
~65	14
~82	15
~94	16
>100	>17

where H represents the weight percentage of hydrophilic groupings in the molecule, such as glycerine, glycols, polyglycols (ethylene oxide polymers), sorbitol, and sorbitan.

Other methods (7-9) for calculating HLB numbers and for approaching emulsion problems revolve around functional group numbers, saturated and unsaturated fatty acids, the HLB numbers of both surfactants and oils, as well as interfacial-tension measurements. However, the most popular approach to resolving emulsion problems appears to be the HLB method described here.

IV. PRODUCT DEVELOPMENT

Just as with any nonemulsified food, the formulator concerned with developing an emulsion-type product should consider such factors as the following:

1. The physical form of the final product (liquid or solid)
2. The pH of the system
3. Normal expected abuse factors, such as light, heat, or cold

In addition, he must consider the type of emulsion desired (oil in water or water in oil).

Furthermore, the formulator must know (or decide on) the desired characteristics of the final emulsified product, since to a large extent these factors dictate the emulsion components and the limits of each.

In general, liquid oil-in-water formulations contain over 70% water. High-viscosity and/or gelled oil-in-water emulsions may contain low solids levels with thickeners added to the aqueous phase, or they may contain high levels of fats, oils, proteins, and carbohydrates. Water-in-oil emulsions are usually higher in

5. FOOD EMULSIONS

oil and solid-fat components and assume the rheology pattern established by the oil phase (see Table 1).

Keeping in mind the desired characteristics, the formulator endeavors to prepare a sample of the product. The method of preparation should employ laboratory equipment that will duplicate the plant facilities as nearly as practical.

In the preparation of an emulsion the required energy input to implement the multifold increase in interface comes from (1) mechanical means and (2) the emulsifier. Thus the choice is a balance of the two sources of energy as is needed to form a suitable initial emulsion. The greater the mechanical-energy input, the less demand there will be on the emulsifier. Conversely, the lower the mechanical-energy input, the greater the demand made on the emulsifier (requiring the maximum reduction of interfacial tension). Usually there is leeway in this stage of the operation, there being more than enough mechanical-energy input and likewise an excess of emulsifier--the emulsifier level usually being fixed at a higher concentration because of stability demands. During formulation studies, with reference to functionality related to emulsion preparation, the best procedure is to utilize in all stages--laboratory, pilot plant, and plant--preparation equipment that is as similar as scaling-up limitations will allow. Attention must be paid to the mechanical-energy input (method and rate), the heating and cooling rates, as well as agitation during cooling.

A time tested product-development technique is as follows:

1. Group the ingredients according to solubilities (aqueous versus nonaqueous).

2. Calculate an approximate required HLB for the combined oil phase based on the type of emulsion desired.

3. Prepare several blends of a low-HLB emulsifier and a high-HLB emulsifier to give HLB values around the calculated value (see Table 5). Choice of ionic type and chemical nature of the emulsifier will be dictated in part by the end use of the emulsion. For

this initial trial the formulator should use a higher concentration of emulsifier than that eventually needed (e.g., 10-30% of the oil phase).

4. Dissolve the oil-soluble ingredients and the emulsifier in oil, using heat if necessary. Maintain at 5 to 10°C above the melting point of the highest melting ingredient unless it is readily soluble in the oil; the maximum convenient temperature is 70 to 80°C.

5. Dissolve the water-soluble ingredients (except acids and salts) in water. If acids and salts are included in the formula, reserve some of the water for later addition with them.

6. If the oil phase is heated, heat the aqueous phase to 3 to 5°C higher than the oil phase.

7. Add the aqueous phase to the oil phase with suitable mechanical agitation. For oil-in-water emulsions propeller stirring is satisfactory. In these instances the initial additions of water are made quite slowly until inversion from a water-in-oil emulsion occurs (evidenced by a decrease in clarity and viscosity); then water addition is made more rapidly. For water-in-oil emulsions homogenization is most satisfactory, and a crude preemulsion is usually prepared and added to the homogenizer.

8. If acids or salts are employed, dissolve in remaining water and add to the cooled emulsion.

9. In the preparation of emulsifiable concentrates the solubility of an emulsifier is of great importance. Here it is usually desirable that the concentrate remain homogeneous for an indefinite period and over a wide range of temperature. The emulsifier must remain dissolved through all storage conditions. It is frequently possible to enhance the solubility of one emulsifier with a coemulsifier. Also, the use of various solvents as couplers or cosolvents is commonplace. Although couplers are highly specific to each system, propylene glycol or even water is frequently used.

5. FOOD EMULSIONS   271

The preliminary formulation should be examined for characteristics other than stability--provided it is stable enough to allow a preliminary examination. [If the formulation is not stable enough, prepare with (a) more emulsifier, (b) an emulsifier with a slightly higher (one unit) HLB number, (c) an emulsifier with a slightly lower (one unit) HLB number, or (d) an emulsifier blend based on a different chemical family.] If it is suitable, a critical selection of the emulsifier comes next. If not, some ingredient changes should be made to create the desired improvement or a modification may be made in the preparation procedure.

To select the best emulsifier, the formulator should (1) establish more exactly the required HLB and (2) establish the best chemical type of emulsifier (specificity factor).

The selection of the best emulsifier involves closely controlled stability testing, both physical and chemical--or ease of preparation if done by the consumer. Here again, based on the nature and use of the final product, decisions must be reached on stability requirements.

Emulsifier selection, with reference to functionality as related to the stability of the emulsion, is handicapped in general by inept definitions of emulsion stability. A stable emulsion shows no indication of coalescence (phase separation) or creaming during the normal expected shelf life, or when frozen and thawed repeatedly, or on exposure to elevated temperatures (40 to 50°C) for various time intervals. Failure under any of these conditions may be allowable in specific instances (e.g., elevated-temperature stability is of no concern with a refrigerated product). As a result for each formulation a set of stability criteria must be assembled for laboratory tests and for quality-control purposes:

1. Minimum room-temperature storage time

2. Number of freeze-thaw cycles

3. Time of storage at 40 and 50°C (or other chosen temperatures

4. The degree of separation or creaming (or other evidence

of instability) that is allowable (e.g., complete separation with good ease of emulsion re-formation on shaking may be acceptable)

For most food emulsions the industry evaluates products at 5, 40, and 50°C. Stability at 5 and 40°C for 3 months is considered minimal, and correspondingly shorter stability is expected at 50°C. Once a product definition of stability has been established, the actual emulsifier selection can begin.

The stability of a dry mix or a fluid that contains an emulsifier or blend of surfactants must also be ascertained. In the former instance performance tests are generally conducted after storage at high and low temperatures. Fluid systems are checked for performance, and if clear as made, are checked for continued clarity or ability to regain clarity after exposure to severe cold.

Pilot-plant manufacturing equipment deserves special consideration because of the need (and tremendous difficulty) of duplicating plant manufacturing techniques. As an example, in preparing an emulsion of moderate viscosity in a typical laboratory beaker (motor-driven-propeller arrangement) the actual work input may be surprisingly high. In equipment scaleup marked differences occur in surface-to-volume rations, peripheral speeds of agitators, tendencies to form a vortex and suck in air or produce foam, rates of heating and especially rates of cooling, and chances for local overheating. Each of these factors serves to increase the complexity of the problem.

Laboratory preparation should endeavor to duplicate plant conditions and, if possible, should err on the side of too little energy input. A batch kettle, either heated or unheated, is approximated by a beaker with a motor-driven propeller of appropriate size (a slow motor speed is preferred, and a simple baffle is usually a desirable improvement, to avoid air entrainment). A plant homogenizer is approximated by a hand homogenizer or by a gear pump with a spring-loaded relief valve. Comparative emulsification of a concentrate is most uniformly handled by motor-timer-controlled shaking.

5. FOOD EMULSIONS

A Waring Blendor, when used to prepare emulsions, imparts relatively fantastic amounts of energy and also incorporates large quantities of air. For this reason it often gives results that are not achieved in subsequent plant scaleup.

An additional major problem in the laboratory occurs with emulsions that are prepared hot and then cooled. A laboratory beaker of emulsion will cool from 60 to 30°C in a few minutes. A 1000-gal tank requires much longer, even with modest forced cooling. For emulsions that contain waxy components the rate of cooling through the melting range can be all important. Hence the best procedure is to determine in the laboratory the best cooling schedule, determine the deviation leeway allowable, and finally set plant conditions to satisfy these criteria. In larger scale equipment fast cooling may be achieved by means of a cooling board or other heat-exchange unit.

When an emulsion system is heated, emulsifier cosolubility is altered, and this may change its behavior characteristics. In addition, care must be taken to avoid unwanted chemical reactions.

It is wise to establish a specific procedure for preparing an emulsion on a plant scale. The order and rate of addition of ingredients are of primary concern. The optimum order and rate are established in the laboratory, but they can seldom be followed exactly under plant conditions. A suitable compromise must be found, and then this order and rate must be followed strictly to maintain quality control. It is possible that apparently small deviations in the order or rate of addition of an ingredient may result in a totally different end product and may in fact govern its acceptance or rejection.

An apt illustration is the preparation of an oil-in-water emulsion by the inversion technique. The oil phase and emulsifier are blended in the tank. Water is then slowly added to the oil phase with stirring. The initially hazy oil-and-emulsifier blend will clarify at first and then will again become cloudy, and the mix will become thick and then suddently thin. Further additions are made rapidly. If the oil is added to the water, a poor emulsion will result. If the water is added to the oil rapidly at first and clearing is not achieved, particle size will be much larger than that obtained by the slow, careful addition procedure.

Specific procedures must be worked out for each formulation. Generally all the oils and oil-soluble ingredients are best combined in the oil phase. Sugars and polyols are usually added with the water, but it is usually preferable, especially for oil-in-water emulsions, to add the salts with the last one-fifth to one-half of the water after a good primary emulsion has been established.

In most instances of liquid-liquid emulsification ambient temperature is to be preferred. With some equipment, heat will be generated during the emulsification step, and this must be removed.

V. EQUIPMENT

The major concerns in the choice of emulsifying equipment are the apparent viscosity in all stages of manufacture, the amount of mechanical-energy input required, and heat-exchange demands.

The choice of emulsification equipment is usually dictated by the application of the resulting emulsions. The purpose of the emulsification equipment, whether it is simple or complex, is to break up or disperse the internal phase into the external phase so that the particle size of the resulting emulsion is sufficiently small to prevent coalescence and the resultant breakdown of the emulsion in the required time of stability.

Hand stirring is the simplest form of agitation. It is reasonably duplicated by slow-speed anchor-type agitators. Depending on the selection of emulsifiers and the ingredients to be emulsified, large- or small-particle-size emulsions that are either stable or semistable may be prepared. Mechanical rotation of the

5. FOOD EMULSIONS

paddles is usually at a slow speed, and unless the emulsion is quite viscous, the agitation efficiency is low. For many viscous emulsions with a high solids content a mechanically rotated paddle or anchor-type agitator is best.

The planetary mixer has actually been developed for one phase of this high-viscosity field, the food industry. In a planetary stirrer the paddle rotates and at the same time the axis about which it rotates follows a circular orbit. In this way a large batch of heavy batter may be intimately mixed. These same mixers are used at higher speeds with a wire whick for aerating and whipping low-viscosity fluids. Because of excessive aeration, they are not adapted to the preparation of low-viscosity emulsions, but are good for high-viscosity systems.

Stirring by means of aeration (bubbling air or gas through a liquid) is not much more efficient than hand stirring unless extremely large volumes of gas are used. A modification of this system, consisting of the injection of live steam into a tank, is usually more efficient because of the condensation of the steam and the resulting cavitation or "steam-hammer effect." The use of air or steam is most practical in low-viscosity systems.

One of the most popular types of emulsification equipment utilizes one or more propellers mounted on a common shaft in a mixing tank. Modifications of this include variation in the location of the propellers in the tank, the use of two or more propeller shafts, and the use of complex propellers. Propeller agitation is most satisfactory for low- and medium-viscosity emulsions. When properly used with adequate emulsifying agents, propeller agitation will result in finer particle size than that obtained with homogenization or milling.

The inclusion of fixed baffles either on the tank wall or adjacent to the propellers, as in a turbine rotor and stator, considerably increases the efficiency of agitation. The use of a turbine agitator is by far the preferred of the two methods since baffle plates in a tank frequently result in areas of little or no agitation (although the general effect is to increase the efficiency of agitation). Turbine-type agitators are available in various sizes, speeds, and rotor-stator clearances, and in many modified designs. Turbine-type systems can be designed to give a very high degree of shearing action. Turbine combinations can be used with higher viscosity fluids than propellers. However, at high viscosities the turbine may not provide sufficient gross agitation of the batch.

The colloid mill can actually be considered to be a modification of a turbine, although in this case the clearance between the rotor and stator is on the order of a few thousandths of an inch. With such small clearances, an extremely high shearing action occurs. The product from a colloid mill usually has a uniform particle size, no doubt due to the fixed clearance between the rotor and stator. Owing to the tremendous shearing forces applied to the emulsion, the temperature rise during emulsification may be from 30°F to as much as 140°F, and in most cases external cooling must be employed. Milling can be done on fluids and pastes, the rate of throughput varying inversely with the viscosity.

In a homogenizer emulsification is effected by forcing the two phases past a spring-seated valve. This is usually done at a relatively high pressure of 500 to 3000 psi. Emulsification occurs not only as the components pass under the valve seat but also when the emulsion impinges against the retaining wall surrounding the valve.

Homogenizers are also built with more than one stage of emulsification, that is, successive relief valves. This is claimed to be of value in some instances wherein the high-pressure homogenization promotes clumping of the fine particles of emulsion that it forms. The second stage of homogenization, at a lower pressure, breaks up the clumps and produces a lower viscosity product. Using comparable ingredients, homogenizers usually give an emulsion of finer average particle size than do colloid mills, though the particle size is not as uniform.

5. FOOD EMULSIONS

A further contrast is that the temperature rise during homogenization (10 to 30°F) is not very large. The actual temperature rise throughout the homogenizer and pump may be only 10 to 30°F, or it may be as high as 50 to 90°F, depending on the type of pump employed. A piston pump gives a smaller temperature rise than does a gear pump. Owing to clearances in the gear pump, a certain quantity of liquid continually bypasses the pump and is partially homogenized before reaching the homogenizer head. Homogenizers will handle liquids or pastes, and the rate of throughput is little affected by viscosity.

A more recent development in the field of emulsifying equipment is the high-frequency, or ultrasonic, oscillator (10). It is possible that a portion of its mechanical action is somewhat similar to that produced by the steam injector--that is, cavitation. Limited data indicate that the ultrasonic emulsifier is best suited for liquids of low viscosity.

As would be expected, many combinations of these devices are employed, and new designs are being explored. Thus for high-viscosity-product manufacture a motor-driven paddle in a jacketed tank is supplemented by the addition of a small high-speed turbine agitator. This is quite satisfactory for the initial emulsification of small quantities of material in the bottom of the tank and assists in emulsification, even at the completion of a batch when the tank is full. Industrial combinations of homogenizers and colloid mills with proportioning scales and pumps for the continuous production of emulsions are available. Means for calculating equipment requirements have been published (11).

The rated power requirements for several different types of agitation are presented in Table 9. The general order of increase in power demand is as follows: propeller mixing--lowest; turbine, homogenizer, and colloid mill--highest.

Laboratory versions of all of these types of emulsification equipment are in common use. Small-scale laboratory and pilot-plant models of planetary mixers, motor-driven propellers,

TABLE 9

Emulsification Equipment

Equipment	Agitator speed (rpm)	Mechanical-energy input	Usable Viscosity range	Heat-exchange demand range[a]
Anchor agitator	Slow	Low	Best for high viscosity	Fair
Wall-scraping anchor agitator[b]	Slow	Low	Best for medium and high viscosity	Good
Propeller mixer[c]	Medium	Low to medium	Best for low to medium viscosity	Fair to poor
Votator	Medium	Low to medium	Best for high viscosity	Excellent
Rotating-cage mill[d]	High	High	Best for low to medium viscosity	Fair
Rotating-disk mill[e]	High	High	Best for low to medium viscosity	Fair
Homogenizer	Slow	High	Low to moderately high	Fair to poor
Colloid mill	High	High	Low to medium	Fair to poor
Mechanical ultrasonic[f]	High	Medium	Low to medium	Fair to good
Electronic ultrasonic	Not critical	Low to medium	Low to medium	Fair

[a] Without auxiliary equipment.
[b] Usually in pairs--counterrotating.
[c] Lightning mixer.
[d] Primier Dispersator.
[e] Cowles Dissolve or Hackmister Disperser.
[f] Rapasonic.

turbines, colloid mills, and homogenizers are available. In addition, use is made of motor-driven egg beaters and shaking machines.

VI. FOOD EMULSIONS

Emulsification occurs in many food applications. One of the simpler foods involving emulsification alone is salad dressing or mayonnaise. In these the emulsifier is generally the added egg yolk, and the product is a high-internal-phase-ratio oil-in-water emulsion, and therefore it gains a structural viscosity because of the phase ratios. The emulsifiers must have a high HLB to promote the desired oil-in-water emulsion type.

An example of the opposite type of emulsion where the added emulsifier also serves almost a single function is oleomargarine. This water-in-oil emulsion contains only a small amount of internal phase, and the product viscosity is essentially that of the blend of partially hydrogenated oils. In addition to emulsion stabilization, the monoglyceride emulsifiers also serve as antispattering agents when the margarine is used for cooking. This effect is probably a result of retaining the water in a high degree of dispersion even at high temperatures.

Emulsifiers promote stable emulsions in canned milk shakes, liquid coffee whiteners, and fluid emulsion shortening; they also maintain stable flavor-oil dispersions in flavor dips like onion dip and in whiskey-sour mixes. Emulsions of antioxidants are also used to promote their distribution on foodstuffs.

Dispersion is, of course, closely akin to emulsification. In this presentation we shall consider that it relates especially to the action of dispersing either solids or liquids in another liquid.

A dry coffee whitener is probably one of the best examples of dispersion control. The product is required to disperse at a controlled rate and give a good bloom over a limited temperature range. The emulsifier used in the formula given in Table 10 ([12]) is a blend of a monoglyceride and polysorbate 60 or polysorbate 65 and is balanced to provide proper manufacturing conditions as well as proper dispersion in the application.

Cake, chocolate-drink, and sauce mixes (e.g., a cheese sauce or gravy mix) all illustrate solids-dispersion situations.

TABLE 10

Typical Coffee-Whitener Formulation[a]

Ingredient	Level (%)
Vegetable fat	35-40
Corn-syrup solids (42 D.E.)	55-60
Sodium caseinate	4.5-5.5
Dipotassium phosphate	1.2-1.8
Emulsifier:	
Mixture of Atmos 150, Span 60, and Tween 60 (60:20:20)	0.3-0.5
or	
Mixture of Atmos 150 and Tween 65 (75:25)	0.2-0.4
Color	q.s.
Flavor	q.s.
Anticaking agent	q.s.

[a] From Ref. [12].
[b] Dry basis.

VII. OTHER USES FOR SURFACTANTS

As more information has been developed on the complex non-Newtonian substances referred to as emulsions, the food scientist has learned that emulsifiers are capable of producing many other beneficial effects in addition to emulsification. These include dispersion, suspension, wetting, aeration, suppression of foaming, lubrication, crystallization modification, and complexing. The purpose of this section is to emphasize the broad functionality of these very versatile materials as well as the fact that they should always be considered as possible aids in solving formulation problems or producing some special effect of interest to the food scientist.

A. Foams

Foams consist of globules of gas dispersed in a liquid. Because the character of the foam governs the appearance and taste appeal of the product, the proper selection of a surfactant is vital.

As in the case of emulsions, the stability of foams varies widely--from a few hours to several months. Similarly the desired density will vary. Although the degree and type of whipping play a major role, a surfactant is almost always required to achieve the desired density and stability.

Surfactant selection varies with product requirements. Usually hard-type lipophilic surfactants, such as a monoglyceride and diglyceride and sorbitan monostearate are blended with polysorbate 60. The concentration will vary from as low as 0.1% to as high as 1.0%. Good foam structure is essential in such food products as cakes, frozen desserts, and shipped toppings (13).

B. Suspensions

Suspensions are stable dispersions of finely divided, insoluble material in a liquid medium. The size of the dispersed particles may vary from 0.1 μ to flocculates or aggregates as large as 100 μ.

Surfactants are used in suspensions to aid the wetting of the insoluble particles. This in turn helps to ensure uniformity of product. In those products in which separation occurs during storage the surfactant helps redisperse the insoluble ingredients when the product is remixed. Generally, hydrophilic surfactants like polysorbate 80 at levels near 0.1% give the best results.

In suspensions surfactants are generally used together with stabilizers or thickeners to produce the desired result. Chocolate drinks are among the most common suspensions in the food industry.

C. Demulsification and Antifoaming

The formulator is generally concerned with making products stable, but there are times, particularly during processing, when he is interested in breaking an emulsion or a foam.

In most cases where demulsification is desired an emulsifier of the opposite type or one that will throw the emulsifier system out of balance is used. Depending on the type of emulsion, a strongly hydrophilic surfactant like polysorbate 80 or a lipophilic surfactant like monoglyceryl or diglyceryl oleate will usually break it.

The best surfactant for breaking foam varies with the foam, but in most cases the suggested products at levels near 0.1% will produce the desired effect.

In foods like ice cream a carefully selected emulsifier produces controlled demulsification. It helps fat particles agglomerate to the size that best produces a dry product (a stiffening

of the structure of the ice cream and a dulling of the surface
caused by proper agglomeration of fat particles).

D. Complexing

In certain applications, such as bread, dehydrated potatoes, and starch-jelly confections, surfactants complex the starch.

1. BAKED FOODS

In bread and rolls the proper surfactant system retards the firming of the crumb that consumers associate with staling. It also conditions the dough and thereby improves machinability while promoting the formation of uniform texture.

Firming of the crumb can be related to starch crystallization. This crystallization is retarded by surfactants, which, it is theorized, enter the helix of the amylose fraction and prevent its leaving the starch granule. This action retards association of the amylose polymers, thus extending the "as baked" quality of the loaf (14).

Monoglycerides and diglycerides have been used for years to retard crumb firming. The composition of this emulsifier is important if ease of handling and good performance are to be obtained. Monoglyceride and diglyceride surfactants are used at levels of 0.25 to 0.50% based on the weight of flour.

Volume and grain are improved in bread apparently by the action of the surfactant on the gluten structure in the dough. It is believed that the surfactant strengthens the gluten, thereby preventing rupture of the cell walls that have entrapped fermentation gases. As the cells form, they tend to form uniformly, giving the finished loaf even grain. In the same manner surfactants produce bread doughs that are dry without being sticky or bucky so that they machine well.

Blending a hydrophilic emulsifier like polysorbate 60 with a carefully selected monoglyceride and diglyceride (15) gives a surfactant that provides both antifirming and dough-conditioning properties.

2. OTHER STARCH-BASED FOOD PRODUCTS

In starch-jelly confections starch crystallization is the major cause of hardening. Although moisture loss may appear to be the cause of staling, tests have shown that firming will proceed even when moisture loss is prevented.

The use of monoglycerides and diglycerides at levels near 0.1% will retard starch crystallization and extend the palatability of this confection.

In other starch-based food products surfactants provide a modifying effect. They reduce the tendency of cooked noodles and macaroni to clump, and they aid the rehydration of dehydrated potato products. Hard-type (highly saturated fat, low iodine number) monoglycerides and diglycerides are also particularly effective in this application.

E. Crystallization Control

The control of crystallization in sugar and fat systems is another function that surfactants perform. Typically surfactants are used for this purpose in chocolate, peanut butter, and sugar-syrup coatings.

1. FAT SYSTEMS

a. Chocolate. Brighter initial gloss and bloom prevention are the two main functions surfactants perform in chocolate and confectionery coatings (16).

Surfactants help form fine fat crystals that create a brighter initial gloss. Compared with chocolate that contains no surfactant at all, the crystals formed in the presence of the

5. FOOD EMULSIONS   285

surfactant system are much finer and more numerous. It is the
extra light reflected from these crystals that provides the brighter
initial gloss.

Bloom is the appearance of solidified fat on the chocolate
surface. It is prevented by a monomolecular surfactant layer that
is formed on the surface of the cocoa fibers. This monolayer prevents the migration of liquefied fat to the surface of the piece
where it solidifies and becomes bloom.

Chocolate and confectionery coatings may contain a combination
of sorbitan monostearate and polysorbate 60 (U.S. Pharmacopoeia)
surfactants at levels of up to 1%. The use of 60 parts of sorbitan
monostearate with 40 parts of polysorbate 60 usually gives the best
results. Confectionery coatings have benefited from the use of
these surfactants for many years. In these coatings a higher ratio
of sorbitan monostearate to polysorbate 60 or sorbitan monostearate
alone is usually more functional.

   b. <u>Peanut butter</u>. Surfactants also perform a crystallization
function in peanut butter. In peanut butter that contains about
50% peanut oil the oil tends to separate during storage. Separation can be prevented by entrapping free peanut oil in the
crystallized surfactant.

Monoglycerides and diglycerides at levels of 1.0 to 2.5% have
proved their ability to overcome oil separation in this manner.
They also improve the spreadability and the palatability of the
product and provide better handling over a wide range of temperatures.

2. *SUGAR SYSTEMS*

The control of crystallization in sugar systems can be
achieved by the use of a small amount of a suitable surfactant. A
hydrophilic surfactant (e.g., 0.1% polysorbate 60) will increase
the rate of crystallization and promote finer crystals.

The most common example of crystallization control via
surfactant is in panned coatings. The use of 0.1 to 0.2% of

polysorbate 60 (based on sugar weight in the coating) reduces panning times by as much as one-third and produces coatings that are more opaque.

F. Wetting

Surfactants are generally effective wetting agents. However, the choice of surfactant is governed by the type of wetting to be accomplished.

To wet a waxy surface with water, the addition of a small amount of a blend of polysorbate 80 and monoglyceryl and diglyceryl oleate surfactants to the aqueous solution will usually produce the desired effect. A typical blend consists of 80 parts of monoglyceryl and diglyceryl oleate to 20 parts of polysorbate 80.

The wetting of powders is a different problem. Too rapid wetting may be undesirable because of clumping or air occlusion, thus making the surfactant level critical. Generally 0.1% of polysorbate 80 added to a powder will produce good results.

G. Lubricating

Monoglycerides and diglycerides possess exceptionally good lubricating qualities that make them useful in food processing. The inclusion of 0.5 to 1.0% of a hard monoglyceride and diglyceride in caramels will reduce their tendency to stick to cutting knives, wrappers, and the consumer's teeth.

VIII. MULTIPLE EMULSIFIER EFFECTS

In several food products surfactants create multiple effects within one system. For example, in ice cream surfactants perform

5. FOOD EMULSIONS

three basic functions: fat dispersion prior to freezing, controlled demulsification during freezing, and aeration during freezing (foam formation). As a result of this triple effect, the ice cream has the desired overrun (aeration), dryness, body, and texture. An emulsifier blend is usually most functional in this case. An 20:80 blend of monoglycerides and diglycerides and polysorbate 80 at a level of 0.1% is reported to be a highly functional system (17). Another example of a food system that utilizes emulsifiers to obtain multiple effects is in the area of whipped toppings. Surfactants help form stable emulsions of the proper particle size. Through their foaming and controlled-demulsification action they produce toppings that aerate easily, are dry, and have good body and texture. Monoglycerides and diglycerides, sorbitan monostearate, and polysorbate are functional products when used alone and in combination. A suggested level is 0.5 to 1%, depending on the composition and properties desired in the final product. Coffee whiteners are another excellent example of a food product that requires the multiple surfactant effect. Although the ingredients of a coffee whitener are similar to those of whipped toppings, the performance characteristics are distinctly different. While emulsion stability is required, aeration and dryness are not. Surfactants are required for these products to aid dispersion and help the whitener achieve its major objective of lightening the color of the coffee.

Liquid whiteners must have good shelf stability, including, in the case of frozen products, freeze-thaw stability. They must disperse readily with no oil separation in hot coffee. Production of a whitener that will meet these requirements demands careful selection of ingredients, including emulsifier. Monoglycerides and diglycerides at a level near 0.6% are recommended for evaluation.

Powdered products do not present the stability problems associated with liquid products. However, proper dispersibility becomes more of a problem with powdered products. For these products blends of lipophilic emulsifiers like monoglycerides and diglycerides and sorbitan monostearate in combination with hydrophilic emulsifiers like polysorbate 60 and polysorbate 65 should be considered. A good starting combination is a blend of a monoglyceride and diglyceride, sorbitan monostearate 60 and polysorbate 60 at a ratio of 60:20:20 and a level of 0.6 to 0.8%.

In chemically leavened cakes surfactants contribute to increased volume, improved texture, tenderness, and extended shelf life. Cake batter is a foam of air bubbles enclosed in films of protein. Surfactants improve the uniformity of the initial foam and the ability of the protein films to entrap the carbon dioxide released by the leavening system during baking. They also aid in dispersing the shortening evenly throughout the batter. This results in a light, palatable cake with good consumer appeal.

Emulsifier selection varies greatly with the type of cake and the results desired. Monoglycerides and diglycerides at a level of about 6% based on the weight of shortening will produce generally good results. Improved results are obtained through the use of sorbitan monostearate and polysorbate 60 surfactant blends in addition to the monoglyceride and diglyceride.

The principal uses of surfactants have been discussed in detail; however, a new area is opening up that will require the multiple effects of emulsifiers more than ever before. The era of convenience foods is here, and they lean heavily on emulsifiers for varied purposes. Coffee whiteners have already been discussed. The pressure-packed and prepared frostings along with cake decorating icings are other examples of products new to the food scene. Synthetic meat and the need for more protein in the diet are only a couple of the challenges of tomorrow that are here today. The emulsion route may provide some of the answers needed to make these concepts commercial realities.

REFERENCES

1. P. Becher, *J. Colloid.*, **19**, 468 (1964).
2. Martin and Cook, in *Remington's Practice of Pharmacy*, Mack, Easton, Pa., 1961, 249.
3. D. S. Kenney, W. E. Grundy, and R. H. Otto, *Bull. Parenteral Drug Assoc.*, **18**(5), 10 (1964).
4. Donnan and Potts, *Kolloid Z.*, **7**, 208 (1910).
5. J. H. Schulman and E. G. Cockbain, *Trans. Faraday Soc.*, **36**, 651 (1940).
6. J. W. McCutcheon, Inc., *Detergents and Emulsifiers*, 1964.
7. J. T. Davies, in *2nd Intern. Congress of Surface Activity, London*, Butterworth's London, 1957.
8. C. D. Moore and M. Bell, *Soap, Perfumery and Cosmetics*, **29**, 293 (1956).
9. H. L. Greenwald, G. L. Brown, and M. N. Fineman, *Anal. Chem.*, **28**, 1693 (1956).
10. R. E. Singiser and H. M. Beal, *J. Amer. Pharm. Assoc. Sci. Ed.*, **49**, 482 (1960).
11. A. Brothman, *Chem. Met. Eng.*, **46**, 263 (1939).
12. Atlas Chemical Industries, Inc., booklet LG-88, p. 5.
13. I. A. MacDonald and G. O. Lensack, *Cereal Science Today*, **12**(1) (1967).
14. Atlas Chemical Industries, Inc., *The Role of Surfactants in Baked Foods*, booklet LG-114.
15. Atlas Chemical Industries, Inc., *New TANDEM 8 Conditioner/Softener*, booklet LG-112.
16. J. W. DuRoss and W. H. Knightly, *Manufacturing Confectioner*, **45**, 50 (July 1965).
17. W. H. Knightly and G. P. Lensack, Assignee, Atlas Chemical Industries, March 10, 1964, U.S. Pat. 3,124,464.

Chapter 6

MEDICINAL EMULSIONS

B. A. Mulley

Postgraduate School of Studies in Pharmacy
University of Bradford
Bradford, Yorkshire, England

I.	INTRODUCTION.	292
II.	EMULSIFYING AGENTS AND OTHER EMULSION ADJUVANTS USED IN MEDICINE.	295
	A. Anionic Emulsifying Agents.	295
	B. Cationic Emulsifying Agents	297
	C. Nonionic Emulsifying Agents	298
	D. Miscellaneous Emulsifying Agents, Thickeners, and Stabilizers	301
	E. Preservatives and Antioxidants.	304
III.	SOME TECHNICAL CONSIDERATIONS IN THE DESIGN AND PREPARATION OF MEDICINAL EMULSIONS.	306
	A. Theoretical Aspects of Preservation	306
	B. Stabilization of Emulsions by Drying.	317
	C. The Multiphase Nature of Some Emulsions	318
	D. Drug Release from Emulsions	320
IV.	PRODUCT REVIEW.	322

A.	Emulsions for Internal Use.	322
B.	Emulsions for External Use.	325
C.	Injectable Emulsions.	334
	REFERENCES.	343

I. INTRODUCTION

Dose forms that are physically emulsions have a long history in medicine and have undoubtedly been used from the very earliest times, before proper scientific investigation of their nature was attempted. It has been said (1) that Grewe, a medical practitioner, made the first mention in Western scientific literature of the application of emulsions in medicine when he reported to the Royal Society in 1674 on the use of oils emulsified with egg yolk. Less than 100 years later forerunners of the pharmacopoeias included descriptions of various emulsions employed for medicinal purposes. For example, an edition of Quincy's *Dispensatory* published in 1770 (2) discussed, among other things, "mixtures of oily, resinous, and other like bodies, with water, in a liquid form, of a white colour resembling milk, and hence called emulsions."

Even in these early records both the basic problems and the possible advantages of presenting medicaments in emulsified form can be clearly seen. For example, in discussing the preparation of emulsions of olive or almond oil the advantages of powdered gum arabic over other emulgents are stated (2) to result in "...the pleasantest form that oils can be given in. The union is also more perfect, and the oil less disposed to separate on standing than in emulsions obtained by other means. Even strong acids added to the emulsion produce no decomposition in it." As another example, the advantage of presenting arum root in emulsified form is described as follows: "In this form arum root, which is normally very pungent to the mouth, does not inconvenience the patient. Used in rheumatic cases it warms the stomach and promotes sweating."

6. MEDICINAL EMULSIONS

At present the basic questions asked at the beginning of the development of medicinal preparations in emulsified form remain those indicated in the examples quoted: (1) the physical attributes required of the product (stability, viscosity, emulsion type, etc.) and (2) the likely relations between the product's physical and biological properties when it is used clinically in man. The main aim of subsequent development must therefore be to devise a formula and a process for preparing the product (together with packing and other details concerned with the final preparation of the finished dose form) that incorporate optimal properties in both respects. In practice thorough investigation of these two aspects has been rarely carried out. Scientific knowledge is still insufficient to predict completely the physical nature and stability of the wide range of emulsion formulas that might be considered as possibilities in any given situation; the study of precise relations between the physical and biological properties of emulsions is even less developed. In many systems the latter may not be an important consideration; in others the problem is so complex that an exact description of the processes involved is not possible at present. However, these considerations are always present, and answers to both problems will progressively emerge with the general development of scientific knowledge in the emulsion field. Interesting examples and partial solutions of both these problems will be shown in the course of this chapter.

The reasons for choosing an emulsion as a suitable dose form for a particular medicinal agent are usually immediately obvious from a cursory examination of its nature and medicinal use, the route by which it is to be used, or the physical properties required of the product. In some cases emulsions are not the only type of product possible, such other dose forms as suspensions or solutions being just as suitable.

Oils have been administered by mouth for a variety of purposes in the past, including the treatment of constipation, where their lubricating or softening action (e.g., mineral oil) or their purgative properties (e.g., castor oil) are of value. Oils as such are unpleasant for a patient to take, but are more palatable when emulsified. In the case of oils containing oil-soluble vitamins administered internally experimental work has shown that emulsification may lead to improved absorption, especially where a patient's digestion of fats is deficient as a result of disease or surgery. Another example where presentation as an emulsion may improve the effectiveness of a medicament is in the administration of heparin. This anticoagulant is normally given by injection because it is destroyed by the oral route in ordinary solution, but it has been shown to be absorbed from emulsions in experimental animals (3).

Emulsions for injection are less widely used, mainly because of the difficulty in finding suitable nontoxic emulsifying agents. Patients undergoing major abdominal surgery have been fed by the intravenous route with cottonseed-oil emulsions supplemented with other nutrients, such as sugars. There is also considerable current interest in immunological preparations in emulsified form because the slow release of the immunizing agent from the emulsion improves the development of antibodies in comparison with other single-dose forms of the antigen.

Many hundreds of drugs have been used for external application to the skin in the form of emulsions and creams. These include antiseptics, steroids, anesthetics, antibiotics, insecticides, and many others. The need to keep the product in contact with the skin and the cosmetic advantages of washability explain their popularity and superiority over aqueous or oily solutions of the drugs. In preparations of this type the biological availability of the medicament may be a problem, and although some investigations have been carried out, the situation is so complex that empiricism is still the rule. Detailed consideration of the range of emulsified preparations in present or former use in medicine is given in Section IV.

Nomenclature in pharmacopoeias and other reference works can lead to some confusion. In the United Kingdom the term "emulsion" is usually reserved for products for *oral* use that are physically

6. MEDICINAL EMULSIONS   295

in this category. Other preparations that are physically emulsions
may have such names as "application" or "liniment," which indicate
their pharmaceutical category. This practice is intended to improve
safety since many products might be mistakenly thought suitable for
oral use simply by virtue of the name "emulsion," which is associated in the lay mind with oral preparations. A further point concerns the physical definition of emulsions. The term is normally
considered to refer to liquid-liquid dispersions, although in fact
many preparations consist of more than two phases, which may or may
not all be isotropic. Two-phase systems in which one of the phases
is in liquid-crystal form are also probably quite common, especially
in systems containing synthetic surfactants. The composition conditions leading to such dispersions are considered in Section III,
but in the chapter as a whole no attempt has been made to exclude
from the discussion products that are generally considered to be
emulsions even though they are almost certainly not simple liquid-
liquid dispersions.

    The following abbreviations are used in this chapter: B.P.,
British Pharmacopoeia; I.P., International Pharmacopoeia; U.S.P.,
U.S. Pharmacopoeia; U.S.N.F., U.S. National Formulary. Information
on the commercial products cited as well as the full names and
addresses of the manufacturers may be found in *Extra Pharmacopoeia,
Martindale* (4).

II. EMULSIFYING AGENTS AND OTHER EMULSION ADJUVANTS USED IN MEDICINE

A. Anionic Emulsifying Agents

    Alkali-metal soaps have long been used in medicine for the
preparation of oil-in-water emulsions, mostly for external use,
as they are too irritant internally. Their deficiencies are well

known: (1) they are incompatible with divalent-metal ions, cationic surfactants, and medicaments; (2) they are stable only at high pH. Soap Liniment, B.P. is a common emulsion product containing an alkali-metal soap and is used as a rubefacient. It contains potassium oleate formed in situ from potassium hydroxide and oleic acid. Similar preparations are described in many other pharmacopoeias.

Divalent-metal soaps are used to prepare water-in-oil emulsions for external use (e.g., oily calamine lotion). Calcium oleate is a very common example and is usually formed in situ by reacting a calcium hydroxide solution with oleic acid.

Amine soaps such as triethanolamine oleate or stearate are used in the preparation of externally applied creams. They give fine-grained emulsions that are more stable than those produced by the alkali-metal soaps.

The sodium salts of sulfated fatty alcohols are very widely used in cream formulas. They are stable over a wide pH range and fairly resistant to divalent-metal ions. Large quantities of fatty alcohols are necessary in the formula if a good emulsion is to be obtained. These emulgents are commonly kept in the form of an anhydrous base (Emulsifying Ointment, B.P.) that will produce good water-washable creams when water is added. Similar products are described in many other pharmacopoeias. In addition to sodium lauryl sulfate and cetostearyl alcohol these bases contain petrolatum, and creams are formed by melting the ointment, dissolving the medicament in the water or the molten oil phase, mixing the two phases at about 70°C, and stirring the product until cold. If the medicament is insoluble in both phases, it is sometimes dispersed in the final cream.

The sulfated fatty alcohols are too toxic for injectable products (5) and are not normally used internally for the same reason.

6. MEDICINAL EMULSIONS

B. Cationic Emulsifying Agents

Cationic surface-active agents have marked bactericidal activity against both Gram-positive and Gram-negative organisms, but are less effective against spores, viruses, and fungi. They are used in a variety of products, including creams, especially for preoperative cleansing of the skin, disinfecting wounds and burns, and removing scabs and crusts from skin. The cationic surfactant usually acts jointly as the emulsifying and antibacterial agent, although fatty alcohols have to be added to improve the stability of the product.

Cationic surface-active agents are incompatible with soaps and anionic agents because the large, oppositely charged ions interact and an inactive precipitate is formed. Their properties are relatively unaffected by metal cations, including calcium ions, and therefore by hard water, but since they combine with proteins, their activity is reduced by serum or organic matter. Their bactericidal activity may be reduced in the presence of nonionic surface-active agents where the proportion of the latter is high (6). In fact high concentrations are used to quench the action of quaternary ammonium compounds in the experimental assessment of the latter by microbiological methods (7).

The antibacterial activity of these compounds is markedly affected by structural factors, particularly the size and length of the alkyl group responsible for the surface activity. The Ferguson principle has been shown to apply if the critical micelle concentration (CMC) is treated as a solubility limit so that the greater activity of the longer chain compounds results from their lower CMC (8). Many of the surfactants used in practice are mixtures of homologs supplied in this form for their improved solubility characteristics and also because individual homologs are more expensive and difficult to manufacture. The Ferguson principle has been shown to extend to multicomponent solutions of quaternary ammonium surface-active agents (9).

Cationic surface-active agents are more toxic than most others. The fatal dose of benzalkonium chloride, cetylpyridinium chloride, and similar compounds is estimated to be 1 to 3 g by mouth (10). Although they are quite widely regarded as relatively nonirritant and of low sensitizing power to human skin (11), this appears to be markedly dependent on the concentration of the compound, and, as is so often found, certain individuals can exhibit serious reactions not found in the majority. For example, Sharvill (12) has reported four cases of extreme sensitivity to cetrimide. Cutler and Drobeck (13) have extensively reviewed the literature on the toxic effects of cationic surfactants.

The important cationic surfactants used in emulsion preparations include Benzalkonium Chloride, I.P., which is also described in many national pharmacopoeias. This has the general formula $C_6H_5CH_2N(CH_3)_2-R$ Cl, in which R represents a mixture of alkyls from C_8H_{17} to $C_{18}H_{37}$. Cetrimide, B.P. is another, consisting chiefly of tetradecyltrimethylammonium bromide with smaller amounts of the C_{12} and C_{16} derivatives. Dequalinium Chloride, B.P. has been used in emulsified preparations and has been claimed to be little affected by the presence of serum, but severe toxic symptoms due to skin reactions have been reported after the use of this compound (14).

C. Nonionic Emulsifying Agents

A very large number of nonionic surfactants have been used or are of potential value in the preparation of medicinal emulsions. Because of their low toxicity it is possible to consider them even for products that are to be injected, and purified grades have been developed for this purpose. The ease with which the proportions of hydrophilic and lipophilic groups can be adjusted to vary the properties within a particular category of surfactant is also attractive for formulation work and further explains the extensive attention that has been devoted to the class.

6. MEDICINAL EMULSIONS

Being un-ionized in solution, nonionic surface-active agents are generally stated to be compatible with both anionic and cationic substances, but they reduce the activity of some preservatives. The mechanism responsible for this has been extensively investigated (6), and it is concluded that interaction of the preservative with the surfactant micelles reduces the concentration free in the aqueous, nonmicellar "phase," which alone is effective as an antibacterial agent. Quantitative study of the interaction has been made by equilibrium dialysis, solubility measurements, and other techniques on a range of preservatives, especially benzoic acid and its derivatives and phenols, and the results correlated in certain instances with the preservative properties of the system. The effect of structural changes in the surfactant has also been investigated. These results and an extension of them to emulsion systems, where partition into the oil phase also affects the preservative activity, are discussed in more detail in Section II.E.

The acute oral toxicities of the sorbitan esters of fatty acids and their polyoxyethylene derivatives are very low in animals. For example, for Tween 60 the LD_{50} in rats is greater than 60 ml/kg (15). In the United Kingdom these compounds are allowed in food products by the "Emulsifiers and Stabilisers in Food Regulations, 1962," made under the Food and Drugs Act of 1955. They are also recognized as safe additives to foods by other national regulatory agencies, including the U.S. Food and Drug Administration, and by the World Health Organization. From the toxicological point of view their use in oral emulsions seems eminently safe.

Fairly extensive study has been made of the oral administration of these compounds in human subjects. For example, sixteen patients ingested 6 g of Span 60 daily for 28 days; the changes found in a wide range of biochemical tests of body functions were insignificant, nor were any general ill effects observed (16). Similar doses of Tween 80 have been ingested for periods of up to 3 to 4 years without ill effects. An accidental dose of 67.2 g of Tween 80 was given to a 4-month-old, 3.5-kg child, virtually without ill effects (17).

Broadly similar results in man have been obtained with polyoxyethylene fatty-acid esters, such as Myrj 52.

Emulsions of coconut oil with nonionic surfactants have been injected intravenously in man, using purified material (18,19). This is discussed in more detail in Section IV.C.

Tween 80 in low amounts has been injected intramuscularly, without any cyst or abcess formation, into patients undergoing treatment for allergies (20).

Investigation of the metabolic fate in man of nonionic emulgents shows that the esters are readily split in the human gut, the liberated fatty acid undergoing normal metabolism, but the fate of the polyethylene glycol moiety depending on its molecular weight. With chort-chain compounds it is largely absorbed and the products are excreted in the urine, whereas with the long-chain compounds it mostly passes through unabsorbed in the feces, the amount absorbed being as low as 2 to 3% when the chain is as long as 40 units (21).

Nonionic surface-active agents like Tween 80 can improve abnormally low fat absorption in man.

Nonionic surface-active agents have been examined for their effect on the oral toxicity of food-coloring agents and butylated hydroxytoluene. No effects (22) were found. In experiments with mice it has been demonstrated that nonionic emulsifiers can promote the action of potent carcinogens on the skin and stomach (23). This effect is not confined to nonionic surface-active agents, but can also be produced by other materials, such as fatty acids. Nevertheless the formulation of potential carcinogens and prolonged use with nonionic surface-active agents is to be avoided (e.g., Coal Tar Solution, U.S.P.). The picture is complex, however. In surface application at high concentrations of carcinogenic hydrocarbon the influence of the solvent is not great, but it can be greater when low concentrations of carcinogen are used.

There are cases in which surfactants reduce the irritancy of certain products; for example, in phenols soaps are used and in iodine preparations, nonionic surface-active agents (24).

6. MEDICINAL EMULSIONS

Officially described emulgents of the nonionic type include the polyoxyethylene fatty-acid esters (e.g., Polyoxyl 40 Stearate, U.S.P), the ethers (E.G., Cetomacrogol 1000, B.P.), glycol and glycerol esters (e.g., Glyceryl Monostearate, U.S.N.F.), and sorbitan derivatives [e.g., Polysorbate 80, U.S.P. (Tween 80)]. Sucrose esters have been used, but have not been officially described. A large number of other nonionic materials have been used less extensively.

The esters are prone to hydrolysis. Their emulsifying properties for pharmaceutical purposes have been studied by Johnson and Thomas (25), and those of the ethers by Hadgraft (26). The ethers are used in oil-in-water creams by incorporating water into a previously prepared anhydrous base, such as Cetomacrogol Emulsifying Ointment, B.P. [cetomacrogol emulsifying wax 30 g, liquid petrolatum (mineral oil) 20 g, and white petrolatum 50 g]. In both cases, as with other emulsifying agents, good emulsions are only formed when large amounts of fatty alcohols are also present. The use of combinations of Tweens and Spans based on the hydrophile-lipophile-balance (HLB) system of designing emulsion formulas is too well known to need further description. A nonionic emulsifying ointment employing this type of surfactant is described in a Belgian formulary (polysorbate 80 7 g, cetyl alcohol 17 g, sorbitol 15 g, white petrolatum 2.5 g, water to 100 g).

Sucrose esters have been investigated as possible emulsifiers for medicinal preparations (27), but have not been widely accepted.

D. Miscellaneous Emulsifying Agents, Thickeners, and Stabilizers

Various cellulose derivatives, which form viscous colloidal solutions, have been used to make oil-in-water dispersions, which are especially suitable for oral use. These materials are not emulgents in the ordinary sense because they are not usually very surface active, but oils may be mechanically dispersed in the viscous solutions they form with water, and the droplets prevented from recombining by the mechanical properties of the gel. The properties of the cellulose derivatives depend on the degree of polymerization (number of anhydroglucose units) in the cellulose chain and the proportion and nature of the groups introduced into the anhydroglucose unit by chemical means.

Methylcellulose, a methyl ether of cellulose containing about 29% of methoxyl groups, is official in many pharmacopoeias, and proprietary preparations include Celacol M and Methocel MC. Various grades are used, the grade being indicated by a number after the name, giving the approximate viscosity in centistokes of a 2% w/v solution at 20°C. The mucilages formed are colorless, odorless, tasteless, neutral, and otherwise inert. They are stable in acid or alkaline solution and in the presence of moderate concentrations of most electrolytes. The viscosity is reduced with temperature rise unless the gel point is reached, but this is reversible on cooling except when heating has been prolonged. Unlike those of natural gums, the viscosities are stable for months at normal temperatures. Preservatives are necessary (e.g., phenylmercuric nitrate, 0.001%), and although interactions have been reported with some preservatives and active ingredients (e.g., phenols and the hydroxybenzoates), they have not been observed with benzyl alcohol, chlorobutanol, or phenethyl alcohol (28). Although widely regarded as safe for oral use, the possibility of toxic effects from the formation of methyl alcohol in the intestine has been mentioned. The seventh report of the FAO/WHO Expert Committee on Food Additives (29) estimates the acceptable daily intake in man as 30 mg/kg, pending further investigation in long-term studies. A typical formula for a methylcellulose emulsion is mucilage of methylcellulose (4%) 27.5 g, fixed oil 40 g, water to 100 g.

Other cellulose derivatives that have been used in emulsions in pharmacy include Hydroxypropylmethylcellulose, U.S.N.F., which has 19 to 30% of methoxyl and 3 to 12% of hydroxy-propoxyl groups attached to the anhydroglucose rings of cellulose by ether linkages. It has the advantage that mucilages are easier to prepare and

6. MEDICINAL EMULSIONS

the gel point is higher. Sodium Carboxymethylcellulose, B.P. is another cellulose derivative used to prepare oil-in-water emulsions, but it is less efficient than methylcellulose.

The relative efficiencies of methylcellulose, ethylmethylcellulose, and sodium carboxymethylcellulose for emulsifying mineral oil and arachis oil have been investigated by Davies and Rowson (30). Emulsifying power increases with concentration, but the low-viscosity grades were found to be better than the high-viscosity ones. The stability of the emulsions was affected by heat: mineral oil emulsions were all stable for 4 to 8 weeks at 40°C, but deteriorated rapidly at 80°C. The effects of acid, alkali, salts, and alcohol on emulsion stability were also studied.

Emulsions stabilized with carrageenan have been investigated by Fitzgerald and Skauen (31), who found mucilage strengths of 0.2 to 0.4% to give the best results.

Gum tragacanth, which swells to a homogeneous, gelatinous mass in water, has been added to emulsions prepared with gum acacia in order to retard creaming, although it has been found not to improve the stability of mineral oil-gum acacia emulsions (32). The activity of preservatives is reduced in the presence of gum tragacanth (33).

Macrogols (polyethylene glycols) are poor emulgents, but are sometimes added as stabilizers. They are also emollient and well absorbed. Officially described macrogols (the numbers indicate the approximate molecular weight) are macrogol 300 (Liquid Macrogol, B.P.), macrogol 400 (Polyethylene Glycol 400, U.S.P.), macrogol 1540 (Polyethylene Glycol 1540, U.S.N.F.), and Macrogol 4000, U.S.P. (Hard Macrogol, B.P.).

Carboxypolymethylene has also been used as an emulsifying agent for preparing oil-in-water emulsions for external use.

Sodium alginate, which forms mucilages with water, has been used to form dispersions of oils in water, but is more commonly used as a thickener. It reduces the activity of many preservatives (34).

Gum acacia has a long history of use as an emulsifying agent for oils intended to be administered by mouth. It contains highly complex glycosidal acids partly present as calcium, potassium, and magnesium salts. Although they are surface active and readily form good oil-in-water (or in certain instances coarse water-in-oil) emulsions with most oils, the high stability of the emulsions seems to be the result of the formation of a microscopic film of considerable thickness round the oil droplets. Study of the formation of these films and other properties of gum acacia have been reported in papers by Shotton and co-workers (see, for example, Ref. 35). Many different procedures have been claimed to give the most efficient use of the gum, and there has been a considerable "mystique" attached to minor aspects of the conditions advocated. Suitable techniques using both "wet" and "dry" gum methods for small-scale (hand) preparation of emulsions are described in standard works on dispensing practice (36).

E. Preservatives and Antioxidants

Among the commonly used preservatives in medicinal preparaion, including emulsions and creams, are phenol (0.5%), chlorocresol (0.1%), phenylmercuric nitrate (0.001%), cetrimide 0.001%), benzalkonium derivatives (0.001%), benzoic acid and the other hydroxybenzoates, and chloroform (0.25%). The hydroxybenzoates are particularly widely used because they are tasteless, odorless, inert, stable, and nontoxic, although they are not such strong bactericides as, for example, chlorocresol.

Of the benzoic acid derivatives, methyl p-hydroxybenzoate has the highest solubility in water and is usually employed in concentrations of 0.1 to 0.2% in aqueous preparations. The higher esters (propyl and butyl) are used as near-saturated solutions. Although their activity is reduced in the presence of nonionic surfactants or when creams containing a lot of oil are to be preserved, they may still be employed if the concentration is raised (37,38). Each ester is active against a different range of organisms, the higher esters being more effective in inhibiting molds and yeasts. All are more effective against Gram-positive bacteria than against Gram-negative ones. Combinations are thought to be better than a single ester since, for solubility reasons, a higher total concentration can be obtained and the combination may be active against a wider range of organisms. A ratio of 2:1 (0.06 and 0.03%) of methyl and propyl esters is often employed. Lower concentrations may not be fully effective (39).

The toxicity of the hydroxybenzoates has been investigated by Matthews et al. (40) and Nathan and Spears (41).

Sorbic acid has also been used as a preservative and is especially useful for products containing nonionic surfactants. At 0.2% it was found more active than 50 other substances and their combinations (42). However, conflicting reports have appeared on its irritant and allergenic properties. Klauder (43) found that even at a concentration of 1% in Hydrophilic Petrolatum, U.S.P. it was almost inert when tested on 50 subjects; Fryklöf (44) found that creams and ointments applied to the face caused erythema and slight itching, and occasionally slight edema in 10 out of 20 subjects. Positive reactions could be obtained down to 0.025% in the most sensitive subjects in a cold cream (water-in-oil emulsion).

Antioxidants are sometimes added to medicinal emulsions in order to reduce the oxidation of active constituents or the oil phase. Even traces of oxidation products from oils are undesirable, especially in orally administered products, because they are so easily noticed by their smell or taste. Oils used in pharmacy often have added antioxidants even before they are incorporated into formulated products. For example, according to the British Pharmacopoeia, mineral oil (Liquid Paraffin, B.P.) is permitted to contain up to 10 ppm of tocopherol or butylated hydroxytoluene; Cod-liver Oil, B.P. may contain dodecyl gallate, octyl gallate, or any mixture of these in a proportion not exceeding 100 ppm.

The most commonly used compounds in pharmacy are the alkyl gallates, ascorbic acid, butylated hydroxyanisole, butylated hydroxytoluene, sodium metabisulfite, and the tocopherols.

Sequestering agents (e.g., citric and tartaric acids), which enhance the effects of antioxidants by reacting with those heavy-metal ions that catalyze ocidation, are also sometimes incorporated. Oil-soluble antioxidants are less effective if the oil is in contact with an alkaline aqueous phase, and this should therefore be avoided.

Birchall and Felix (45) found that the propyl, octyl, and dodecyl esters of gallic acid, alone or in combination with citric acid or acetonedicarboxylic acid, were the most effective preservatives of arachis oil. Extensive studies on oxidation and on the protective action of antioxidants for oils in the solubilized and emulsified state have been made by Carless et al. (46) and others (47,48).

Little has been published in the pharmaceutical literature on the toxicity of antioxidants. Most of them appear to be safe under the conditions in which they are normally used in medicine. Consideration of their possible toxic hazards in food products has received much attention (see, for example, Ref. 49), and provisional estimates have been made of an acceptable daily intake for man of the commonly used compounds. This is about 0.2 to 0.5 mg/kg for such compounds as the gallates. Butylated hydroxytoluene has been reported to have a lower safety margin than other antioxidants, and recommendations have been made that it should no longer be permitted in foods (50).

III. SOME TECHNICAL CONSIDERATIONS IN THE DESIGN AND PREPARATION OF MEDICINAL EMULSIONS

A. Theoretical Aspects of Preservation

Emulsions of the water-in-oil type do not cause much difficulty because most likely contaminants do not grow in nonaqueous media, but products in which water is the continuous external phase often provide good growth conditions for bacteria, molds, and fungi. Prevention of contamination depends partly on the use of uncontaminated raw materials, correct processing and packaging, and the use of packaging materials that are themselves uncontaminated. However, unless the preparation is produced in a sterile condition and then used aseptically, a chemical preservative is necessary to maintain the relative freedom from microorganisms normally aimed at during manufacture and to prevent the growth of those that unavoidably come in contact with the product during storage or use.

The increasing use of many new ingredients with unknown effects on the efficacy of the preservative has in recent years given rise to some difficulty in satisfactorily formulating products with respect to the preservative, and failures probably occur quite widely. There is also at present considerable concern about the microbial contamination of a wide range of pharmaceutical preparations since the discovery that many were unsatisfactory (51). Some official monographs for pharmaceutical materials and formulated products now lay down standards for the maximum permitted levels of microorganisms of various kinds and tests for the absence of certain others that are pathogenic (52). Preservation is thus of considerable importance. Quite a lot of work, both theoretical and practical, has been reported in the literature and provides a good guide to the principles involved, although the range of possible product variations is such that a definitive answer cannot be given to every possible case. There are also some aspects of the physical principles involved that are as yet unresolved.

Basically the problem divides into (1) the biological one of defining the concentration of preservatives in aqueous solution necessary to prevent the growth of likely contaminants and (2) a physicochemical one, because in an emulsion system the preservative added may not all remain in the aqueous phase (53,54). Some will dissolve in the oil and thus lower the concentration in the water, which in many cases may be sufficiently reduced to makde the preservative ineffective. More specific incompatibilities between the preservative and the constituents of the product may reduce the concentration free in the aqueous phase. For example, part of the preservative may be bound to any macromolecular excipients added for stability reasons (55), or it may be solubilized in the micelles formed by most emulsifying agents (56,57). Preservatives that are weak acids or bases are affected by pH changes in the product because in some cases only the un-ionized molecules are active (58). In many practical instances more than one of these factors may reduce the concentration of preservative free in the aqueous phase, but it is convenient and simpler to discuss them individually, building the theoretical treatment in stages.

1. PARTITION OF PRESERVATIVE INTO THE OIL PHASE

It has long been realized that oil-water partition can affect the activity of both preservatives and drugs. For example, Husa and Radin (59) attributed the improved antiseptic value of phenol in certain ointment bases to the lowering of its solubility in the nonaqueous phase, which favors an increased concentration in the water. Regarding the preservative action of the hydroxybenzoates, Atkins (53) reported that the partition of the propyl p-hydroxybenzoic acid ester in certain oil-water systems so favored the oil phase that only about 10% remained in the aqueous phase. The methyl ester was considered more satisfactory because about two-thirds remained in the aqueous layer. Biological tests made on creams with similar oil and water phases showed that even for deliberately contaminated creams 0.1 to 0.15% of the methyl ester was effective as a preservative agent, whereas 0.25% of the propyl ester was unsuitable.

Garrett and Woods (54) investigated the distribution of benzoic acid between peanut oil and water, studied its dependence on pH, and pointed out its significance to the preservative action. Further experiments on the distribution of methyl p-hydroxybenzoate between oils and surfactants and water have been reported by Hibbott and Monks (60). The low partition coefficients found for mineral oil and almond oil were thought to account for the successful use of the preservative for these materials, but difficulties

6. MEDICINAL EMULSIONS

with newer oil phases like isopropyl myristate and formulations containing nonionic surfactants were not considered surprising because these all had much higher coefficients. The addition of such solvents as propylene glycol and glycerin to the aqueous phase was shown to reduce the loss of preservative to the oil phase. For example, 20% propylene glycol in the aqueous phase reduced the partition coefficient for the ester from 31 to 16 for oleyl alcohol and water.

Hibbott and Monks (60) also reported the incidence of mold growth in formulated products containing a nonionic surfactant, cetyl alcohol, and oil phases of mineral oil, isopropyl myristate, and diethyl sebacetate or mixtures of these. The overall concentration of preservative was from 0.25 to 0.15% and its concentration in the aqueous phase, calculated from the partition experiments, was 0.0071 to 0.0443. Creams with more than 0.018% preservative in the aqueous phase were uncontaminated, but below this concentration mold growth was found.

Although the calculations made from the distribution measurements are subject to some criticism because a single distribution coefficient is unlikely to fully describe the amount of preservative in the phases in such complex formulated products, the results clearly show how important it is to consider the reduction in the effective concentration of preservative due to partition into the oil phase.

A series of papers have been published since 1962 by H. S. Bean and co-workers on the activity of preservatives in simple oil-water systems and in more complex emulsion systems containing a surfactant. Physical measurements on the distribution of the preservative between the phases and the effect of formulation factors such as phase-volume ratio have been made and correlated with the effectiveness of the preservative against organisms dispersed in the system. In the first paper (61), in which phenol was examined in mineral oil-water or arachis oil-water systems, it was reported that the bactericidal activity of the phenol, although mainly dependent on its concentration in the water, was also affected by the phase-volume ratio. The activity of the bactericide increased at high ratios of oil. This was considered to be an interface effect because the microorganisms tended to concentrate at the oil-water interface, which was thought to have a high concentration of adsorbed preservative. The interface effect was not great except at high oil-water ratios. An alternative explanation, in the case of the arachis oil systems, could be that this oil contains extractives that increase the action of the phenol. In a more recent paper (62), the activity of phenol and p-chloro-m-cresol in petrolatum-water systems emulsified with polysorbate 80 was attributed to the concentration of the phenols free in the aqueous phase.

A paper by Garrett (63) gives a theoretical account of a model for predicting the action of preservatives in various systems including emulsions. He considered the constitutional factors that can reduce the thermodynamic activity of the preservative, especially those that are weak acids, and developed a mathematical model accounting for each. Briefly, this gives a number of enhancement factors by which it is necessary to multiply the biologically effective concentration μ of the preservative to give the total concentration P_T needed to maintain the inhibitory concentration. The basic expression is

$$P_T = \mu(f_1 \times f_2 \times f_3) \qquad (1)$$

where f_1, f_2, and f_3 are the necessary enhancement factors arising from binding of the preservative to macromolecular constituents of the product (including surfactant micelles), partition into an oily phase or ionization of the preservative, and its chemical instability, respectively. This approach will be further considered later.

The following theoretical description is derived from the work cited here, but it uses a somewhat different approach to account for the effect of micellar solubilization of preservative. This treatment is possibly oversimplified, but it will probably prove

6. MEDICINAL EMULSIONS

satisfactory until the theory of solubilization is further clarified experimentally.

2. SIMPLE PARTITION INTO THE OIL PHASE

For an un-ionized preservative, which is unassociated in either phase, the final concentration free in the water, C_w, after equilibrating an aqueous phase of initial concentration C with the oil phase can be derived as follows:

$$CV_w = C_w V_w + C_o V_o \qquad (2)$$

where V_w and V_o are the volumes of water and oil phases, respectively, and C_o is the final concentration in the oil phase.

Dividing by V_w and substituting $\phi = V_o/V_w$ (oil-water phase-volume ratio), we obtain

$$C = C_w + C_o \phi \qquad (3)$$

If the oil-water partition coefficient $k_w^o = C_o/C_w$, then

$$C = C_w + k_w^o C_w \phi \qquad (4a)$$

$$\text{or} \quad C = C_w(1 + k_w^o \phi) \qquad (4b)$$

Thus, if the concentration of the preservative usually regarded as satisfactory in an aqueous solution, the partition-coefficient, and the phase-volume ratio are all known, C can be calculated.

Preservatives that are capable of ionizing in the aqueous phase (e.g., organic acids) need a slightly more complicated treatment. The quantity C_w has two components, the concentration of un-ionized acid, $[HA]_w$ and the ionized acid $[A^-]_w$. Normally only the un-ionized fraction has significant biological activity. For example, Rahn and Conn (58) found that the inhibitory concentration of benzoic and salicylic acids for a yeast was almost independent of pH if the undissociated acid alone was considered to be the active species. The difference in activity between undissociated molecules and ions of the same acid is usually attributed to the poor penetration of ions through living-cell membranes.

Returning to the use of preservatives that can ionize in oil-water systems, it is necessary to consider the distribution of the total preservative among the various species and phases of the system. This will depend on k_w^o, K_a (the dissociation constant of the acid in the aqueous phase), the phase-volume ratio, and the pH.

Substituting $[HA]_w + [A^-]_w$ for C_w in Eq. (3) gives

$$C = [HA]_w + [A^-]_w + C_o \phi \tag{5}$$

But $C_o = k_w^o [HA]_w$ and $[A^-]_w = K_a [HA]_w / [H^+]$. Substituting in Eq. (5), we get

$$C = [HA]_w + K_a [HA]_w / [H^+] + k_w^o [HA]_w \phi \tag{6}$$

or

$$C = [HA]_w (1 + K_a / [H^+] + k_w^o \phi) \tag{7}$$

The concentration of preservative to be added initially to the aqueous phase can therefore be calculated from this equation.

A further complication occurs if the preservative can associate in one of the phases, because the simple distribution law is no longer valid. Benzoic acid, which dimerizes in many nonpolar solvents, can be cited as an example. The modified distribution coefficient is $k'' = {}^2\sqrt{C_o}/[HA]_w$, or $C_o = k''^2 [HA]_w^2$. Substituting in Eq. (7), we obtain

$$C = [HA]_w (1 + K_a / [H^+] + k''^2 [HA]_w^2 \phi) \tag{8}$$

6. MEDICINAL EMULSIONS

3. EMULSIFIED OIL-WATER SYSTEMS CONTAINING A SURFACTANT

Many oil-in-water emulsions contain quite a high concentration of surfactant in the aqueous phase, and here the concentration of the preservative free in the water depends not only on partition into the oil but also on the interaction between the preservative and the surfactant micelles. This interaction has been extensively studied by workers in the pharmaceutical field (see, for example, Refs. 64 through 67) as well as many authors in other fields, but there are major differences in the interpretation of the results as well as direct conflicts in experimental data. Some of these may result from the use of poorly characterized surfactants of mixed and variable composition. The simplest description of the systems is to treat the micelles as a phase with solutes partitioned between the water and micellar phases. Although this is reasonably satisfactory, it may be necessary to invoke other mechanisms in order to account for some of the data. For example, according to Mitchell and Brown (66), a single partition coefficient cannot be used to describe the solubilization of chloroxylenol in nonionic surfactant micelles.

If the simple pseudophase theory of micellar solubilization is used, the initial concentration needed in the aqueous phase, before partition into the oil and micellar phases, in order to achieve a final concentration free in the water, C_w, can be derived as before:

$$CV_w = C_w V_w + C_o V_o + C_m V_m \tag{9}$$

where C_m is the concentration of preservative in the micelles and V_m is the micellar volume. If k_w^m is the partition coefficient between the water and micellar phases, then

$$CV_w = C_w V_w + C_w k_w^o V_o + C_w k_w^m V_m \tag{10}$$

Dividing by V_w, using $q = V_m/V_w$, and rearranging gives

$$C = C_w (1 + k_w^o \phi + k_w^m q) \tag{11}$$

There is a difficulty in defining and obtaining values for the micellar volumes, but this can be partly overcome by defining all concentrations and phase ratios in weight terms (68) and assuming that the micellar phase contains no water molecules. This is almost certainly an approximation, but the errors involved in the assumption are small and not important in the present context.

To use these equations without further experimental work it is necessary to know the values for the various constants. Both ϕ and q can be calculated for particular formulations. Some of the necessary partition coefficients for a few common preservatives have been calculated from literature data and collected in Tables 1 and 2. Although the data in these tables will be satisfactory for many practical purposes and certainly an improvement over the use of preservative concentrations that neglect the factors considered here, it must be clearly stated that the values quoted are subject to a number of limitations. For example, oil-water partition coefficients may have been obtained at a single solute concentration. The micelle-water partition coefficients are subject to the same difficulty. However, even when the partition isotherm is curved, as it may be in cases where the whole solubility range is considered, it is often linear over a reasonable range and in this region adequately described by a single coefficient (67). It should also be noted that the theory assumes that oil dissolved in the micelles and surfactant in the oil phase does not affect the partition of the preservative.

Inspection of the values in the tables shows a number of interesting features. The micelle-water distribution coefficients are not very sensitive to changes in the structure of the surfactants unless there are large changes in the proportions of the lipophilic and hydrophilic groups. This agrees with treatments of micellar solubilization in which it is regarded as a nonspecific solution process. It is also apparent that structural changes in preservatives favoring oil solubility result in very high coefficients favoring the micellar phase, especially for phenolic

6. MEDICINAL EMULSIONS 315

materials like chlorocresol. Preservative concentrations that are ordinarily effective in aqueous solutions are thus unsatisfactory in many modern emulsified systems, as has been found in practice. As an example, substitution of values from Table 2 for methylparaben in an emulsion of almond oil (20%) with Tween 80 (5%) gives [Eq. (11)] the preservative concentration required as about 0.5%, that is, five times that normally used. Although values for distribution coefficients in many systems are not available, for practical purposes interpolation from the results in Tables 1 and 2 may be reasonably satisfactory. Alternatively values may be determined experimentally, with sufficient accuracy in most cases, by single-point solubility measurements.

TABLE 1

Micelle-Water Distribution Coefficients for Benzoic Acid[a]

k_w^m	Surfactant	Temp. (°C)	Ref.
46	Polysorbate 80	30	68
40	Cetomacrogol 1000	30	68
50	$C_{16}E_{16}$	25	69
46	$C_{16}E_{30}$	25	69
33	$C_{16}E_{40}$	25	69
27	$C_{16}E_{96}$	27	69
77	$C_{12}E_{7}$	20	67
70	$C_{12}E_{8}$	20	67
70	$C_{12}E_{9}$	20	67
67	$C_{12}E_{10}$	20	67
73	$C_{12}E_{11}$	20	67
73	$C_{12}E_{12}$	20	67
44	Cetomacrogol 1000	25	70
47	Cetomacrogol 1000	25	65
45	Myrj 49	25	71
22	Myrj 59	25	71

[a]Calculated from literature data.

TABLE 2
Oil-Water and Micelle-Water Distribution Coefficients for Some Common Preservatives[a]

Preservative	k_w^o	Oil	Temp. (°C)	Ref.	k_w^m	Surfactant	Temp. (°C)	Ref.
Methylparaben	0.3	Mineral oil		59	57	Tween 80	27	74
	6	Almond oil		59	43	Myrj 52	27	74
	18	Isopropyl myristate		59				
	47	Castor oil		59				
	56	Diethyl phthalate		59				
	204	Diethyl adipate		59				
Ethylparaben					84	Myrj 52	27	74
Propylparaben					326	Myrj 49	25	73
					230	Myrj 52	25	73
					261	Myrj 52	27	74
Sorbic acid					38	Tween 80	30	72
					29	Myrj 52	30	72
Phenol[b]	0.067	Mineral oil	25	60	20	Polysorbate 80	25	61
	5.6	Arachis oil	25	60				
Chlorocresol[b]	1.7	Mineral oil	25	61	450	Polysorbate 80	25	61

[a]Calculated from literature data.
[b]Nonlinear distribution isotherms can be obtained for phenols in micelles formed by nonionic surfactants.

6. MEDICINAL EMULSIONS   317

B. Stabilization of Emulsions by Drying

Investigations have been made into the possibility of overcoming the inherent stability problems of emulsions by drying them, storing in the dried form, and reconstituting the product when required. This process is only possible if a solid support for the oil phase is added to the product.

Richter and Steiger-Trippi (75) studied mineral oil emulsions prepared with 0.9% of sodium oleate dried by a spray-drying process. The solid supports were Tylose SL 400 (1%) and Aerosil Standard (2%). The solids were essential if the dried emulsion was to be successfully reconstituted. An upper limit of 20% of mineral oil was the maximum proportion of oil phase that could be satisfactorily dried and reconstituted. Between 20 and 30% the emulsion could be dried, but it broke immediately on reconstitution. When the proportion of mineral oil was above 30%, the emulsion could not be dried.

A more extensive investigation of solid supports in dried emulsions was reported by Lladser, Medrano, and Arancibia in 1968 (76). The oil phase was again mineral oil, mainly restricted to 10% of the emulsion volume. The emulgent was a mixture of polysorbate 80 and sorbitan monooleate (5:3), 20% of the volume of the oil phase. A range of solid supports were studied. Also investigated were the so-called crystalline materials, such as manitol, urea, glycine, sorbitol, glucose, sucrose, and lactose (13.3%), as well as a number of hydrophilic colloids at concentrations of 1 to 2%, including sodium alginate, polyvinylpyrrolidone, bentonite, gum acacia (5%), Aerosil, and hydroxyethylcellulose. In some cases mixed crystalline and colloid-type supports were studied. The emulsions were freeze dried, and the products were assessed for ease of reconstitution, particle-size distribution of the emulsions, and creaming rate. The effect of storing the freeze-dried product for 20 and 40 days at 0°C, room temperature, and 40°C were investigated. Redispersible products were obtained with the crystalline supports manitol, glycine and urea, but not with the others. Of the colloidal supports, only sodium alginate gave a redispersible product and that with difficulty. Mixtures of supports--for example, Aerosil and manitol--gave good results with 5 to 10% of oil, but not with 15 to 20%. Ease of redispersion was in some cases so good as to be almost instantaneous, for example, with manitol and glycine as solid supports. The mixed support Aerosil and manitol also gave good results even with 15% of oil phase. All the reconstituted emulsions had particle-size distributions displaced to larger sizes, but the emulsions were in many cases still satisfactory. The creaming rate of the products was also increased after freeze drying. There were no great differences between products stored for 20 or 40 days at any of the temperatures investigated, although those kept at 0°C were on the whole slightly better. Dried emulsions stored at 40°C tended to turn yellow during storage.

C. Multiphase Nature of Some Emulsions

Although emulsions are commonly treated as two-phase systems in which adsorption of the emulgent at the interface is responsible for the stability of the dispersion, more than two phases are often present, particularly in systems containing synthetic surfactants. The solubility of many surfactants is low and is often limited by separation, not of a solid phase, but of a liquid-crystal phase. Some of the most important features of emulsions (emulsion type, stability, and viscosity) may be related to the multiphase character of the dispersion, and the gellike physical properties of liquid crystals appear to be especially important in emulsion stability, since this phase often tends to surround the drops of the disperse phase. In the pharmaceutical and related fields a number of authors have published work on this aspect of emulsions (77-86).

The simplest emulsions have three components: "oil," water, and an emulsifying agent. At low emulgent concentrations, from the phase-rule point of view, the systems belong to those categories where pairs of partially miscible liquids have their miscibility increased by the presence of the third component. In cases where the third component is a surfactant, its distribution between the phases is unusual in that it is controlled by the formation of micelles (for a detailed investigation of a nonionic system and literature review, see Ref. 87). The characteristics of the systems depend on the nature of the components, particularly the chemical type and molecular weight of the surfactant, and on the polar/nonpolar character of the oil phase. For example, the short-chain nonionic surfactant $C_8H_{17}(OCH_2CH_2)_6OH$, although miscible with water, octanol, and dodecane, dissolves almost exclusively in the octanol phase in water-octanol mixtures, but in water-dodecane mixtures, almost entirely in the aqueous phase. These remarkable properties are related to micelle formation of the surfactant in the two systems. In both cases, however, high concentrations of the surfactant do not induce the formation of other than two isotropic liquid phases. When the alkyl-chain length of the surfactant is raised slightly to 10 carbon atoms, however, the two-liquid-phase region in water-dodecane mixtures is confined to a small area on the ternary phase diagram up to 1 to 2% of the surfactant (88). At higher concentrations a narrow three-phase region occurs where a liquid-crystal phase as well as two isotropic liquid phases are present. In this region very stable emulsions are formed. Higher surfactant concentrations still give dispersions that have the appearance of emulsions, although only one of the two phases present is liquid, the other being liquid crystal. From these results and general phase-rule studies on surfactant systems (65) it seems likely that many emulsions used in practice are not simple liquid dispersions.

Emulsions often contain more than three components. Normally the emulsifying system contains, in addition to the surfactant, a long-chain alcohol or other stabilizer (e.g., Tween-Span mixtures). Such phase-rule studies as have been made show that the stabilizers tend to increase liquid-crystal formation is surfactant systems, so that the liquid-phase regions are even further reduced. The ratio of surfactant to stabilizer also exerts interesting effects on the distribution of the surfactant between the phases. In the short-chain nonionic surfactant system already mentioned (87) the surfactant is almost completely removed from the aqueous phase into the organic phase when sufficient octanol is added to the dodecane and there is an intermediate three-liquid-phase region where multiple-drop dispersions readily form (85). The phase relations therefore seem to be of considerable import in a number of emulsion systems and may help to explain many of their physical properties. The distribution of the active ingredient between the phases in medicinal emulsions, which is probably affected by micelle formation, must also influence such properties as drug release, so that investigations of this aspect of emulsions should lead to interesting new results and principles.

D. Drug Release from Emulsions

It has long been realized that the composition and properties of emulsions have an effect on the release of incorporated drugs and that the solubility of the drug itself in the oil and water phases or its interaction with the emulsion constituents also has significance. Many biologically active molecules are organic bases that form water-soluble salts with inorganic or organic acids, whereas the base is usually insoluble. The distribution of the medicament as base in the oil and water phases of an emulsion is thus likely to be very different from its distribution as a salt. The emulsion type (oil-in-water or water-in-oil), viscosity, and transport across internal or external interfaces when surfactants are present would also be expected to have an influence on the passage of drug molecules from the dispersion. Examples, with qualitative descriptions of such effects, in other sections of this chapter include the differences in release of various salts of acridines from oil-in-water and water-in-oil emulsions, the effect of viscosity on the

6. MEDICINAL EMULSIONS   321

activity of antigenic materials dispersed in emulsions, prolongation of the hypnotic effect of barbiturates in injected emulsions compared with the sodium salts of the drugs in aqueous solution, and many others. Penetration of materials into emulsions, especially where decomposition of the active ingredient is involved, may also be a consideration.

Quantitative physical description of drug release from various semisolid dose forms including emulsions has received some attention, especially in regard to preparations applied externally to the skin, where the drug cannot act until it is released and absorbed through the skin barrier. In many systems transport across the skin barrier, rather than release from the base, is the rate-controlling process for the drug's reaching the site at which it acts. In such cases the rate of release from the base is not important, but where the skin is damaged or where there are moist lesions, release of drug from the base may be rate limiting (89).

In a general paper on the formulation of dermatological preparations Wurster discusses (89) a theoretical equation derived by W. I. Higuchi relating the "effective diffusion coefficient" of a drug in emulsion bases where the volume fraction is low (~0.1) to the diffusion coefficients in the internal and external phases, their volume fractions, and the partition coefficient. For most drugs there is little difference between the diffusion coefficients in either phase, and therefore drug release depends mainly on the partition coefficient. The effect of the internal phase will be very marked where distribution is in its favor. This equation agrees with practical experience.

Koisumi and Higuchi (90) studied the release of an organic base from emulsions when the partition of the base between the phases is changed by altering the pH of the aqueous phase. They investigated the release of pyridine from water-in-oil emulsions in vitro, using hexadecane as the oil phase and sorbitan sesquioleate, HLB number 3.7 (Arlacel C), as the emulgent, with varying concentrations of acid in the aqueous phase. Under these conditions increasing acidity reduces the amount of pyridine released because ionization of the pyridine inhibits partition into the continuous oily phase. These workers found experimentally that the amount released (as measured through a cellophane membrane) was linearly related to the square root of the time allowed for diffusion to take place, showing that, even if the diffusion coefficient was concentration dependent, the square-root relationship held, as for previously studied systems where the diffusion coefficient was independent of concentration. They were able to obtain good correlation with a theory of release based on the concept of an "effective" permeability constant for the hetero-geneous emulsion system which is related to the permeability constants of pyridine in hexadecane and water, and the volume fractions of the phases.

In a study of the influence of the HLB numbers of surfactants on ephedrine release from emulsified systems, Waggoner and Fincher (91) measured release from both water and oil alone as controls and from emulsified systems containing nonionic surfactants of varying HLB numbers by a dialysis method. The HLB number of the surfactant affected the release rates, the highest occurring in the upper HLB range. The surfactant did not impair the release of ephedrine from the oil control system, but an inhibitory effect was noted with the water control system. The release from emulsions indicated that an interfacial barrier was present, and it was concluded that release of the drug from the oil phase was the rate-limiting step in its release from the emulsion.

IV. PRODUCT REVIEW

A. Emulsions for Internal Use

Emulsions used internally need to be prepared with nontoxic emulgents, and the naturally occurring gums and thickening agents have long been the most important materials used in pharmacy for this purpose. Mineral oil emulsions, where a liquid hydrocarbon is

6. MEDICINAL EMULSIONS

the disperse phase of an oil-in-water emulsion, are used as a treatment for constipation. The beneficial properties of the emulsion result from the fact that saturated hydrocarbons in the oil phase modify the consistency of the stool because they are not absorbed, thus making defecation easier. Prolonged use of such products may be undesirable for two reasons. First, the emulsions may reduce the absorption of fat-soluble vitamins, probably by their being partitioned into the nonabsorbable mineral oil. Second, there have been reports that very small mineral oil particles *can* be absorbed into the bloodstream (92,93) and from there deposited in various tissues, with adverse effects. For example, if sufficient particles are trapped in the fine capillaries of the lungs, symptoms resembling pneumonia occur, and there is also a risk that deposition of mineral oil in any tissue might induce the formation of cancerous cells. Emulsions produced mechanically with efficient homogenizers are particularly likely to have fine particles that can pass through the intestinal wall, and therefore the use of such equipment in the preparation of mineral oil emulsions needs careful consideration.

Many of the official formulas for mineral oil emulsions are designed for extemporaneous preparation of small quantities by hand. Gum acacia is usually the emulgent, and in the British Pharmacopoeia the "dry gum" method of preparation is described. Creaming is reduced by the incorporation of gum tragacanth into the formula to increase the viscosity of the disperse phase. The British Pharmacopoeia also describes an emulsion with methylcellulose 20 as the dispersing agent. This probably produces a coarser emulsion, which should reduce the absorption of mineral oil. Similar emulsions (e.g., Mineral Oil Emulsion, U.S.N.F.) are described in U.S. compendia. Agar is used as the dispersing agent in some proprietory preparations. Mineral oil emulsions with other medicaments, such as phenolphthalein, which is a purgative acting on the bowel by its irritant properties, and magnesium hydroxide, which has purgative properties owing to its ability to induce water retention in the gastrointestinal tract, are also commonly used. It is possible to disperse mineral oil with hydrated magnesium hydroxide, without other emulsifying agents, the hydroxide acting as a solid emulgent. The properties of various formulas of this type have been reported (94). Castor oil, a purgative acting by its irritant qualities, has also been given in emulsified form.

A number of other oils are given internally. For these gum acacia and gum tragacanth are commonly used as the emulsifying and stabilizing agents, respectively. Examples include fish oils that contain vitamins (e.g., Cod-Liver Oil Emulsion, B.P., 1959). A formula for a halibut-liver-oil emulsion has been particularly recommended (95).

The hypnotic paraldehyde, which is usually given as a coarse dispersion in water with gum tragacanth as the dispersing agent, has been dissolved in oil, and the solution emulsified with Tween and Span (paraldehyde 40 ml, mineral oil 44 ml, Tween 60 5 g, Span 80 5 g, saccharin sodium 300 mg, peppermint water to 120 ml) (96) has been administered in this form. It is claimed that it is more pleasant to take because it has no aftertaste.

Emulsified oils intended as liquid food replacements or supplements are used orally where ordinary solid food cannot be readily ingested by the patient, either from general weakness or because the mouth (especially the teeth) or esophagus is damaged or has been subjected to surgery. Such emulsions need not be sterile, but the emulsifying agents should be suitable for oral use. A product containing arachis oil (dextrose 25 g, arachis oil 50 ml, Span 60 1.1 g, Tween 60 1.1 g, methyl p-hydroxybenzoate 100 mg, propyl p-hydroxybenzoate 100 mg, lime oil 0.13 ml, citric acid 750 mg, green or yellow coloring and water to 100 ml), which is claimed to be more stable and palatable, and to have a higher caloric content (510 calories in 100 ml) than earlier formulas, has been described by Richards and Whittet (97). Similar coconut-oil (98) and cottonseed-oil emulsions have also been devised.

Silicone emulsions of liquid silicones or liquid silicones containing finely powdered silica (4 to 7% w/w), which have antifoaming properties, have been used for the treatment of flatulence

6. MEDICINAL EMULSIONS   325

and meteorism; for the elimination of gas, air, or foam before
radiography; and for the relief of abdominal distension and
dyspepsia.

A number of flavoring adjuvants are kept in concentrated form
as emulsions and used as required in medical preparations, thus
avoiding certain storage and measuring problems. These products
are frequently relatively unstable because the emulsion is only
required to be sufficiently stable for an aliquot to be obtained.
Examples of this type of product are Peppermint Emulsion, B.P. and
Anise Emulsion, B.P., 1954. The emulsifying agent is often an extract of quillaia bark containing saponins, which are highly
surface-active. Chloroform Emulsion, B.P. is a similar product,
but is also used for its preservative as well as its flavoring
properties.

B. Emulsions for External Use

A very wide range of antibacterial agents have been formulated
as creams for external application and used for the treatment of
small wounds and other minor injuries where infection would delay
healing. For more extensive wounds dressings impregnated with
solutions of the drug are more suitable.

Proflavine and other acridines, and propamidines are all
potent antiseptics effective against Gram-positive and Gram-negative bacteria, but not against *Proteus vulgaris, Pseudomonas
aeruginosa*, and some strains of *Escherichia coli* in the strengths
usually employed. Having high-molecular-weight cations, these
materials are inactivated by soaps and other anionic emulsifying
agents, and it has therefore been stated that creams should be prepared with nonionic surface-active agents (99). Older formulas
(e.g., Proflavine Cream, B.P.) produced water-in-oil dispersions
because yellow beeswax and wool fat were used as the emulgents.
Such preparations have been shown on the basis of invitro tests to
have little activity because the acridines are dissolved in the

aqueous phase and not released through the oil phase. Other salts (e.g., n-valerates) that are soluble in both oil and water have been found to be the most active.

Aminacrine hexylresorcinate in the form of 0.2% water-washable oil-in-water cream has been claimed to be an antimycotic agent that is almost specific against *Malassezia furfur*, although less active against other fungi (100). Aminacrine is also used as the hydrochloride and is sometimes preferred to proflavine, as it is nonstaining. As with other acridines, the base is precipitated at pH > 8. Acriflex is a widely used oil-in-water cream containing 0.1% of aminacrine hydrochloride.

Propamidines and related derivatives also have the advantage of being nonstaining. They are used for the treatment of minor burns and skin abrasions. The derivative dibromopropamidine isethionate is used similarly and is also active against infections due to penicillin-resistant staphylococci and some strains of *Pseudomonas aeruginosa*. It has also been used in the treatment of impetigo (101). No official formulas have been devised, but an early "ethical" formula (i.e., Brulidine Cream) was of the castor oil-water emulsion type. This product has also been reported (102) to be effective in promoting the healing of leg ulcers. The report considered that, although better than other antiseptics, it was not as good as some traditional remedies.

Phenolic and chlorinated antiseptics are widely used in medicine for a variety of purposes. Phenol itself and other low-molecular-weight phenols are toxic when used internally or externally and are mostly applied as general purpose disinfectants for rooms, linen, instruments, etc., or used in preparations as preservatives rather than as active medicinal agents. Higher phenols, which are less toxic, have been used externally as antiseptics, and certain chlorinated compounds, such as chlorhexidine dihydrochloride, are also suitable for external use. Although presented in many forms, a number of emulsions or creams containing these materials have been used medicinally.

6. MEDICINOL EMULSIONS

In the case of phenolic antiseptics in emulsified or solution form, inactivation by nonionic surfactants is an important consideration. The effect of partitioning into the micelles, normal with any surfactant, is markedly in favor of the micellar "phase" in undersaturated systems, for example, with chloroxylenol (66), so that the antimicrobial activity, which depends on the concentration free in the aqueous "phase" is very much reduced. The significance of this in the use of phenolic preservatives for emulsions prepared with nonionic surface-active agents has already been discussed, but the same principle also applies to preparations in which a phenol is used as the medicament rather than simply a preservative.

For cheap general purpose disinfection the so-called Black and White Fluids (British Standard Specification, B.S. 2462, 1961) are often used. Black Fluids are solutions of coal-tar acids or similar acids derived from petroleum, or mixtures of these, with or without hydrocarbons, and contain emulsifying agents. For use they are diluted with water, when an emulsion is formed. There is a range of grades available, and tests for activity and requirements for stability of the emulsions are specified in the Standard. White Fluids are similar, but are in emulsion form even when undiluted.

Hexachlorophene is widely used for application to the hands for the reduction of bacterial flora and helps prevent cross-infection via hospital personnel. It has a persistent action due to accumulation in the skin and retains its activity in the presence of soap, unlike many other antiseptics. It is often used in the form of a medicated soap, although creams are also common; an example is pHisoHex (hexachlorophene 3% in an emulsion containing lanolin, white petrolatum, and an anionic sulfonated ether detergent). Several applications are necessary before the skin flora are substantially reduced (103). It has been reported that pHisoHex is effective in the treatment of acne (104). Hexachlorophenes in general, and pHisoHex in particular, have been given a favorable assessment in comparison with other antiseptic preparations or measures of skin disinfection (see, for example, Ref. 105). Reports of brain damage to animals following experiments with this agent have raised doubts about the safety of products containing it (106) and have resulted in various restrictions on its use.

Chlorohexidine is a potent topical antiseptic, cationic in nature, although not surface active like the other antibacterial cationic surfactants; it is effective in high dilutions against a wide range of Gram-positive and Gram-negative bacteria. It is often presented in cream form prepared with nonionic emulsifying agents [chlorohexidine gluconate solution (20%) 5 ml, mineral oil 10 g, cetomacrogol emulsifying wax 25 g, water to 100 g). Many of its salts are of low solubility (e.g., the sulfate has a solubility of 1 in 20,000), and it is therefore incompatible with solutions containing such ions. The gluconate, acetate, and hydrochloride are the most commonly used salts. Having a high molecular weight, the antiseptic is incompatible with soaps and other anionic emulsifying agents. Apart from its general use for minor skin abrations, it has been successful in obstetrics creams, as it is claimed to be entirely bland to the vaginal mucosa and very effective, in comparison with other antiseptics, against *Pseudomonas aeruginosa*, *Escherichia coli*, *Staphylococcus pyogenes*, and group α,β hemolytic streptococci (107).

Another use of such creams occurs in the treatment of certain skin conditions, such as recurrent boils, and also in nasal application (usually combined with neomycin) where nasal carriers of stephylococci often infect themselves or others postoperatively. Conflicting reports on the value of such treatments have appeared (108,109).

Preparations for the treatment of parasitic infections of the skin and hair include those containing benzyl benzoate, lindane, and DDT. The first is an oil that forms the disperse phase of an oil-in-water emulsion used to treat scabies. This infection can extend over considerable areas of skin, and a fluid preparation that spreads easily is most suitable. A low concentration (2%) of

6. MEDICINAL EMULSIONS

an anionic emulsifying wax, usually containing sodium lauryl sulfate, produces a good emulsion (e.g., Benzyl Benzoate Application, B.P.), although because of the low viscosity, precautions against foaming and the incorporation of air into the final product are necessary during its preparation. In Benzyl Benzoate Lotion, U.S.P. the emulgent, an anionic organic soap, is formed from triethanolamine and oleic acid. In the case of DDT preparations used against parasites in the hair, it is necessary to dissolve the active ingredient in a suitable solvent before producing the oil-in-water emulsion that is commonly used. This is more acceptable because it is eventually readily washed from the hair. The British Pharmacopoeia gives a formula containing DDT, xylene, and citronella oil, the emulgent being Emulsifying Wax, B.P. Lindane is more rapid in action than DDT and effective at lower concentrations, but it is more volatile and less persistent. Similar formulations of this agent have been devised and used as for DDT.

Antihistamines are quite widely used for local application to the skin where reactions of an allergic nature are involved, but many dermatologists think that local applications are inadvisable and that oral use is preferable because of the high incidence of skin sensitization. Ellis and Bundick (110) reported, however, that 25% of 200 dermatologists thought their use was of value in localized neurodermatitis and anogenital pruritus. The activity of the antihistamines, which are usually dissolved as salts in the aqueous phase of oil-in-water creams, may be affected by binding to nonionic emulgents (111). A variety of antihistamines have been presented in the form of creams, including antazoline hydrochloride, chlorcyclizine hydrochloride, diphenhydramine hydrochloride, and mepyramine maleate. Of these, antazoline has some local-anesthetic properties and is said to be less irritating to tissues than most other antihistamines. These preparations are also used for the relief of insect bites and sunburn.

Corticosteroids are commonly applied externally dispersed in semisolid preparations and often produce dramatic effects in skin diseases in which inflammation is a prominent feature: eczema, dermatitis of various kinds, psoriasis, and intertrigo. Both greasy ointment-type and cream-type (oil-in-water) bases have been used, each having its own characteristic advantages and drawbacks. Traditional ointment bases are emollient, occlusive (retarding loss of moisture from the skin), and have shown satisfactory release of corticosteroid in clinical tests. The occlusive character has often been considered a disadvantage, but it is now thought that retention of water in skin is an important factor in maintaining its condition. Constituents of natural origin that have an emollient action, although beneficial, may not be as important a factor as was once thought in skin care. This aspect of the properties of bases may be particularly important in the type of skin conditions treated by corticosteroids. The disadvantages of fatty bases are that they do not mix with exudate and are thus unsuitable for moist lesions and that, being greasy and sticky, they do not spread well. This may cause the patient either to apply excess material, thus possibly creating side effects like collagen atrophy or adrenal suppression, or to find subjective reasons for rejecting the treatment altogether. Cream bases are more suitable for moist lesions, being miscible with water, and since they are less sticky, they also spread more readily. Release of steroid can be capricious, possibly owing to partition into the oil phase or to an effect of the surfactant; another disadvantage is the need for a preservative. This is undesirable because the disease conditions in which steroid therapy is indicated increase the possibility of sensitization effects.

For external use steroids with pronounced antiinflammatory, rather than mineralocorticoid, action are required. Hydrocortisone is an early example of this type, but more active compounds, such as betamethasone, fluocinolone and triamcinolone, have since been developed.

Toxic effects, such as inhibition of pituitary and adrenal function in response to stress, can be produced if excessive

6. MEDICINAL EMULSIONS

amounts are used and large areas treated, especially if the technique applied involves the use of an occlusive polyethylene dressing. Covering the treated area with such a dressing has been found to increase steroid activity many times, possibly by modifying the transmission characteristics of the skin by increasing its water content. Reports of such effects have been published when as much as 45 g a day of triamcinolone cream (0.01 or 0.1%) has been used on patients with extensive psoriasis (112). Another danger with corticosteroid preparations is the possibility of suppressing the warning symptom of inflammation where the condition is complicated by the presence of infection. Combination preparations in which an antibiotic or other antibacterial agent is added to the corticosteroid are therefore common in this field.

Often a comparison between formulations is made and release of the active ingredients is assessed by the vasoconstricotr test. This is based on the blanching effect of corticosteroids when applied to the skin (113). A paper by Collard (114) discusses the problems of formulating steroid preparations.

Hydrocortisone as such and its acetate ester were the first steroids commonly used in external preparations. They were usually presented in cream form, and since they were of lower activity than more recently developed compounds, 1% of the active agent was normally used. Nonionic surfactant creams have been widely used as vehicles (Hydrocortisone Cream, B.P.), and despite the reported interaction between phenolic preservatives and nonionic surfactants, chlorocresol has been used as the preservative.

Of the more active steroids, betamethasone as the valerate or sodium phosphate derivative (0.1%) has been applied in cream form in similar bases to those just described. Antifungal or antibacterial agents (e.g., clioquinnol or neomycin) have been added for use in dermatoses complicated by infections. Fluocinolone acetonide is used in similar cream bases at even lower concentrations (0.025%), also with or without the same antiinfective agents Triamcinolone acetonide is another steroid used especially in external preparations, although the official B.P. formula has an anionic-surfactant cream base. There are many reports on the effectiveness of all these steroids in a variety of skin conditions, although there is also a considerable proportion of relapses when treatment is discontinued in some skin diseases.

Recently new formulas have been developed in attempts to overcome the problems associated with such potent drugs as the corticosteroids, in particular the difficulty of ensuring that the low concentrations used (often less than 0.1%) are evenly dispersed in the vehicle and adequately in contact with the skin so that absorption can readily take place. This is thought to be more satisfactorily accomplished by dissolving the steroid in a suitable solvent and dispersing the solution in the base, rather than by attempting to disperse a microfine powder, with its attendant difficulties of reproducing the grinding processes necessary. In one formulation (115) the steroid is dissolved in the minimum amount of propylene glycol (approximately 5% of the total amount of the final product), and the solution is dispersed as the internal phase of an anhydrous organic liquid-in-oil emulsion in molten petrolatum with wool fat as the emulsifying agent. Comparison of this formulation with steroid dispersed in five other ointment or cream bases of the British Pharmacopoeia, with or without occlusive wrapping, by clinical test or the vasoconstrictor assay already mentioned, has shown it to be superior (116,117). This improved activity has been ascribed to the wetting of the keratin layer of the skin by the propylene glycol solution, thus carrying the steroid into the dermis in concentrated form. The fact that the drug is already in solution may also explain the superiority over solid dispersions where the drug must presumably dissolve before being absorbed.

Another technique in which the steroid is also dissolved in propylene glycol, and which has other advantages, has recently been described (118). Long-chain saturated fatty alcohols such as stearyl alcohol, glycerin, and polyethylene glycol are mixed hot with the steroid solution in propylene glycol, and the mixture is

6. MEDICINAL EMULSIONS

"snap-cooled" under high shear. The product is a soft white semisolid in which the liquid phase is held in the spaces between a "crystalline network of hydrogen-bonded molecular chains;" the whole may possibly be described as a macrogel. Presumably the gel network consists mainly of the fatty alcohol, and although it has been described as crystalline, it could possibly be liquid crystal. After clinical assessment of the product it has been claimed that fluocinolone acetonide is released and absorbed from it better than from traditional emulsified bases (119). It has also been claimed that in addition to the usual cosmetic advantages of a cream (misible with exudate and easily washed off), it is resistant to bacterial and mold growth; hence no preservative is necessary, and there is no consequent liability to sensitization by such. One possible disadvantage is that it softens above 35°C, and some propylene glycol may separate at this temperature. It melts at 52°C and in cooling forms a granular cream.

Barrier creams and other skin protectives have become quite widely used against irritants due to incontinence, colostomy discharges, or in the prevention and treatment of bed sores, napkin rash, and other skin conditions. Many formulas, particularly those meant to protect against water-soluble irritants, contain water-repellant silicone fluids often emulsified to give semisolid creams for ease of application and removal while still retaining their protective properties.

Early reports emphasized the advantages of silicone creams over traditional methods. For example, Bateman (120) described a regimen in which the cream was smeared on twice a day for 1 week and then once a day, with as little washing of the skin as possible. A controlled trial on various formulas reported by Roberts (121) showed that a content of at least 20% of dimethicone 200 was necessary for an effective product. Comparison with traditional skin care with soap, water, spirit, and powder showed, according to Roberts (122), that a silicone cream was more effective. But other reports comparing various regimens and preparations for the care of normal and diseased skin conclude that traditional soap and water is better than silicone preparations, including creams (123), for normal skins, although a spray of silicone in alcohol was thought to be particularly good for abnormal skin.

A wide range of formulas has been described. The official B.P. dimethicone cream uses a low-viscosity silicone (dimethicone 20) in relatively small proportion, but the oil phase also contains 40% of liquid petrolatum. Despite the fact that the emulgent is a mixture of a cationic surfactant and a long-chain fatty alcohol that also possesses antibacterial activity, a phenolic preservative is used. The use of other dimethicones of different viscosity has been shown to make no difference to the ease of preparation or to the appearance and viscosity of this cream (124). Another formulation containing a much higher proportion of dimethicone, but no petrolatum, and a dispersing agent of methylcellulose is claimed by the developers (125) to be unlikely to act as a sensitizer, although preservatives were necessary (benzoates or phenethyl alcohol). Proprietary preparations containing silicones appear to be all of the oil-in-water cream type. A review of barrier creams is given by Alexander (126).

A better degree of skin protection to the area round ileostomy, cecostomy, or colostomy, where it is necessary to prevent irritation due to proteolytic or other discharges, may be obtained with creams containing high proportions of aluminum powder and titanium dioxide (127).

C. Injectable Emulsions

Where extensive surgical operations are carried out on the gastrointestinal tract or in certain clinical conditions where the digestive processes are upset, patients are often unable to take food by mouth, and it is necessary to maintain them on material that can be introduced directly by the intravenous route (128).

6. MEDICINAL EMULSIONS

This is relatively simple in the case of carbohydrate constituents because sugars can be supplied in aqueous solution, but is not possible for fats, without which a sugar-based diet is unbalanced and inadequate for anything but short-term use. The potential value of products for intravenous use containing emulsified fats has therefore led to the development of satisfactory preparations, and patented proprietary products are now available and widely used.

Such information as has been published in the open literature and disclosed in patents suggests that the main problems in development were the provision of nontoxic emulsifying agents and the production of emulsions of small particle size in a process that involved a sterilization stage. Most emulsions are not very stable when subjected to autoclaving, and since it has been found that emulsions with particles larger than 7 to 8 µ (most should be much smaller) are unsuitable for intravenous use, as such particles may be trapped in the fine capillaries of the lungs causing pneumonia-like symptoms, processing techniques that avoid the production of coarse particles even with autoclaving have had to be developed.

According to the patent (129) on which the Upjohn Company's Lipomul appears to be based, a purified soya lecithin is a suitable emulsifying agent. Lipomul itself contains cottonseed oil 15%, dextrose 4%, a coemulsifier of the polyethylene oxide-polypropylene type (Pluronic F68) 0.3%, purified soya phosphatide fraction 1.2%, and water.

The purified phosphatide is prepared by dissolving Lecithin RG in about 10 times its volume of ethanol and stirring the solution with 2 parts by weight of aluminum oxide as adsorbent at room temperature for 30 min. The adsorbent is removed, and the purified soya phosphatide fraction is isolated by removing the alcohol under reduced pressure. It is tested before use for absence of a blood-pressure-depressant effect in cats, by a method similar to that published by Geyer and co-workers (130). According to claims made in the patent (129), emulsions prepared with the purified emulsifier produce significantly fewer undesirable reactions than earlier intravenous fat preparations, which produced reactions in as many as 30% of patients. Emulsions containing higher proportions of oil tend to become too viscous for intravenous use, and it is also stated that particles should be smaller than 2.5 μ in diameter and preponderantly below 0.5 μ.

Further descriptions of other emulsifier systems, purification methods for the individual components, and aspects of the formulation and assessment of intravenous fat emulsions that seem to be satisfactory in dogs have been published by Singleton and coworkers (131). The primary emulsifier was polyethylene glycol monopalmitate, and procedures designed to remove diester and unreacted glycol from laboratory and commercial products (e.g., Lipal 4P) were described. Two secondary emulsifiers were investigated, the tartaric acid ester of monoglycerides of cottonseed-oil fatty acids (Drewmulse 5998A), which was freed from unreacted tartaric acid, and Pluronic F68, which was used as received. Variations in Lipal 4P, Drewmulse, and Pluronic concentrations were studied for 15% cottonseed-oil emulsions with respect to the stability of the emulsions and the particle size of oil droplets. The best formulation with the minimum concentration of emulsifiers was (in parts by weight) cottonseed oil 15, Lipal 4P (filtrate fraction) 1.2, Drewmulse 5998A 0.3, Pluronic F68 0.3, and isotonic glucose solution 83.2. Preparation procedures involving homogenization, sterilization, and subsequent methods of cooling the product, which considerably affect the quality of the product (size of particles), are described. The best method of cooling after sterilization was found to be mechanical rolling of the container in a cooler at 4°C.

In clinical practice where large volumes of intravenous fluids are being administered it is convenient to add any other drugs being given to the one fluid. Lynn (132) studied the stability of disodium carbenicillin (15 g/500 ml) and sodium cloxacillin (5 g/500 ml) in Intralipid. No loss of potency of the semisynthetic penicillins was found after 24 hr at 23°C, but some aggregation of emulsion particles was observed immediately after

6. MEDICINAL EMULSIONS

mixing the emulsion with the former, although no further changes were noted after 6 and 24 hr. Cloxacillin produced similar aggregation effects, but also some coalescence after 24 hr. Sodium ampicillin (10 g/500 ml) did not appear to change the microscopic appearance of the emulsion. The last two drugs maintained their potency for 6 hr. Lynn concluded that in spite of these results the manufacturer's advice that drugs should not be added to the product should probably be respected.

Reviews of emulsifier systems of possible application for intravenous emulsions have been published by Geyer (133) and by Lambert and co-workers (134).

Vitamin K in oil-in-water emulsion form is sometimes given by intramuscular or intravenous injection when an antidote to anticoagulant therapy is necessary. The injection usually contains 10 mg of the vitamin per milliliter, and it is important to inject it very slowly and to inspect the product carefully for separated droplets before use. Monographs are given in the British and U.S. pharmacopoeias (phytomenadione injection and sterile phytonadione emulsion) for products, but no details of the emulgents to be used are described.

Ljimberg and Jeppson (135) have described some experiments on the release of oil-soluble drugs dissolved in soya-bean oil, after emulsifying the product and injecting it intravenously into mice as test subjects. They found that the hypnotic effect of emulsified phenobarbital and thiopental was prolonged in comparison to that of aqueous solutions of the sodium salts of these drugs. No noticeable side effects were produced by the emulsified products.

Immunological preparations are commonly presented as simple aqueous solutions or sometimes in forms in which the antigenic material, which is usually of fairly high molecular weight, is adsorbed onto freshly precipitated inorganic material such as aluminum hydroxide. Many preparations in the solution form are of low or negligible activity if given as a single dose. Freund and Bonanto (136) demonstrated that enhanced and sustained antibody titers could be obtained in animals with water-soluble antigens like diphtheria toxoid if the aqueous antigen solution was emulsified with mineral oil to give water-in-oil dispersions. Since the early reports some dozens of papers have been published describing experiments with a variety of antigenic materials in emulsified form and assessment of their biological effectiveness in man and animals. It seems that these preparations may have advantages over other forms and will thus become of some medical significance.

An example is a fairly recent study by Hobson et al. (137) on a large number of adult volunteers which demonstrated the value of emulsified Asian-influenza vaccine. Hobson et al. studied a vaccine prepared from formalin-inactivated A/Singapore/1/5 virus blended with an equal volume of Drakeol 6R containing 10% Arlacel A. Although this preparation was not directly compared with conventional vaccine saline solution, the investigators found that the emulsified preparation induced higher peak antibody titers in a larger proportion of subjects than would be expected for saline vaccine and that the persistence of circulating hemaglutination-inhibiting antibody was considerably prolonged even at a virus-antigen dose of only one-tenth that customarily used in the form of saline vaccine. Titers 1 month after vaccination were as high as or higher than those after 3 months, which suggested that the combination of vaccine with oil had not impeded the rate of antigen release so that early immunization was not adversely affected. This is an important property required of any vaccine, and emulsified formulations that do not give a rapid buildup of antibody would have a serious disadvantage compared to preparations with more immediate effects, even if the emulsified preparations had other desirable properties. In addition to the biological properties of emulsified vaccines, their potential would obviously be affected if they did not retain their potency on storage. Hobson and co-workers found that the potency of the Asian-influenza vaccine was not significantly affected by 1 year's storage at 4°C.

We shall review some of the literature reports on vaccines in the emulsified form, paying particular attention to such links between the formula and physical properties of the preparations and their medical and pharmaceutical characteristics as have been suggested as being significant. Many references to the literature appear in the review by Lazarus and Lachman (138) and in the paper by Berlin (139), which gives 24 references to work before 1960.

The earlier workers (136,140) were able to demonstrate that mineral-oil emulsions of such antigenic materials as horse serum and diphtheria toxoid gave enhanced and sustained antibody titers in animals in comparison with more conventional preparations. An additional synergistic effect was found if killed tubercle bacillae were added to the aqueous phase of the emulsions. The oil phase was usually mineral oil, and the emulgent was Falba. The latter is of indefinite chemical composition, but consists of a mixture of beeswax, mineral oil, and oxycholesterols presumably derived from lanolin.

Although excellent antibody responses were obtained with preparations using Falba as emulgent, occasionally nodules were found in the subcutaneous tissues at the site of injection, and in some cases abcesses also formed. This was thought to be partly due to the nature of the emulsifying agent, and a number of others have been investigated, of which Arlacel A appears to have been found the most satisfactory (141). This emulgent has been used not only because it gives suitable emulsions but also because it appears to enhance lymphatic proliferation at the site of injection, thus increasing the antibody response. There have been reports that certain batches of this emulgent are toxic, and it is claimed that a purer form, now available as "Specially Treated Arlacel A," is free from these effects (138). Influenza vaccines prepared with this surfactant normally have an oil phase of Drakeol 6 (9 parts) and Arlacel (1 part) in an emulsion containing viral constituents in the aqueous phase, the oil-water phase-volume ratio being 1:1.

Berlin (139) has published extensive results that attempt to link the physical properties, especially viscosity, of emulsified influenza vaccines and the antigenic response. He studied changes

in emulsion viscosity produced by alterations in the viscosity of
the oil phase of the emulsion. Eighteen petroleum-oil fractions of
varying viscosity were studied, and changes in their antigenic
effectiveness were measured in mice. In general the less viscous
emulsions were found to give higher antibody responses. The effect
of altering emulsion viscosity by adding 12-hydroxystearin (Thixcin)
to the oil phase was also studied. Again the less viscous emulsions
were more effective. The adjuvant effect was completely abolished
with the highest concentration of Thixcin (400 mg/ml of oil).
Similar results were obtained in experiments in which the viscosity
of the emulsions was altered by changing the proportion of aqueous
phase or by varying the amount of mechanical agitation used to
prepare it. Berlin stated that earlier unpublished observations
lead him to believe that poorly stable emulsions do not give an
adjuvant effect and that very stable emulsions might also be less
effective. This suggests that those of intermediate stability might
be the best formulation to aim at. Berlin (139) reported results
showing that emulsions of intermediate stability (produced by varying the proportion of aqueous phase) gave geometric mean hemoglutination-inhibition titers double the mean for those of low and
high stability.

Berlin (139) considered two possible modes of action to explain
the increased effectiveness of antigenic material in water-in-oil
emulsion form. In the first the mineral oil of the emulsion is
considered to function passively by attracting cells that form a
local antibody-producing organelle surrounding the emulsion. An
alternative mechanism considers the emulsified antigen as a depot
that retards destruction and elimination of the antigen by body
fluids, but gradually releases antigen for transport to antibody-forming tissues. Berlin suggested that the second mechanism is
consistent with the fact that the effectiveness of the emulsion is
not related to the volume so long as a constant amount of antigen
is present. This is supported by the observation that no difference was found between antibody responses in mice in which the

6. MEDICINAL EMULSIONS 341

emulsion is found in a few large deposits and mice in which the
emulsion is found in numerous small droplets near the site of inoculation. Berlin argued that these observations indicate that the
surface area of emulsion retained at the site of inoculation and
presumably the surface area of antibody-forming organelle does not
influence the amount of antibody produced, and this favors the
second mode of action. He also considered that his results show
that there is an optimum rate at which antigen is released. If
the emulsion is very unstable, the antigen is released so quickly
that little response is observed, as for antigen in saline solutions or in the form of oil-in-water emulsions. If the emulsion is
too stable, the antigen may be released too slowly to produce an
optimum response.

Somewhat similar results have been found in comprehensive investigations on the nature of the oil phase by Wilner and coworkers (142). In order to define the chemical nature of the oil
phase clearly, they investigated a range of pure hydrocarbons.
Straight-chain compounds with carbon chains of 13 to 18 units were
unsatisfactory. The lower members were toxic; the higher were
solids at room temperature. Branched-chain alkanes with 16 to 30
carbon atoms were found to be suitable. As in Berlin's experiments, Wilner et al. showed that activity is inversely proportional
to emulsion viscosity, and it was their lower-molecular-weight oils
that produced less viscous emulsions. However, the lower-molecular-weight oils gave side reactions in which marked neutrophilic or
granulomatous infiltration occurred at the injection site. Branch-chain hydrocarbon oils with 24-carbon-atom skeletons or Drakeol 6VR
were less reactive in this respect.

Several reports have shown that low-viscosity emulsions are
more effective and that these will be produced by low-viscosity
oils and an aqueous-phase volume that is not too high. Such emulsions have other practical advantages in that they are easier to
inject in conventional hypodermic syringes. Lazarus and Lachman
(138) stated that the viscosity should be not more than 10,000 cP

and preferably lower if a continuous filling syringe is to be used in, for example, the mass injection of chickens or turkeys with fowl-cholera vaccines in emulsified form.

A formulation that produces a free-flowing emulsion and diffuse depots on injection has been described by Herbert (143). He made a water-in-oil emulsion with an aqueous phase containing ovalbumin in physiological saline and an equal volume of oil phase (Drakeol 6VR) emulsified with Arlacel A. This product was then used as the oil phase in an oil-in-water emulsion produced with an aqueous phase of 2% Tween 80 in saline. The emulsion was formed by shaking and a short burst of ultrasound. The product was a multiple emulsion in which the antigen solution formed a secondary aqueous disperse phase in a primary disperse oil phase of Drakeol with a continuous phase of the saline solution containing the Tween. When ovalbumin in the multiple emulsion was compared with ovalbumin in a water-in-oil emulsion for relative effectiveness in producing antibody response in mice, it was found to give an earlier and increased response. The multiple emulsion was stable when stored for 5 months at 56°C and a further 7 months at room temperature.

There is a patented formula (144) that has been claimed to be effective for many types of vaccine (145). The oil phase is peanut oil containing 4% aluminum monostearate, the emulgent is Arlacel A (5%), and the oil-water phase-volume ratio is 1:1. Although this incorporates a number of features that appear to be desirable from the literature, Fox and Wittner (146) have found that streptococcal antigens in water-in-oil emulsions were less stable than vaccines prepared by adsorption onto freshly precipitated aluminum hydroxide. This indicates that the same adjuvant may not be satisfactory for all vaccines and that each system needs independent study for effectiveness and stability.

Although it is too early to say what position emulsified antigenic preparations will eventually hold, it can be safely stated that single-dose emulsions have proved effective in allergic conditions and that immunization has been successfully established in man and animals with a variety of biological materials.

REFERENCES

1. Anon., *Chemist Druggist*, *132*, 475 (1940).
2. *The New Dispensatory*, 3rd ed., J. Nourse, Bookseller in Ordinary to His Majesty, London, 1770.
3. R. H. Engel and S. J. Riggi, *J. Pharm. Sci.*, *58*, 1372 (1969).
4. R. G. Todd, ed., *Extra Pharmacopoeia, Martindale*, 25th ed., Pharmaceutical Press, London, 1967.
5. A. F. Cascorbi, F. G. Rudo and G. G. Lu, *J. Pharm. Sci.*, *52*, 803 (1963).
6. D. L. Wedderburn, in *Advances in Pharmaceutical Sciences* (H. S. Bean, A. H. Beckett, and J. E. Carless, eds.), Vol. 1, Academic Press, London, 1964, pp. 195-264.
7. British Standard method for "Laboratory Evaluation of Disinfectant Activity of Quaternary Ammonium Compounds by Suspension Test Procedure," *B.S. 3286*, 1960.
8. N. D. Weiner, F. Hart, and G. Zografi, *J. Pharm. Pharmacol.*, *17*, 350 (1965).
9. H. H. Laycock and B. A. Mulley, *J. Pharm. Pharmacol.*, *22*, 157S (1970).
10. J. M. Arena, *J. Amer. Med. Assoc.*, *190*, 56 (1964).
11. D. L. Wedderburn, *Brit. J. Ind. Med.*, *17*, 125 (1960).
12. D. Sharvill, *Lancet*, *1*, 771 (1965).
13. R. A. Cutler and H. P. Drobeck, in *Cationic Surfactants* (E. Jungermann, ed.), Dekker, New York, 1970, p. 527.
14. R. B. Coles, W. T. Simpson, and D. S. Wilkinson, *Brit. Med. J.*, *2*, 688 (1964).
15. E. Eagle and C. E. Poling, *J. Food Sci.*, *21*, 348 (1956).
16. S. S. Waldstein, H. M. Schoolman, and H. Popper, *Amer. J. Digest. Dis.*, *21*, 181 (1954).
17. E. Chusid and J. Diamind, *J. Pediat.*, *46*, 222 (1955).
18. W. R. Waddell, R. P. Geyer, F. R. Olsen, and F. J. Stave, *Metab. Clin. Exptl.*, *6*, 815 (1957).
19. C. E. Mayer, J. A. Francher, P. E. Schurr, and H. D. Webster, *Metab. Clin. Exptl.*, *6*, 591 (1957).

20. E. A. Brown, *Rev. Allergy Appl. Immunol.*, *15*, 325 (1961).
21. P. J. Culver, C. S. Wilcox, C. M. Jones, and R. S. Rose, *J. Pharmacol. Exptl. Therap.*, *103*, 377 (1951).
22. P. H. Elworthy and Joseph F. Treon, in *Nonionic Surfactants* (M. J. Schick, ed.), Dekker, New York, 1967, pp. 923-970.
23. K. Setala, L. Merenmies, L. Stjernvall, Y. Aho, and P. Kajanne, *J. Natl. Cancer Inst.*, *23*, 969 (1959).
24. N. A. Allawala and S. Riegelman, *J. Amer. Pharm. Assoc., Sci. Ed.*, *42*, 267 (1953).
25. C. A. Johnson and J. A. Thomas, *Pharm. J.*, *175*, 51 (1955).
26. J. W. Hadgraft, *J. Pharm. Pharmacol.*, *6*, 816 (1954).
27. H. Hopkins and L. D. Small, *J. Amer. Pharm. Assoc., Sci. Ed.*, *49*, 220 (1960).
28. C. K. Bahal and H. B. Kostenbauder, *J Pharm Sci.*, *53*, 1027 (1964).
29. Anon., *WHO Tech. Rept. Ser.*, No. 281 (1964).
30. R. E. M. Davies and J. M. Rowson, *J. Pharm. Pharmacol.*, *12*, 154 (1960).
31. B. W. Fitzgerald and D. M. Skauen, *J. Amer. Pharm. Assoc., Sci. Ed.*, *44*, 358 (1955)
32. R. A. Anderson and M. J. Woollard, *Aust. J. Pharm.*, *43*, 213 (1962).
33. P. C. Eisman, J. Cooper, and D. Jaconia, *J. Amer. Pharm. Assoc., Sci. Ed.*, *46*, 144 (1957).
34. G. Richardson and R. Woodford, *Pharm. J.*, *192*, 527 (1964).
35. E. Shotton and K. Wibberly, *J. Pharm. Pharmacol.*, *11*, 120T (1959).
36. C. Gunn and S. J. Carter, *Cooper and Gunn's Dispensing for Pharmaceutical Students*, 11th ed., Pitman, London, 1965.
37. F. D. Pisano and H. B. Kostenbauder, *J. Amer. Pharm. Assoc., Sci. Ed.*, *48*, 310 (1959).
38. H. Sokol, *Drug Std.*, *20*, 89 (1952).
39. J. H. S. Foster, *Pharm. J.*, *192*, 429 and 461 (1964).
40. C. Mathews, J. Davidson, E. Bauer, J. L. Morrison, and A. P. Richardson, *J. Amer. Pharm. Assoc., Sci. Ed.*, *45*, 260 (1956).

41. P. W. Nathan and T. A. Spears, Nature (London), 192, 658 (1961).
42. M. Barr and L. F. Tice, J. Amer. Pharm. Assoc., Sci. Ed., 46, 445 (1957).
43. J. V. Klauder, J. Soc. Cosmetic Chemists, 11, 249 (1960).
44. L. E. Fryklöf, J. Pharm. Pharmacol., 10, 719 (1958).
45. G. A. Birchall and R. I. Felix, J. Pharm. Pharmacol., 12, 186T (1960).
46. J. E. Carless and J. R. Nixon, J. Pharm. Pharmacol., 12, 348 (1960).
47. J. Swarbrick and C. T. Rhodes, J. Pharm. Sci., 54, 903 (1965).
48. A. G. Mitchell and L. S. C. Wan, J Pharm Sci, 54, 699 (1965).
49. Anon., Sixth Report of FAO/WHO Expert Committee on Food Additives, WHO Tech. Rept. Ser., No. 228 (1962).
50. Anon., Food Standards Committee Report on the Review of the Antioxidant in Food Regulations, 1958, H. M. Stationery Office, London, 1963.
51. L. O. Kallings, F. Ernerfeldt, and L. Silverstople, Microbiological Contamination of Medicinal Preparations, Report to the National Board of Health, Sweden, 1965.
52. The U.S. Pharmacopoeia, 18th ed., Washington, D.C., 1970.
53. F. Atkins, Mfg. Chemist, 21, 51 (1950).
54. E. R. Garrett and O. R. Woods, J. Amer. Pharm. Assoc., Sci. Ed., 42, 736 (1953).
55. G. A. Nowak, Soap, Perfumery, Cosmetics, 36, 914 (1963).
56. A. E. Alexander and A. R. Trim, Proc. Roy. Soc. (London), B133, 220 (1946).
57. H. S. Bean and H. Berry, J. Pharm. (London), 3, 639 (1951).
58. O. Rahn and J. E. Conn, Ind. Eng. Chem., 36, 185 (1944).
59. W. J. Husa and J. M. Radin, J. Amer. Pharm. Assoc., 21, 861 (1932).
60. H. W. Hibbott and J. Monks, J. Soc. Cosmetic Chemists, 12, 2 (1961).
61. H. S. Bean, J. P. Richards, and J. Thomas, Boll. Chim. Farm., 101, 339 (1962).

62. H. S. Bean, G. H. Konning, and S. A. Malcolm, *J. Pharm. Pharmacol.*, *21*, 173S (1969).
63. E. R. Garrett, *J. Pharm. Pharmacol.*, *18*, 589 (1966).
64. N. K. Patel and H. B. Kostenbauder, *J. Amer. Pharm. Assoc., Sci. Ed.*, *47*, 289 (1958).
65. B. A. Mulley, in *Advances in Pharmaceutical Sciences* (H. S. Bean, A. H. Beckett, and J. E. Carless, eds.), Vol. 1, Academic Press, London, 1964, pp. 87-194.
66. A. G. Mitchell and K. F. Brown, *J. Pharm. Pharmacol.*, *18*, 115 (1966).
67. M. Donbrow, E. Azaz, and R. Hamburger, *J. Pharm. Sci.*, *69*, 1427 (1970).
68. B. A. Mulley and A. J. Winfield, *J. Chem. Soc.*, 1459 (1970).
69. N. G. Patel and N. E. Foss, *J. Pharm. Sci.*, *54*, 1495 (1965).
70. K. J. Humphreys and C. T. Rhodes, *J. Pharm. Sci.*, *57*, 79 (1968).
71. M. Donbrow and C. T. Rhodes, *J. Chem. Soc.*, 6166 (1964).
72. S. M. Blaug and S. S. Ahsan, *J. Pharm. Sci.*, *50*, 138 (1961).
73. F. W. Goodhart and A. N. Martin, *J. Pharm. Sci.*, *51*, 50 (1962).
74. S. M. Blaug and S. S. Ahsan, *J. Pharm. Sci.*, *50*, 441 (1961).
75. A. Richter and K. Steiger-Trippi, *Pharm. Acta Helv.*, *36*, 322 (1961).
76. M. Lladser, C. Medrano, and A. Arancibia, *J. Pharm. Pharmacol.*, *20*, 450 (1968).
77. R. Salisbury, E. E. Leuallen, and L. T. Chavkin, *J. Amer. Pharm. Assoc., Sci. Ed.*, *43*, 117 (1954).
78. R. J. James and R. L. Goldenberg, *J. Soc. Cosmetic Chemists*, *11*, 461 (1960).
79. B. A. Mulley, *J. Pharm. Pharmacol.*, *13*, 205T (1961).
80. B. W. Burt, *J. Soc. Cosmetic Chemists*, *16*, 465 (1965).
81. F. Lachampt and R. M. Vila, *Amer. Perfumer Cosmet.*, *82*, 29 (1967).
82. S. Friberg and L. Mandell, *J. Pharm. Sci.*, *59*, 1001 (1970).
83. F. A. J. Talman and E. M. Rowan, *J. Pharm. Pharmacol.*, *19*, 417 (1967).

6. MEDICINAL EMULSIONS 347

84. B. W. Barry and E. Shotton, *J. Pharm. Pharmacol.*, *19*, 110S (1967).
85. B. A. Mulley and J. S. Marland, *J. Pharm. Pharmacol.*, *22*, 243 (1970).
86. J. S. Marland and B. A. Mulley, *J. Pharm. Pharmacol.*, *23*, 561 (1971).
87. J. S. Marland, Ph.D. thesis, Bradford University, England, 1970.
88. A. A. A. Ali and B. A. Mulley, unpublished work, 1971.
89. D. E. Wurster, in *Safer and More Effective Drugs*, American Pharmaceutical Association, Washington, D. C., pp. 111-140.
90. T. Koisumi and W. I. Higuchi, *J. Pharm. Sci.*, *57*, 87 (1968).
91. W. G. Waggoner and J. H. Fincher, *J. Pharm. Sci.*, *60*, 1830 (1971).
92. L. Schneider, *New Engl. J. Med.*, *240*, 284 (1949).
93. G. Forbes and A. Bradley, *Brit. Med. J.*, *2*, 1566 (1958).
94. Anon., Pharm. Soc. Lab. Rept. No. P/64/31 (1964).
95. Anon., Pharm. Soc. Lab. Rept., *Pharm. J.*, *185*, 454 (1960).
96. W. H. Hassler, G. Stahl, and L. Cross, *Bull. Amer. Soc. Hosp. Pharmacists*, *11*, 124 (1954).
97. R. M. E. Richards and T. D. Whittet, *Pharm. J.*, *175*, 141 (1955).
98. J. Tober and T. W. Autian, *J. Amer. Pharm. Assoc., Sci. Ed.*, *47*, 422 (1958).
99. S. E. Wright, *Pharm. J. N. Z.*, *22*, 180 (1950).
100. E. H. Zimmerman, *J. Amer. Med. Assoc.*, *176*, 23 (1961).
101. L. G. Tullock and R. P. Warin, *Practitioner*, *187*, 827 (1961).
102. A. G. Fergusson and J. C. P. Logan, *Brit. Med. J.*, *1*, 871 (1961).
103. Report by the Public Health Laboratory Service Committee on the Testing and Evaluation of Disinfectants, *Brit. Med. J.*, *1*, 408 (1965).
104. I. E. D. McLean, K. T. Graham, and M. O. East, *Practitioner*, *189*, 82 (1962).

105. E. J. L. Lowbury, H. A. Lilly, and J. P. Bull, *Brit. Med. J.*, *1*, 1251 (1963).
106. Anon., *Pharm. J.*, *208*, 53 (1972).
107. R. M. Calman and J. Murray, *Brit. Med. J.*, *2*, 200 (1956).
108. L. G. Tullock and R. P. Warin, *Practitioner*, *187*, 827 (1961).
109. R. J. Henderson and R. E. O. Williams, *Brit. Med. J.*, *2*, 330 (1961).
110. F. A. Ellis and W. R. Bundick, *J. Amer. Med. Assoc.*, *150*, 773 (1952).
111. A. R. Hurwitz, P. P. Duluca, and H. B. Kostenbauder, *J. Pharm. Sci.*, *52*, 893 (1963).
112. K. S. Taylor, F. D. Malkinson, and C. Gak, *Arch. Dermatol.*, *92*, 174 (1965).
113. A. W. McKenzie and R. B. Stoughton, *Arch. Dermatol.*, *86*, 608 (1962).
114. R. E. Collard, *Pharm. J.*, *186*, 113 (1961).
115. N. Senior, *Pharm. J.*, *206*, 99 (1971).
116. I. Sarkany, J. W. Hadgraft, G. A. Caron, and C. W. Barrett, *Brit. J. Dermatol.*, *77*, 569 (1965).
117. J. Tissot and P. E. Osmundsen, *Acta Dermato-Venerol.*, *46*, 447 (1966).
118. J. P. Garnier, *Pharm. J.*, *207*, 475 (1971).
119. J. P. Garnier, *Clinical Trials Journal*, *8*, 55 (1971).
120. F. J. A. Bateman, *Brit. Med. J.*, *1*, 554 (1956).
121. G. W. Roberts, *Lancet*, *1*, 283 (1960).
122. G. W. Roberts, *Lancet*, *1*, 1207 (1959).
123. J. P. W. Hughes and G. S. Wigley, *Brit. Med. J.*, *2*, 153 (1962).
124. Anon., Pharm. Soc. Lab. Rept., *Pharm. J.*, *187*, 187 (1961).
125. M. A. Cooke and A. P. Launchbury, *Brit. Med. J.*, *1*, 292 (1961).
126. P. Alexander, *Mfg. Chemist*, *36*, 55 (1965).
127. W. Swallow, *Pharm. J.*, *187*, 407 (1961).
128. H. B. Lehr, J. E. Rhoads, O. Rosenthal, and W. S. Blakemore, *J. Amer. Med. Assoc.*, *181*, 745 (1962).

129. C. E. Meyer, J. A. Fancher, and P. E. Schurr (to Upjohn Co.), U.S. Pat. 2,945,869 (1960).
130. R. P. Geyer, D. M. Watkin, L. W. Matthews, and F. J. Stare, J. Lab. Clin. Med., 34, 688 (1949).
131. W. S. Singleton, J. L. White, L. L. Di Trapani, and M. L. Brown, J. Amer. Oil Chemists' Soc., 39, 260 (1963).
132. B. Lynn, J. Hosp. Pharm., 28, 71 (1970).
133. R. P. Geyer, Physiol. Rev., 40, 150 (1960).
134. G. F. Lambert, J. P. Miller, and D. V. Frost, J. Amer. Pharm. Assoc., Sci. Ed., 45, 685 (1956).
135. S. Ljimberg and R. Jeppson, Acta Pharm. Suecica, 7, 435 (1970).
136. J. Freund and M. V. Bonanto, J. Immunol., 48, 325 (1944).
137. D. Hobson, C. A. Lane, A. S. Beare, and C. P. Chivers, Brit. Med. J., 2, 271 (1964).
138. J. Lazarus and L. Lachman, Bull. Parenteral Drug Assoc., 21, 184 (1967).
139. B. S. Berlin, J. Immunol., 85, 81 (1960).
140. J. Freund and A. W. Walter, Proc. Soc. Exptl. Biol. Med., 56, 47 (1944).
141. J. E. Salk, M. L. Bailey, and A. M. Laurent, Amer. J. Hyg., 55, 439 (1952).
142. B. I. Wilner, M. A. Evers, H. D. Troutman, F. W. Trader, and I. W. McLean, J. Immunol., 91, 210 (1963).
143. W. J. Herbert, Lancet, 2, 771 (1965).
144. M. E. Hilleman, U.S. Pat. 3,149,036 (1964).
145. M. R. Hilleman, Amer. Rev. Resp. Dis., 90, 683 (1964).
146. E. N. Fox and M. K. Wittner, J. Immunol., 97, 86 (1966).

Chapter 7

EMULSION PAINTS

Gerould Allyn

Rohm and Haas Company
Philadelphia, Pennsylvania

I.	INTRODUCTION.	352
	A. History	352
	B. Advantages.	354
	C. Disadvantages	355
	D. The Nature of the Paint Business.	356
	E. Sales Position of Emulsion Paints	356
	F. Definitions	357
II.	LATEX-PAINT VEHICLES.	360
	A. Manufacture	360
	B. Types .	362
III.	PAINT FORMULATION	367
	A. Manufacturing Processes	367
	B. Pigments.	369
	C. Extenders	371
	D. Additives	371
IV.	TRADE-SALES COATINGS.	374
	A. Applications.	374

	B. Types of Latex Paint 376
V.	LATEX MAINTENANCE PAINTS 379
VI.	INDUSTRIAL LATEX PAINTS. 380
	A. Coil Coating 381
	B. Electrocoating 382
	C. Curtain Coating. 383
	D. Flood Coating. 383
	REFERENCES 384

I. INTRODUCTION

A. History

The past two decades have witnessed tremendous growth in the use of latex (water-emulsion) paints. Introduced in their present form shortly after World War II, they now dominate the market for interior flat wall coatings and account for more than half the exterior-house-paint market. They are also making steady inroads into the maintenance and industrial paint markets. Sales of water-emulsion paints have soared from less than 4×10^6 gal ([1]) in 1950 to more than 21×10^7 gal in 1968 ([2]).

Water-thinned paints are the oldest type of coating known to man. The ancient Egyptians decorated their dwellings, coffins, and tombs with several types of aqueous coating systems, including gum arabic, egg white and yolks, animal blood, milk, berry juices, glue, and lime plaster ([3]). Until the latter part of the 19th century, whitewash (slaked lime and water) was used extensively as interior and exterior house paint.

Although these water-base paints were effective as decorative coatings, they offered little protection to the surfaces coated. With the introduction of the more durable oil-base paints and varnishes, the use of water-thinned paints declined markedly.

7. EMULSION PAINTS

In the last quarter century, however, the trend has reversed; water-base paints are a major factor in the paint industry once more. This reversal stems from development of several new water-emulsion vehicles for making tough, durable, and easy-to-apply paints.

Emulsion paints go back to the early 1930s. Oil-base emulsion paints were introduced then, some 20 years before modern synthetic latex paints became available. The earlier emulsion paints were based on dispersions of drying oils in water containing casein and other emulsifying agents. Later oil-modified alkyd emulsion paints having improved properties were developed. They were used for painting both interior walls and exterior masonry. Coatings based on an alkyd-resin emulsion were used on masonry buildings at the 1939 New York World's Fair.

The age of synthetic latex paints began shortly after World War II with the development of butadiene-styrene latices for interior paints. The original butadiene-styrene latex systems were recommended primarily for use on interior walls, wallboard, plaster, and cinderblock. With the introduction of poly(vinyl acetate) homopolymers in 1953, latex paints began to be used on exterior masonry. Poly(vinyl acetate) wall primer coatings and sealers also became popular due to their fast drying speed.

At about the same time paint makers introduced acrylic-resin emulsion systems and suggested these for use on interior wall surfaces. Because of their durability and excellent alkali resistance, the use of acrylics spread to exterior coatings for stucco, masonry, and cinderblock.

From this point on, the use of both poly(vinyl acetate) and acrylic emulsion paints for exterior masonry climbed steadily. In addition, both moved rapidly into butadiene-styrene's share of the interior market for plaster, wallboard, and masonry. By 1957, latex house paints for primed exterior wood surfaces were in use.

Today latex emulsions are sold to make not only the foregoing latex paints but also paints for concrete floors, gloss and semigloss interior wall and trim coatings, exterior gloss trim coatings, as well as maintenance and industrial finishes.

Half of all interior latex flat paints now produced are made with poly(vinyl acetate) emulsions. Acrylic-resin emulsions hold first place among exterior latex house paints, with about half that market. There is also large use of vinyl acetate copolymers and some use of highly polymerized, solubilized linseed oils. Butadiene-styrene's share of the latex-house-paint market has slowly dwindled and will likely continue to do so, although it is still the least expensive of latex resins. Only about 20% of interior latex flat paint now is based on this resin.

B. Advantages

Latex paints have gained wide acceptance by both the consumer and industry. The popularity and explosive growth in the use of latex paints are due to their unique properties. For one thing, latex paints are easy to apply and give professional appearing results. For another, they are quick drying (within an hour or two) and make the completion of a job much faster than do the older oil-base paints. Moreover, equipment used in applying latex paint can be conveniently washed out with water.

Other advantages, particularly of the acrylic and some vinyl acetate types, include the following:

1. Little or no odor

2. Excellent resistance to alkalies in fresh plaster, stucco, or masonry

3. Long-term outside durability

4. Applicability in damp weather or on wet surfaces

5. Excellent resistance to blistering on wood surfaces

7. EMULSION PAINTS

The growing use of latices in household and industrial finishes is also due to elimination of the fire, explosion, toxicity, and organic solvent hazards that are commonly associated with the application of solvent-base paint. Moreover, latex paints reduce air-pollution problems resulting from the volatile organic solvents used in conventional solvent-base paint.

C. Disadvantages

Latex paints do have some disadvantages in comparison with the old solvent-base systems. They spread easily, and because of this there is a tendency for painters to spread latex paints too thinly, giving poor protection and appearance. Thus two coats of latex paint may be needed to hide the surface as well as one coat of an alkyd-oil type. In addition, water-emulsion paints usually must be protected against freezing and cannot be applied at temperatures much below $50°F$.

Another disadvantage is that latex-paint films are more permeable to moisture vapor than is oil-base paint. Although this is helpful in overcoming the blistering of paint films, it does not prevent the rusting of metals.

Emulsion paints do not penetrate porous surfaces as well as organic solvent-base paint. This can lead to poor adhesion on improperly prepared surfaces. For example, painting over old chalky paint surfaces can cause premature failure due to lack of penetration at the chalky interface. Besides, latex paints do not adhere well to greasy or oily surfaces.

Finally, the water in emulsion paint has a high heat of vaporization that slows evaporation in baked or heat-cured coatings; its fixed boiling point makes it difficult to control the evaporation rate as is possible with organic solvents.

D. The Nature of the Paint Business

The paint business is divided into two broad categories: trade sales and industrial coatings. Trade-sales paint includes shelf goods sold through retail or wholesale outlets to do-it-yourself buffs who do most of the repainting of residential interiors and to painting contractors for use on new construction, repainting, refinishing, and general maintenance. Total trade-sales paints account for about 54% of the paint industry's dollar volume. Perhaps 80% of the trade-sales volume is paint for general use on homes, offices, and other buildings (4).

On the other hand, industrial coatings take in a variety of products. They are sold by the paintmaker to other manufacturers for industrial maintenance and protection as well as for coating products like automobiles, furniture, appliances, and machinery.

Shipments of all trade-sales paint (including emulsion systems) totaled close to $1.5 billion in 1969 and accounted for just under half the total gallons of all paint sold that year. In the same year the sales of latex emulsion house paint reached $600 million (4), or 200 million gal (2). Total industrial coatings shipments amounted to some $1.3 billion in 1969 (4).

E. Sales Position of Emulsion Paints

As already noted, emulsion paints dominate the market for interior flat wall house paints, accounting for approximately 80% of the total sales. The swing to the use of latex paint on outside surfaces has boosted its sales to 46% of the total market for exterior house paints (2).

The most dramatic growth in recent years has come in latex gloss and semigloss interior paints. Their sales shot from about 3×10^6 gal in 1966 to some 13×10^6 gal in 1969.

In contrast to the rapidly increasing sales of trade-sales paints, latex paints now account for only 5 to 10% (2) of all industrial or product finishes. They have been slow to catch on. This stems in part from the generally higher costs of latex paints. Other disadvantages in the use of latex paints in industrial application include the following:

1. Usually painting facilities must be revamped to handle latex coatings.
2. Clean, grease-free surfaces are required to ensure good adhesion.
3. Water requires more heat and is more costly to evaporate than organic solvents; its evaporation rate cannot be controlled.
4. The very-high-gloss finishes required in many industrial applications are difficult to achieve.

Nevertheless, this picture is rapidly changing, as coil or strip coatings (e.g., thermosetting acrylic emulsion systems) are coming up fast in this $50-million-a-year market.

F. Definitions

1. EMULSION

An emulsion is a two-phase system containing two immiscible or sparingly soluble liquids to which an emulsifier, buffers, catalysts, and protective colloids are added. With oil-in-water emulsions, like the latices, the external phase is water containing salts, buffers, catalysts, and emulsifier. The internal phase comprises swollen micelles of the emulsifier and monomer droplets at the start of polymerization.

In the final product the internal phase is made up of polymer particles on whose surfaces the emulsifier is adsorbed.

Technically the latex paints discussed in this chapter are classified as dispersions, and not emulsions as the name implies. Latex paints consist of finely divided particles dispersed in water. True emulsions, on the other hand, are comprised of a liquid dispersed in an immiscible liquid.

2. PAINT

Paint is used to protect, decorate, and obscure a surface. The two basic ingredients in paint are the pigment and the vehicle. Pigment, small particles of opaque materials, gives paint its color and hiding power, or ability to obscure a surface.

The vehicle (carrier for the pigments) is the liquid part of the paint and consists essentially of two components: the thinner and the binder. The former is the volatile part of paint that evaporates, leaving behind a solid paint film. Nevertheless, it plays a very important role not only in the application properties of paint but also in the physical and chemical properties of the paint itself.

The binder remains in the paint film to bind the pigment particles into a uniform film after the volatile liquid evaporates. Binder type and amount largely determine most of a paint's properties--durability, washability, toughness, and adhesion.

In short, latex paints are those in which the thinner for the paint vehicle is water instead of mineral spirits, turpentine, or some other organic solvent used in oil- or alkyd-base paints.

Another class of water-thinned paints is of the solution type. These are homogeneous water-soluble systems. The binders, or polymers, such as oils and alkyds that are used in these systems, usually contain carboxyl or hydroxyl groups to render them water soluble. The binders of most modern water-thinned paints are water emulsions of polymers or copolymers of acrylic esters, poly(vinyl acetate), or butadiene-styrene. This chapter deals only with the emulsion systems.

3. FILM FORMATION

The chemical and physical reactions involved in film formation are not fully understood. This is particularly true of the part taken by some of the additives. In general, however, after latex paint is applied to a surface, water evaporates from the emulsion, forcing the dispersed polymer particles together. As the water

7. EMULSION PAINTS

continues to evaporate--packing the particles tightly--surface-tension forces act strongly to squeeze the individual polymer particles to a near-homogeneous film (5). Figure 1 shows butadiene-styrene particles in the last stages of film formation.

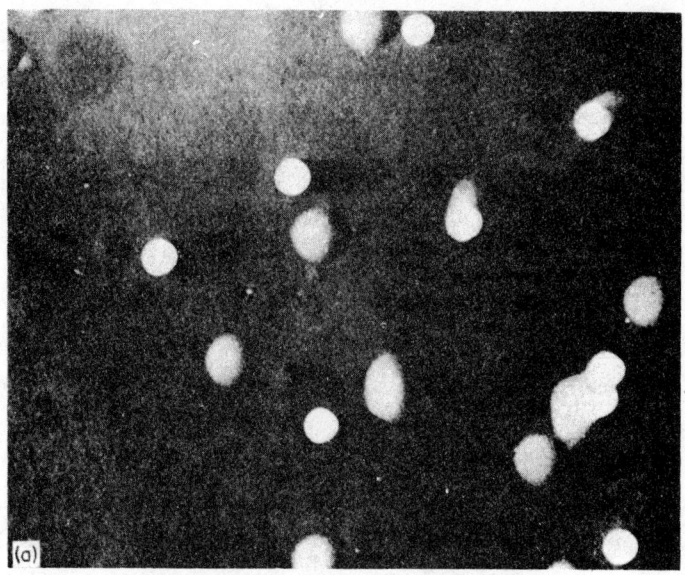

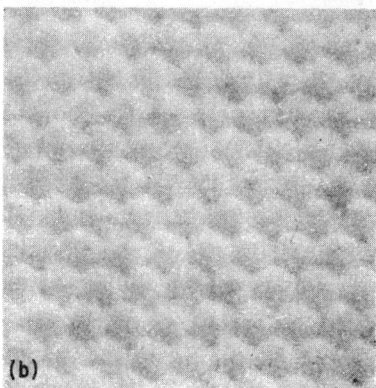

FIG. 1. Butadiene-styrene latex in the last stages of film formation. Notice the tendency of the soft butadiene-styrene particles (left) to coalesce. The white particles were chemically treated to provide contrast. The particles (right), as the water evaporates, group together and gradually lose their identity in a nearly homogeneous film. (Courtesy of the Dow Chemical Co.).

To get a continuous film, the latex particles must be slightly soft at the lowest temperature at which paint is applied, but hard enough to form a tack-free film when the water evaporates. Most latex paints are formulated to have good film formation down to 50°F.

Time, temperature, relative humidity, porosity of the substrate, particle size, binder type, and quality of dispersion all affect film formation, and paintmakers must consider all of these factors in formulating latex paint.

II. LATEX-PAINT VEHICLES

A. Manufacture (6,7)

1. POLYMERIZATION

Latex-paint vehicles are usually prepared by the emulsion polymerization of monomers like vinyl acetate, acrylic esters, or butadiene and styrene. With this rapid, but readily controlled technique, the insoluble monomer is emulsified with the aid of suitable emulsifying agents (surfactants) and agitation. The addition of sufficient emulsifier permits dispersion of the stirred monomer phase into a stable system. On the other hand, the addition of electrolyte or water-miscible solvents, evaporation, or severe mechanical stress can cause demulsification and gelation.

Surfactants, among which emulsifiers are a special class, reduce the interfacial tension between the monomer and water phases, so that with agitation, the monomer is dispersed or emulsified. The surfactant, which may be anionic, cationic, or nonionic, provides sites where polymerization occurs. Although emulsifiers are only slightly soluble in water, they are capable of forming hydrated aggregates called micelles within which the polymer particles are formed. These aggregates often comprise 50 to 100 surfactant molecules and are capable of soaking up water and monomer.

They can form because surfactant molecules are composed of mutually repellent polar, or ionic, parts and nonpolar, or nonionic parts. In these aggregates the hydrophilic polar parts are oriented toward the external aqueous phase while the hydrophobic nonpolar parts are oriented inwardly. When a monomer (or other water-insoluble liquid) is present, the micelles are somewhat swelled by imbibing monomer, thus increasing the solubility of the latter.

The polymerization of monomers can be initiated by heat, ultraviolet light, or high-energy radiation--or by the action of suitable chemical catalysts. In making latex vehicles, catalysts or initiators are generally used. This is mainly because reactions with catalysts are rapid and easy to control.

By varying the ingredients and conditions of polymerization, a wide variety of properties in the resulting polymer emulsion can be obtained.

2. COPOLYMERIZATION

Most commercial latex vehicles used today are copolymers made from two or more monomers--for example, the copolymerization of vinyl acetate and ethylene to form a vinyl acetate-ethylene emulsion system. This polymerization process makes good use of the monomers' individual characteristics in the resulting copolymer. Hence, by a wise choice of monomers for copolymerization, it is possible to custom-make a vehicle for specific end uses and with specific properties.

The performance properties of these polymers can also be varied by incorporating plasticizers and other additives. However, these external modifiers, though useful, frequently lose their efficacy because of their tendency to leach out of a paint film during weathering or aging.

The strength of a polymer is directly proportional to its molecular weight; thus the toughest paint films are formed from latex polymers having the highest molecular weights. Among the other process variables that affect film properties are the proportion of monomers in a particular latex, the number of side chains, and the extent of crosslinking between chains.

B. Types

As late as 1965 styrene-butadiene copolymers were the most widely used latices for the trade-sales market. Since then, however, tough, water-resistant paints based on vinyl acetate and on acrylic esters have captured the major share of this market. This happened mainly because butadiene-styrene films continue to cure after application and become brittle and yellow on aging. Furthermore, they do not have the flexibility and color retention of the acrylics and vinyls. Alkyd- and oil-emulsion paints and vinyl chloride copolymer systems also vie for trade-sales business.

As shown in Table 1, the trade sales of the "big four" in the latex-paint market totaled about 15×10^7 gal in 1967, a 10% increase over the preceding year ([8]).

TABLE 1

Distribution of Trade Sales Latex Paints (10^6 gal)[a]

Latex paint	1966	1967	Change (%)
Interior:			
Poly(vinyl acetate)	45.0	45.5	0.0
Poly(vinyl acetate) acrylic	7.0	10.0	43.0
Butadiene-styrene	29.0	26.0	-10.0
Acrylic	11.5	16.5	43.0
Total	92.5	98.0	5.0
Exterior:			
Poly(vinyl acetate)	10.5	10.5	0.0
Poly(vinyl acetate) acrylic	11.0	16.5	50.0
Acrylic	21.5	24.0	12.0
Total	43.0	51.0	18.0
Grand total	135.5	149.0	10.0

[a] From Ref. [8].

7. EMULSION PAINTS   363

1. ACRYLIC EMULSIONS

The production of acrylic latex paint jumped from about 10^6 gal in 1954 (9) to some 4.5×10^7 gal in 1967 (8). In 1969 acrylic emulsion paints accounted for nearly half the market for exterior latex paints and 15 to 20% of the market for interior flat wall paints (4). Currently they have little competition in the gloss- and semigloss-latex-enamel market.

Introduced in 1953, acrylic-resin emulsions are generally noted for their toughness, water-white color, outdoor durability, and good adhesion to a wide variety of surfaces. Although they cost more than the three other widely used latex emulsions (vinyl acetate, vinyl acetate-acrylic, and butadiene-styrene), acrylic emulsion paints offer an excellent combination of performance properties and greater latitude in formulating.

Acrylic-latex paints are defined as those that are predominantly the polymers and copolymers of the esters of acrylic or methacrylic acid. Chemically the acrylics are long-chain polymers of methacrylate and acrylate esters

$$\begin{pmatrix} & CH_3 \\ & | \\ (-CH_2-C-)_n \\ & | \\ & C=O \\ & | \\ & OR \end{pmatrix} \qquad \begin{pmatrix} & H \\ & | \\ (-CH_2-C-)_n \\ & | \\ & C=O \\ & | \\ & OR \end{pmatrix}$$

polymethacrylate   polyacrylate

where R is an alkyl radical such as ethyl, butyl, or 2-ethylhexyl. The length of the alkyl radical in the ester grouping has a marked influence on the physical properties of the polymer. Short radicals (e.g., methyl and ethyl) result in hard, tack-free, and tough polymers. The substitution of longer chains in the ester grouping results in soft, tacky polymers with low tensile strength.

The methyl groups on the α-carbon of the methacrylate polymers also have an important influence on polymer properties. Generally methacrylate polymers are harder and less flexible than the corresponding acrylate polymers.

Thus a wide spectrum of polymers, from very hard, relatively inflexible to very soft, highly flexible can be prepared, allowing chemists to custom-make resins for specific end uses.

Thermoplastic acrylic resins, used mostly in air-drying applications, were the first to be used in acrylic paints. However, increased interest in water-base systems for industrial baking finishes sparked the development of thermosetting emulsion polymers. These consist of acrylic copolymers and crosslinking monomers that contain functional reactive groups such as carboxyl and hydroxyl. Thermosetting emulsion-base paints cure when baked to form tough, hard finishes. They are especially suitable for industrial applications that require finishes of maximum hardness, toughness, adhesion, gloss, durability, and chemical resistance.

2. ALKYD EMULSIONS

As already noted, alkyd-resin emulsions pioneered the way to modern latex emulsions. However, with the introduction of the new latex paints in 1949 and their subsequent wide acceptance, interest in alkyd-emulsion paints dwindled (10).

The declining use of alkyd-emulsion paints, like that of butadiene-styrene coatings, resulted partly from the fact that such paint films harden to become brittle and yellow with age.

Nevertheless, alkyd emulsions are widely used as additives to latices for exterior coatings to improve film adhesion as well as increase the solids content and reduce costs.

Alkyds are polyesters of a polyhydric alcohol (e.g., glycerol or pentaerythritol) and a polybasic acid or anhydride (e.g., phthalic anhydride) modified with a fatty acid or oil.

3. Butadiene-Styrene

Spurred by the availability of this low-cost copolymer shortly after World War II, paint chemists developed butadiene-styrene

7. EMULSION PAINTS

latices for interior paints. Huge quantities of butadiene and styrene copolymers were used to make synthetic rubber during World War II. Hence the term "latex" as applied to paints based on such copolymers. Actually, latex paint do not contain rubber.

Butadiene-styrene's share of the $600-million-a-year latex trade-sales market has slowly tapered off since the introduction of paints based on poly(vinyl acetate) and on acrylic resin. About 20% of latex interior flat wall paint is now based on the butadiene-styrene copolymer (4). This copolymer is not used much in exterior latex paint.

Interestingly, shipments of latex interior emulsion paints in 1958, the bulk of which was used as wall coatings, amounted to roughly 5×10^7 gal. Butadiene-styrene emulsion paints accounted for about 66% of this total (11). In 1963 shipments of butadiene-styrene emulsion paints dropped to about 35% of a total of 7.9×10^7 gal of interior latex paints (11).

Used alone, styrene resins yield water-white, hard, and chemically resistant films when coalesced by baking at high temperatures or when externally plasticized. However, styrene is generally copolymerized with about 35% butadiene as an internal plasticizer (part of the copolymer chain), as represented by the following skeleton diagram:

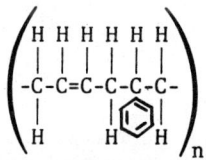

butadiene-styrene copolymer

The butadiene-styrene copolymer is still the least expensive of the "big four." Paints based on it cure slowly to form tough films with excellent resistance to water and alkalies. However, they generally do not have the good adhesion of the other latices. In addition, butadiene-styrene-based paint films harden by oxidation and polymerization (these latices are only partly polymerized during manufacture) similar to the way air-drying organic solvent-based paints do. This postapplication reactivity causes the film to become brittle and yellow with age.

4. OILS

Highly polymerized linseed-oil emulsions and solubilized oils are used to make paints having good exterior durability (12). Although not as widely used as the other latices, they compete in the same market. These penetrating finishes give good adhesion to rough, porous surfaces. Moreover, the formulation of high-gloss finishes is relatively easy with solubilized oils.

5. VINYL CHLORIDE (13,14)

Poly(vinyl chloride) emulsion paints produce films that are hard, tough, and chemically resistant. However, they do not form films at room temperature and are susceptible to decomposition by ultraviolet light and heat. To overcome these deficiencies, poly(vinyl chloride) is usually combined with a small amount of another monomer such as acrylic or vinyl acetate. These latices, which combine the good properties of vinyl chloride with the combined monomer, are used to make paints having excellent durability, chalk resistance, and tint retention.

6. VINYL ACETATE

Just a bit more expensive than butadiene-styrene systems, vinyl acetate polymer and copolymer systems are widely used by the latex industry. A skeletal representation of poly(vinyl acetate) is shown below.

$$\left(\begin{array}{c} \text{H H H H} \\ \text{-C-C-C-C-} \\ \text{H O H O} \\ \text{C=O C=O} \\ \text{CH}_3 \text{ CH}_3 \end{array} \right)_n$$

poly(vinyl acetate)

7. EMULSION PAINTS

About 50% of the interior flat wall paints now on the total latex market are made with poly(vinyl acetate) emulsions (4). Most of the vinyl acetate used in latex paints is copolymerized with other materials, such as dibutyl maleate or fumarate, vinyl stearate, as well as ethyl or butyl acrylate. Vinyl acetate-acrylic copolymer systems are moving in on the all-acrylic trade-sales latex market with more than 10% of the interior market and about a third of the exterior latex business (2).

Poly(vinyl acetate) systems are popular because they outperform butadiene-styrene latices and are less expensive than straight acrylics. They produce films with excellent flexibility, durability, and adhesion.

III. PAINT FORMULATION

A. Manufacturing Processes

Proper paint-formulation techniques are just as important as the choice of emulsion type in getting desirable film properties. The procedure for making latex paint varies with the equipment available. Basically the process involves pigment dispersion (called grinding), followed by the addition of emulsifier and other additives, adjustment of pH and viscosity, and tinting.

Much of the equipment used in paintmaking is for mixing and dispersing pigments. The most critical part of paintmaking is incorporation of the pigment with the binder (Fig. 2). Frequently used for these operations are high-speed impeller mills like the Cowles Dissolver (made by Morehouse-Cowles, Inc.) and sand mills. Other equipment, such as ball and roller mills, is also used.

The impeller mill is a disk-type, high-speed mill that operates on the same principle as a milk-shake mixer. All dispersers of this kind have the same basic mixing mechanism, though machines vary widely in size and horsepower. The impeller, a high-speed

FIG. 2. Worker at a modern paint plant disperses pigments for latex paints in a disk-type impeller mill.

disk, performs three basic functions: wetting, circulating or pumping, and dispersing (15). Impeller disks are furnished in various diameters, each one carefully tailored to handle a specific job.

Sand milling is also widely used for pigment dispersing. The process involves agitation (with sand) of a crude slurry of pigment and emulsion (16). Dispersion is accomplished by means of the fluid shear developed between rotating disks and adjacent layers of sand. Paint ingredients are premixed before passing them through the sand mill. After grinding to the desired degree of dispersion, the paste is filtered through a carefully sized screen, which retains the sand and allows the dispersed material to pass through (16).

With the impeller mill, the production of latex paint is divided into two operations: pigment dispersing at about 5400 ft/min (disk peripheral speed) and mixing at about 1500 disk peripheral speed (the letdown). Paint ingredients are usually added as follows:

Stage 1. Pigment Dispersion:
  Water
  Dispersant
  Pigments:
    Titanium dioxide
    Extenders
    Colors

Stage 2. Letdown:
  Thickener
  Latex
  Preservative
  Additives:
    Wetting agents
    Defoamers
    Optional ingredients for specific needs

Usually the dispersing and wetting agents, coalescent, antifoamer, water, and thickener solution are charged to a 500- to 5000-gal tank and mixed. The pigments and extenders are then slowly added as the mixer speed is increased. The pigment paste is then mixed at a high impeller speed until all aggregates are broken down and pigments are homogeneously dispersed.

After this, the dispersed pigment paste is diluted by adding the bulk of the emulsion vehicle, defoamer, and additional water or thickeners as necessary to adjust viscosity. Mixing continues at a low impeller speed until the paint batch is thoroughly blended. Finally, the pH of the finished latex paint is adjusted if necessary. The product is then ready for packaging.

B. Pigments (17,18)

1. *WHITE PIGMENTS*

White pigments provide color and hiding power. The major white pigment used is titanium dioxide. Other functions of the pigments are to help impart chemical resistance, good application properties, and various decorative effects. For example, by manipulating the type and amount of pigment incorporated into the binder, paint chemists can produce paints whose dried films are high-gloss enamels or flat, low-sheen finishes. Falling in between are semigloss and many other finishes.

Whether a dried film is glossy or flat is primarily a function of the pigment volume concentration, which is the ration of the volume of solid pigment to the total volume of solid pigment plus latex solids in percent and is calculated as follows:

$$\text{pigment volume concentration (\%)} = \frac{V_{ps}}{V_{ps} + V_{ls}} \times 100$$

where V is volume and the subscripts ps and ls stand for pigment solids and latex solids, respectively.

Glossy films have lower pigment volume concentrations than flat ones. The film thickness of the dry paint film, important for maximum durability, depends in part on the solids content.

2. COLOR PIGMENTS

The colored pigments used in latex paints are numerous and contain many color variations. Basically, they fall into two categories: organic pigments, which are usually hydrophobic, and inorganic pigments, which are usually hydrophilic. The hydrophilic types are relatively easy to incorporate into latex paints. Although the hydrophobic pigments are more difficult to incorporate, the correct blend of surfactants for a particular pigment makes the job a bit easier. The organic colors commonly used are phthalocyanine blue and green, toluidine red, carbon black, and Hansa yellow. The inorganic ones include iron oxide red, yellow, brown, and black; green chrome oxide; zinc yellow; and chromium oxide green.

C. Extenders

Extenders are inert materials used in latex paints to regulate flow properties, to extend hiding properties, and to help cut raw-material costs without sacrificing quality. Extenders also contribute solids to the paint and aid in film build. A wide variety of inert pigments can be used as extenders in latex paints. Selection depends on the performance properties desired. Each extender --clay, talc, silica, calcium carbonate, or mica--has some unique physical characteristic. Each imparts different performance properties to a given paint.

D. Additives (17,18)

A wide variety of raw materials--often in relatively small amounts--are combined with latices to make paints. A typical 1-gal white-topcoat-paint formula is given in Table 2.

TABLE 2

Typical 1-Gallon White-Latex-Topcoat Formulation

Ingredient	Pounds
Water	0.86
Wetting agent	0.03
Dispersing agent	0.15
Antifoamer	0.02
Freeze-thaw stabilizer	0.60
Preservative	0.01
Thickener (2.5% solution)	0.85
Titanium dioxide	2.50
Extender	2.04
Latex emulsion (45-55% solids)	4.60
Coalescing agent	0.12
Buffer (28% NH_4OH)	0.02
Total	11.80

Besides pigment and vehicle, several additives are used in latex-paint manufacturing: dispersing and wetting agents, antifoamers, freeze-thaw stabilizers and wet-edge agents, preservatives, thickeners, coalescent agents, and buffers (Table 2).

1. DISPERSING AND WETTING AGENTS

To obtain a stable latex paint, it is necessary to disperse most pigments and extenders in water before they are added to the emulsion. Dispersing and wetting agents (a combination of surfactants and pigment deflocculants) provide the best stability.

In an aqueous suspension of pigments there is a strong tendency for the pigments to adhere to each other when they collide. Because the attractive forces between the surfaces of the particles are large, flocculation of particles may result. To defloculate such suspensions, surface-active materials--which, in part, overcome the strong interparticle attraction--are added.

The types of compounds used as dispersing agents include sodium polyphosphates, the sodium salts of lignin sulfonic acids, aryl alkyl sulfonic acids, and carboxylated polyelectrolytes. They work by imparting like charges to pigment particles, which repel one another.

Nonionic wetting agents are also used along with pigment dispersants to wet pigment particles. These are mainly based on octylphenoxy polyethoxy ethanol and alkyl aryl ethers.

2. ANTIFOAMERS

Foam in latex paint may result in short weight in the package. It may also affect application properties, resulting in a rough, pockmarked film. To minimize foaming during manufacture and to prevent foam formation during application, antifoaming agents are added to emulsion paint systems.

3. FREEZE-THAW STABILITY AND WET-EDGE AGENTS

One disadvantage of latex-emulsion paints is their susceptibility to freezing. To enhance the freeze-thaw stability of emulsion

7. EMULSION PAINTS

paints, paintmakers use such materials as ethylene glycol, propylene glycol, and diethylene glycol to depress the freezing point of water. These, however, remain in the film temporarily as a water-soluble plasticizing material and thus soften it and impair its early water resistance.

Freeze-thaw stabilizers also help to improve wet-edge time, that is, the ability of a newly applied paint to stay wet long enough to blend in with a subsequent lapping during application.

4. PRESERVATIVES

Such compounds as the organomercurials and the newer nonmercurial preservatives are widely recommended for use in latex paints to retard bacterial growth in the can and mildew formation on exterior paint finishes. The amounts to be used are determined by the particular vehicle used, the surface, and the geographical location. The warm, humid climates of the South are especially conducive to mildew growth.

Preservative concentration can influence the ultimate film performance. Preservatives that are relatively water soluble are detrimental, as they tend to wash out of the film on aging. Water-solubilized grades of mercurial preservatives may also react with other ingredients in paint (e.g., dispersants). Interaction between these two materials reduces both the dispersant concentration, leading to pigment flocculation, and the paint's resistance to microbial and fungus attack.

5. THICKENERS

Thickeners help control viscosity and provide protective colloid action between the dispersed pigment particles. They are also used to adjust the flow characteristics of latex paints to provide the proper application consistency. The most popular protective colloids are methylcellulose, hydroxyethylcellulose, sodium and ammonium polyacrylates, and poly(vinyl alcohol).

6. COALESCING AGENTS (FUSION AIDS)

Coalescents are added to an emulsion system in order to soften the resin for improved film formation, especially under borderline application conditions, such as low temperatures or high winds. As the water evaporates from a film, resin particles are forced closer together until they fuse. However, because most latices are not soft enough to fuse at low temperatures, plasticizers are added to aid low-temperature coalescence. Poor coalescence may cause premature drying, resulting in milky or crumbly films. Such materials as pine oil, glycol ethers, and tributyl phosphate are often used as coalescents.

Coalescing agents also help to improve such film properties as leveling, scrub resistance, and adhesion. However, excessive quantities will impair the paint's package stability, drying speed, and freeze-thaw stability.

7. BUFFER

Each emulsion paint has a pH level at which it is most stable. Batches of paint are usually adjusted to a certain pH range during manufacture. Volatile buffers are generally preferred because they evaporate from the drying paint film and do not reduce the film's water resistance.

IV. TRADE-SALES COATINGS

A. Applications

1. GENERAL USES

Latex emulsions offer paint manufacturers wide latitude for producing various types of latex paint for both interior and exterior applications depending on the ratio of pigment to binder. Whether the surface is wood, cement, stucco, or brick, there is a

7. EMULSION PAINTS

specific latex paint designed for its maximum protection and beauty. For example, masonry surfaces contain alkali and require a paint that will not be adversely affected by this condition.

Latex paints are well suited for use on porous surfaces like masonry wallboard and plaster as well as wood surfaces. Indeed, the widest use of latex paint has been for interior wallboard and exterior masonry surfaces. However, many of these paints have been formulated as coatings for exterior house surfaces, both wood and metal.

Besides the standard flat topcoat paints, tint bases, and primer coatings that were the first developed, the improved properties of the newer latex emulsions make possible one-coat topcoats as well as gloss trim paints. These may be applied by brush, roller, or spray gun.

2. SURFACE PREPARATION (19)

In addition to proper formulation, latex paints deliver good service only if they are applied according to instructions. Surface preparation is one of the key steps in applying paint. Of course, the amount of preparation necessary depends on the condition and nature of the surface to be painted. Sound unpainted surfaces may need no more than a washing or a thorough brushing to remove dust, chalk, dirt, grease, or other surface contaminants. However, if a previously painted surface is partially peeling, flaking, or blistering, the damaged areas should be sanded smooth before recoating.

After cleaning, holes and cracks should be filled with mortar or caulking compound before paint is applied. Relatively sound painted surfaces require only a thorough washing with clean water or with a detergent like trisodium phosphate solution, followed by a thorough clean-water rinsing. Stubborn dirt, rust, or excessive roughness can be removed with sandpaper or steel wool; grease and oil can be removed by washing with mineral thinner or turpentine. Mildew can be scrubbed off exterior surfaces with a dilute sodium hypochlorite bleach solution, followed by rinsing with water.

If previously applied paint on exterior wood is blistering or peeling severely, it must be sanded down to the wood or removed with paint remover. Careful burning with a blowtorch is helpful in removing badly peeling paint. Heavily chalked masonry, concrete, or stucco may require sandblasting or power wire-brushing. Flaking or peeling may be caused by excessive moisture coming from within or outside sources. The source of this moisture should be eliminated before repainting.

B. Types of Latex Paint

1. PRIMERS (10)

Unfinished surfaces and bare spots often are coated with a primer before painting. Primers help to secure a reliable bond between the topcoat and its substrate. They also contribute other properties to the surface, such as enamel holdout (uniformity of finish), sealing of surfaces, and corrosion resistance.

a. Masonry primers. The most widely used latex primers for masonry surfaces are simply topcoats thinned with water. They are resistant to water and alkalies, and provide excellent sealing of masonry surfaces.

b. Plaster and wallboard primers. Quick-drying interior latex primers, or undercoats as they may be called, are especially effective on these substrates that have high absorption properties and alkalinity. Unlike the masonry primers, which are water-thinned topcoats, primers (or primer-sealers) for plaster and wallboard are specially made. They have pigment volume concentrations of 35% compared to 45 to 55% for topcoats.

c. Wood primers. Latex primers can be used on any type of bare wood to provide a bond for topcoats and help seal wood pores. Some woods, such as cedar and redwood, contain water-soluble

7. EMULSION PAINTS

stains that can bleed through a latex coating. To control the leaching of these stains, painters use stain-resistant latex wood primers containing basic lead silicate or dibasic lead phosphite. In such systems the lead-containing pigment apparently reacts with the stains leached from the wood to form insoluble substances.

2. FLAT TOPCOATS AND TINT BASES (17)

A major use of latex emulsions has been in the formulation of a variety of flat wall paints for wood, masonry, plaster, and wallboard. These make use of latex paints' excellent adhesion, chalk resistance, wet adhesion, formulation latitude, durability, as well as ease of application and cleanup.

The pigment volume content of white topcoats usually ranges from 35% to as high as 55% for masonry topcoats. On the other hand, tint bases (a way of providing thousands of color shades) usually have a pigment volume content of 35 to 45%. Basically, a tint base for making custom-colored paints is a white paint that can be tinted to a desired color by adding small amounts of colorants. The colorants are usually alkyd- or glycol-based. Desirable tint bases accept and give good dispersion of the added colorants and still have good performance properties. Most latex systems have excellent color acceptance and performance.

3. SEMIGLOSS AND GLOSS PAINTS (20)

Development of emulsion-base gloss paints was slow until 1965 because of the difficulty in formulating finishes with suitable gloss, wet-edge retention, leveling (i.e., lack of brush marks), and adhesion to glossy surfaces. Once these obstacles were overcome, latex gloss paints (which are currently based primarily on acrylic emulsions) seized a big share of the market formerly dominated by solvent-thinned enamels.

The use of interior semigloss enamels has shown dramatic growth, climbing from about 3×10^6 gal in 1966 to some 13×10^6 gal, or more than 20% of the entire interior gloss market, in 1969.

The degree of gloss shown by latex gloss paints depends to some extent on the pigment volume concentration and drops with increasing pigment volume content. For this reason latex gloss paints are made within a pigment volume concentration range of 20 to 30%. Figure 3 shows the effect of binder level on degree of gloss of a typical latex paint.

The outstanding features of gloss and semigloss latex enamels made with acrylic emulsions include the following:

1. Excellent leveling

2. Outstanding adhesion to old glossy oleoresinous undercoats, even under the humid conditions existing in bathrooms and kitchens

3. Very good gloss retention, plus resistance to yellowing and embrittlement with age--properties not shown by the once-unchallenged solvent-thinned gloss paints

4. Excellent wet-edge retention--easily controlled by the use of propylene glycol in the formulation

5. High gloss and hiding power

6. Low odor, easy application, and fast cleanup

Although latex gloss paints were initially designed for interior use and were so sued until 1968, many companies are now manufacturing high-quality latex gloss trim paints for exterior applications.

4. FLOOR PAINTS (21)

Some latex emulsions made from hard polymers have been specifically developed for use in paints to coat concrete floors. These paints display the ease of application, adhesion, durability, and quick-drying properties typical of latex-emulsion systems. Dampness from areas such as basements can cuase some other floor finishes to blister and lose adhesion, but latex floor paints resist moisture quite well.

7. EMULSION PAINTS

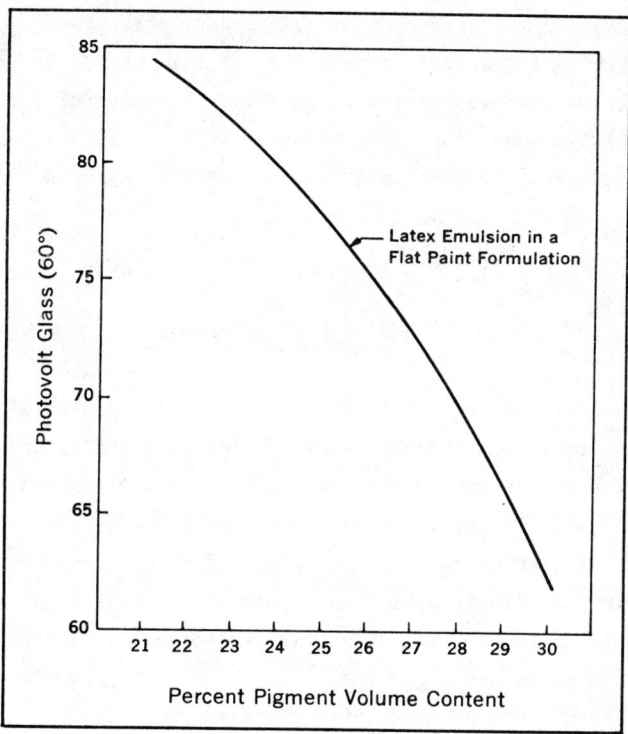

FIG. 3. Effect of pigment volume content on gloss at a constant titanium dioxide level (275 lb/100 gal paint). As expected, gloss drops with increasing pigment volume content.

V. LATEX MAINTENANCE PAINTS

Latex maintenance paints are used on such surfaces as tanks, piping, and asphalt, and must perform under various severe conditions: heat, moisture, and chemical fumes and solvents. Because of the attractive application and performance features characteristic of latex paints, they have recently gained a toehold in this market. In addition, latex emulsions, which may be used to formulate both primers and topcoats, have good adhesion and corrosion resistance over clean metal. Freedom from noxious odor and the fire hazard associated with organic solvents during application also contributes to the growing acceptance of latex maintenance emulsion paints.

The rigid demands imposed on maintenance coatings have, in the past, prevented large-scale penetration of this market by latex paints. Latex systems, nevertheless, offer a number of attractive features that account for the continuing interest in developing suitable coatings for both interior and exterior maintenance paint.

VI. INDUSTRIAL LATEX PAINTS

Another growing use of latex emulsion is in industrial finishes suitable for products ranging from automobiles and airplanes to furniture and building products (e.g., factory-prefinished aluminum and mineral fiberboard exterior siding). Industrial latex finishes, which find use as primers and topcoats, now account for 5 to 10% of this $1.3-billion-a-year market dominated by solvent-base finishes. The industrial market has just begun to be tapped by latex paints.

The steady inroads that latex coatings are making in the solvent-base market can be largely attributed to the valuable performance advantages associated with latex-emulsion systems. The film toughness, mar resistance, excellent flexibility, superior durability, and good chemical resistance are characteristics that make coatings based on high-molecular-weight latex-emulsion polymers attractive to industrial coaters.

Moreover, the application advantages of emulsion systems are helping to boost sales in the industrial market. These advantages include low odor as well as elimination of fire hazards and solvent fumes, all of which help to improve the working environment.

Industrial latex finishes may be applied by a number of techniques: coil coating, electrocoating, curtain coating, or flood coating.

7. EMULSION PAINTS

A. Coil Coating

A major outlet for certain latex emulsions, principally thermosetting acrylics, is in coil coatings for building products. Coil coating is a high-speed efficient factory prefinishing technique in which huge rolls of aluminum or steel strips are treated and coated on one or both sides by rollers for subsequent fabrication into finished end products (22) (Fig. 4). Advantages include the following:

1. Low operating costs--about half the cost of applying paint by spraying or one-fifth the cost of hand painting
2. Better control of film thickness
3. Elimination of nearly all paint waste

FIG. 4. A paint line operator checks the automatic recoiling of finished aluminum at Alcan Aluminum Corporation's Warren, Ohio plant. Coil coating is a fast and efficient way for prefinishing steel or aluminum. (Photo courtesy Amchem Products, Inc.)

With a coil-coating line, the entire series of finishing steps (coating, baking, and quality-control checks) are consolidated in one area and into one continuous process.

This technique for applying latex paints has potential for sizable growth. Some industrial coaters, however, are reluctant to switch over to coil coating partly because of the high cost of installing a high-speed line, running up to $1 million (4).

B. Electrocoating

Latex emulsions are still attempting to carve a foothold in the relatively limited, but growing, electrocoating market. Electrocoating is a finishing technique similar to the electroplating of metallic films. The object to be coated (aluminum, steel, or copper) acts as one of the electrodes, usually the anode. The tank containing the paint usually serves as the cathode, while the paint particles act as the electrolyte (22). As current is applied across the tank, the resin and pigment particles are deposited on the object being coated as a continuous film. This process is followed by removal of the object from the bath and by rinsing with a water spray. With emulsions, each of the dispersed polymer particles carries an electrical charge, stemming from either a polar group on the polymer itself or from the surfactant that is adsorbed on the particle surface (23). Film thickness depends primarily on the current applied.

Several latex paint and emulsion manufacturers are investigating this area. One thermosetting acrylic emulsion has been designed for electrocoatings and is also suggested for use as a clear finish, as a high-gloss decorative enamel, and as a primer (24). More are on the way.

The most marked advantage of electrocoating is the ability to form a uniform film on surfaces--such as inaccessible or hard-to-reach corners and sharp edges--that are difficult to coat by

7. EMULSION PAINTS

conventional methods of application. Furthermore, the film thickness can be controlled, and sagging is eliminated. In addition, paint loss is negligible and operating costs are low.

Despite these many advantages, electrocoating has not made a strong showing--the present market is $7 to $10 million (4)--partly because of high installation costs. For coating small objects in tanks of, say, 5000 gal, costs might run from $25,000 to $250,000 (4). A tank large enough for coating automobile bodies, holding roughly 60,000 gal of paint, can cost $1.5 to $2 million.

C. Curtain Coating

Another industrial finishing technique is curtain coating for factory-prefinishing flat or curved objects like mineral fiberboard, composition board, and ceiling tile.

With this technique, the object to be coated is placed on a conveyor and passed through a curtain of paint on to a second conveyor. Paint is pumped from a reservoir to a tank that has an adjustable slot at the bottom. Paint falls through the slot in a thin curtain on the object; the overflow drains to a trough and is pumped back to the reservoir. Film thickness depends on conveyor speed (which can travel up to 450 ft/min), paint solids content, and the thickness of the paint curtain.

D. Flood Coating

This application technique for factory-prefinishing large, complex objects is a relatively simple operation. The equipment consists of a pool of paint fed through a hose to a nozzle that floods paint over the entire object. Excess paint is allowed to drain off or is removed with an air knife and is recirculated. Problems with foaming have limited the use of latex paints in flood coating.

REFERENCES

1. E. Schulte, *Official Digest* (now *Journal of Paint Technology*), April 1961.
2. Charles H. Kline & Co., Inc., *Marketing Guide to the Paint Industry* (IMG-1-69) 1969.
3. J. J. Mattiello, ed., *Protective and Decorative Coatings,* Vol. 1, Wiley, New York, 1941.
4. D. M. Kiefer, *Chem. Eng. News*, December 22, 1969.
5. Dow Chemical Co., *Dow Latexes for Interior Paints*, July 1966.
6. Rohm and Haas Co., *Emulsion Polymerization of Acrylic Monomers*, Bulletin CM-104, May 1969.
7. Celanese Corp. of America, Chemical Div., *Vinyl Acetate Monomer ... in Emulsion Technology*, Bulletin No. S-56-1.
8. A. Banov, *American Paint Journal*, September 9, 1968.
9. Anon., *Chem. Eng. News*, February 15, 1960.
10. A. G. Roberts, *Organic Coatings*, Building Science Series 7, U.S. Department of Commerce, National Bureau of Standards, February 1968.
11. *1963 Census of Manufacturers*, Industry Series, Paints and Allied Products, SIC Code 2851, U.S. Department of Commerce, Bureau of the Census.
12. Spencer Kellogg Div., Textron, Inc., *Linaqua, Water-Soluble Linseed Oil for Exterior House Paints*, 1962.
13. O. Z. Tyler, B. F. Goodrich Chemical Co., unpublished work. 1960.
14. Anon., *Chem. Eng. News*, August 29, 1966.
15. Rohm and Haas Co., *Resin Rev.*, *18*, No. 2 (1968).
16. E. I. du Pont de Nemours & Co., Inc., *Sand Grinding Process for Dispersing Pigments*, 1961.
17. Rohm and Haas Co., *Rhoplex Acrylic Emulsions for Outdoor Paints*, Progress Report 12, C-266, 1969.
18. Air Reduction Chemical & Carbide Co., *Airco Polyvinyl Acetate Latices for Interior and Exterior Paints*.

19. National Lead Co., *Surface Preparation and Product Information*, 1969.
20. Rohm and Haas Co., *Resin Rev.*, *18*, No. 2 (1968).
21. Rohm and Haas Co., *Resin Rev.*, *18*, No. 1 (1968).
22. Rohm and Haas Co., *Resin Rev.*, *19*, No. 1 (1969).
23. J. W. Hagan, Paper for Presentation at New York Society For Paint Technology Meeting, March 1965.
24. Conchemco Corp., Product Bulletin on Acrylic Resin 311-203, Jan. 1972.

Chapter 8

ASPHALT EMULSIONS

R. L. Ferm

Chevron Research Company
Richmond, California

I.	INTRODUCTION.	389
II.	ADVANTAGES OF ASPHALT EMULSIONS	390
III.	PRODUCT TYPES AND NOMENCLATURE.	391
IV.	EMULSIFIERS FOR ASPHALT	392
	A. Structure and Function.	392
	B. Anionic Emulsifiers	393
	C. Cationic Emulsifiers.	395
	D. Amphoteric Emulsifiers.	397
	E. Nonionic Emulsifiers.	397
	F. Clay Emulsifiers.	398
V.	IMPROVEMENT OF ASPHALT EMULSIFIABILITY.	399
VI.	ADDITIVES IN EMULSIONS.	399
	A. Viscosity Control	400
	B. Adhesion Improvement.	400
	C. Stability Improvement	402
	D. Workability Enhancement	402
VII.	MANUFACTURE OF EMULSIONS.	403

	A. Emulsification	403
	B. Plant Design	404
	C. Storage and Handling	405
	D. Waste Disposal	407
VIII.	EMULSION PROPERTIES AND TESTING	407
	A. Total Asphalt and Solids Content	407
	B. Viscosity	408
	C. Particle Size and Distribution	408
	D. Particle Charge	409
	E. Storage Stability	410
	F. Mechanical Stability	410
	G. Dehydration Rate	410
	H. Stability in the Presence of Mineral Aggregates	411
IX.	THE TECHNOLOGY OF PAVING WITH EMULSIONS	412
	A. Seal Coating	412
	B. Pothole Patching	414
	C. Emulsion Paving	415
	D. Soil Stabilization	416
	E. Slurry Sealing	417
X.	INDUSTRIAL EMULSIONS	419
	A. Roof and Surfacing Coatings	419
	B. Paper and Building-Board Sizes	421
	C. Adhesives	422
	D. Metal-Coating Uses	422
	E. Applications in the Mineral Industries	423
	F. Concrete Additives	423
	G. Soil Building Blocks	424
	H. Oil-Well Treatment	424
XI.	AGRICULTURAL AND WATER-CONSERVATION USES	425
	A. Agricultural Mulches	425
	B. Soil-Erosion Control	427
	C. Canal and Reservoir Sealants	427
	D. Water Harvesting	428
	E. Underground Sealing of Sandy Fields	429

8. ASPHALT EMULSIONS

    F. Climate Modification. 429
    G. Fertilizer Component. 430
    REFERENCES. 430

I. INTRODUCTION

Asphalt emulsions continue to grow in importance for road building and for a wide variety of industrial uses. Improved formulations are increasing the versatility of these products and are bringing their use into new fields, such as agriculture and water conservation. The objective of this chapter is to summarize background information and recent developments that will further aid in exploiting the useful properties of these materials. Other reviews on asphalt emulsions have been published by Day and Herbert [1], Mertens and Borgfeldt [2], and Drukker [3]. The background information presented here is largely supplemental to these.

The advantages of using asphalt in emulsified form were first demonstrated for road construction in the period 1906-1914. Substantial use for paving purposes did not occur until the end of the 1920s. The history of this development has been summarized by Forrest [4]. Ross [5] has described early road-construction methods with emulsions. Only anionic emulsions were manufactured until the early 1950s. The performance advantages of cationic emulsions were first demonstrated in Europe at that time. Because of their rapid setting and improved adhesion to a wider variety of aggregates, the cationic emulsions have been steadily displacing the anionic asphalt emulsions.

The usefulness of emulsified asphalt for protective coatings and as a means of making mixtures of asphalt with fibers and other particulate matter led to a wide variety of industrial applications. These employ mainly anionic emulsions. The volume of these emulsions is much smaller than that of paving emulsions.

The total production in the United States of asphalt emulsions was 2,157,415 tons in 1969 (6). Of this amount 2,056,716 tons were used in paving.

II. ADVANTAGES OF ASPHALT EMULSIONS

The main advantage of emulsified asphalt over other asphalt forms is fluidity at ambient temperatures. This feature may also be obtained by dissolving asphalt in hydrocarbons to obtain the "liquid asphalts," or "cutbacks." However, these products can be fire hazards if the solvent is highly volatile, and they are slow curing if the solvent has reduced volatility. Furthermore, many liquid-asphalt products must be applied at temperatures of 180 to 200°F for satisfactory spraying characteristics.

The ability to mix with wet aggregates is an important advantage of emulsions over other asphalt products. An aggregate-drying operation is needed to make hot asphalt mixes. It is feasible to carry on road work with certain emulsions under threatening weather conditions. Cationic emulsions may have particularly good early resistance to rain damage (7, 8).

Asphalt emulsions are particularly well suited for the surface treatment of worn and raveled pavements by the application of chip seals or aggregate-slurry mixtures. It is difficult to apply hot asphalts as a thin overlay in rock-chip seal-coat construction, as exact coordination with the spreading of the rock chips is required to obtain good adhesion. Emulsions are considerably more useful for this type of road surfacing. Liquid asphalts are used also for chip sealing, but require long curing times. Maintenance and resurfacing techniques that can be carried out quickly with minimum shutdown time for the roadway are becoming more important as urban density increases. The slurry-sealing technique satisfies this requirement and also provides a low-cost resurfacing method. A slurry seal is a fluid mixture of a fine, specially graded aggregate, asphalt emulsion, and water (9), spread in a thin layer to restore pavement thickness, fill cracks, and smooth out irregularities. Properly applied slurries give a very durable, skid-resistant surface. Recently developed "Quick-Set" slurry-seal systems (10) allow the road to be opened to traffic 15 to 20 min after laydown.

III. PRODUCT TYPES AND NOMENCLATURE

Asphalt emulsions are classified according to their relative dehydration, or setting, rates: rapid setting (RS), medium setting (MS), and slow setting (SS). Cationic emulsions were first indicated by adding the letter K as a suffix (e.g., SS-K). More recently cationics have been designated by using C as a prefix (e.g., CRS or CSS). The latter system has been adopted by the American Society for Testing and Materials (11). This system also includes provision for classifying products according to viscosity grades. Thus the Saybolt Furol viscosities at 122°F of CRS-1 and CRS-2 emulsions are in the ranges of 20 to 100 and 100 to 400, respectively. An emulsion whose designation does not contain the letter K or C is always assumed to be anionic. The letter h is added as a suffix (e.g., CSS-1h) to indicate that a relatively hard asphalt (i.e., penetration at 77°F of 40 to 90) has been used in the emulsion. Otherwise, asphalt in the penetration range 100 to 200 is generally used.

Penetration macadam, formed by spraying an emulsion onto a base of uniformly graded aggregate, or mixtures of such an open-graded aggregate with an emulsion that are bladed or otherwise spread to form a pavement require the MS-type products. Road bases or pavements made with aggregates consisting of a wide range of particle sizes (i.e., dense-graded aggregates) require the SS-type emulsions.

This is because the preparation of such aggregate-emulsion mixtures requires extensive mixing action and hence additional emulsion stability. The slurry seals described in the preceding section also require SS emulsions. Furthermore, the SS emulsions have sufficient stability to permit mixing with soils for roadbed stabilization or dilution with water for use as dust-laying agents.

No general designations have evolved for industrial asphalt emulsions.

IV. EMULSIFIERS FOR ASPHALT

A. Structure and Function

Almost all asphalt emulsions used for road work are of the oil-in-water type. Water-in-oil, or inverted, emulsions have been used for some stockpile mixes, for patching, and in some industrial products. Thus it is necessary to choose an emulsifier whose hydrophobic-hydrophilic balance is conducive to forming the desired emulsion type.

Emulsifiers used in making RS and CRS emulsions generally contain linear alkyl groups as the hydrophobic components. These surfactants are characterized by their ability to pack together and form monomolecular films. Such films either totally or partially surround the emulsion droplets and provide the electrically charged double layers that give stabilization. Emulsions of this type have limited stability when diluted with water. Water dilution results in breakup of the emulsifier films and transfer of the emulsifier to the bulk water phase. Naphthenic acid salts, which can be used to make RS emulsions, represent an exception to the foregoing picture of stabilization. Possibly these surfactants, which may occur naturally as the acids in asphalt, may be partially solubilized in the asphalt-emulsion droplet after the salt is formed.

8. ASPHALT EMULSIONS

The SS and CSS emulsions are usually made with emulsifiers that contain cyclic hydrocarbon groups, such as lignin or rosin derivatives. These "bulky" emulsifiers cannot orient themselves in orderly, close-packed molecular films around the emulsion droplets. They probably stabilize the emulsion by clumping together as aggregates that act as barriers between the asphalt droplets. The emulsifier concentrations required are generally above the critical micelle concentrations which suggests this mode of stabilization. This would also explain the usual ability of SS and CSS emulsions to withstand high dilution with water accompanied by good retention of stability.

B. Anionic Emulsifiers

Almost every known type of anionic surfactant appears in the asphalt-emulsion patent literature. Petroleum-derived materials include naphthenates, sulfonates, cresylates, and wax acid soaps; all are generally used as the sodium slats. Of these, the naphthenates are the most important, as they occur naturally in many asphalts in concentrations that allow emulsification by contacting the molten asphalt with a caustic solution. Some commercial emulsions are manufactured by this means. The acid number of the asphalt must be about 0.75 mg KOH per gram or higher to give satisfactory emulsions by simple caustic emulsification. Generally sodium hydroxide or potassium hydroxide is the caustic. The acid numbers of asphalts from various sources were reported by Caro (12). These ranged from 0 to 3.

Naphthenic acids can be extracted from petroleum and mixed with low-acid-number asphalt or added as the alkali-metal salts to the aqueous phase in the emulsification process (13). The acid numbers of naphthenic acids used as emulsifiers should be in the range 75 to 175 (14). Carboxylic acids in a California petroleum have been structurally characterized by Seifert and Teeter (15).

Sulfonic acid salts resulting from the acid treatment of petroleum were claimed as asphalt emulsifiers in a number of early patents. However, these tend to be oil-soluble soaps and thus are not effective for making the oil-in-water emulsions desired in present-day products. The development of water-soluble detergent sulfonates has provided more useful asphalt emulsifiers (10, 16, 17). Sodium bis(dodecylphenyl) ether disulfonate is useful in making emulsions whose breaking rate can be controlled by mixing with coagulation agents, such as calcium chloride or cement (18). Monoalkyl and dialkyl benzene sulfonates of average molecular weight (400 to 500) are claimed to give emulsions of superior storage stability (19). Emulsions made with oil-soluble sodium petroleum sulfonates in the asphalt phase and a water-soluble sodium alkylbenzene sulfonate in the aqueous phase are claimed to give unusually water-resistant bituminous coatings on mineral aggregates (20).

Tall-oil soaps are useful asphalt emulsifiers (21). Sulfonated tall oil is reported to improve the adhesion of the deposited asphalt to aggregate (22). Prior to the advent of the lower cost tall-oil-soap emulsifiers, oleic acid soaps were widely used.

Other fatty-acid derivatives used as emulsifiers include hydroxystearic acid (23), fatty-acid pitch (residue from fatty-acid distillation) (24), and residues from the refining of various oil seeds (25).

Rosin soaps were used to make early emulsions (26) and were usually incorporated as additives to the asphalt phase prior to emulsification. Polymerized rosin is claimed to improve emulsion stability (27). A pine-chip-resin extract (Vinsol resin, Hercules Company) has been widely used in making SS-type emulsions (28). Vinsol resin emulsions are characterized by small particle size and good storage stability.

Lignin sulfonates obtained from the paper-pulping industry as the sodium ammonium, and calcium salts can be used to make SS-type emulsions (29, 30). However, such emulsions tend to have only fair storage stability. Combinations of lignin and saponified-rosin derivatives give improved performance (31). Lignite and humic acid derivatives have also been used to disperse asphalt, but are not commercially important.

A variety of protein-containing or derived products have been used as asphalt emulsifiers. Blood in coagulated liquid or dried form makes very stable emulsions. Formaldehyde (32) or thymol (33) must be added to prevent microbial degradation during storage. Gelatin in the presence of various inorganic salts to prevent gel formation has been used as an asphalt emulsifier (34). Casein has been used in combination with saponified rosin (35) or potassium tannate (36). Lecithin and its derivatives have been shown to be useful asphalt emulsifiers (37-39). "Glue" has also been used (40). Alkali-hydrolyzed yeast is an effective asphalt emulsifier (41). Other polymeric, nonprotein asphalt emulsifiers include maleic anhydride-ethylene copolymer derivatives (42) and the condensation products of acrylic acid reaction with halogenated polyisobutylene (43).

C. Cationic Emulsifiers

Simple C_{12} to C_{18} monoalkylamine hydrochloride or acetate salts were used to make the first cationic asphalt emulsions (44, 45). Emulsions of improved stability result with N-alkylpoly-methylenediamine salts having C_{12} to C_{22} alkyl groups and two to four methylene groups (46). Secondary alkyl polymethylenediamines, in which the alkyl group contians 13 to 19 carbon atoms, are claimed to be superior for asphalt emulsification to primary fatty alkyl polymethylenediamines (47). Mixtures of C_8 to C_{22} alkyl or alkenyl monoamine salts and the corresponding triamine salts are also claimed to have superior emulsifying properties (48).

Polyethoxylated derivatives of N-fatty alkyl polymethylene-diamine salts give emulsions that are outstanding in their ability to coat a wide variety of mineral aggregates (49-51). Polyamine-ethylene oxide condensates further reacted with tall-oil fatty acids have been used in combination with oxyethylated fatty alcohols, such as oleyl or cetyl alcohols reacted with 10 moles of ethylene oxide per mole of alcohol (53).

The reaction products of fatty acids, particularly tall-oil acids, with polyamines like diethylenetriamine or tetraethylenepentamine are a valuable class of asphalt emulsifiers. The fatty amidoamine products can be used in the form of their mineral acid salts (53-55) to give emulsions with improved storage stability and adhesion to a wide variety of aggregates. Fatty imidazoline amines are easily formed by further heating the fatty amidoamines; these, as salts of mineral acids, are also good asphalt emulsifiers (56). The sulfamic acid salts of imidazoline surfactants have been proposed as asphalt emulsifiers (57). Combinations of fatty imidazolines and nonionic emulsifiers, such as ethoxylated octylphenol, have been used to give emulsions of improved mixing stability with electronegative aggregates (58).

Alkyl quaternary ammonium salts, particularly the halides, have the advantage as emulsifiers in CRS asphalt emulsions of reducing runoff from the pavement during seal-coating operations (59). These include such surfactants as tallow trimethylammonium chloride, alkylpyridinium halides, and alkylimidazolinium halides. Small amounts of sterols in emulsions made with quaternary ammonium emulsifiers further reduce pavement runoff and improve adhesion to the aggregate (60). A study of the effect of structure of alkylpyridinium emulsifiers on asphalt-emulsion stability was made by Dobozy (61). A unique feature of quaternary ammonium emulsifiers is their ability to form alkaline cationic emulsions; the pH of such emulsions may range from 7 to 11 (62). Quaternary ammonium emulsifiers containing an aromatic group like phenyl or naphthyl attached to a C_8 to C_{22} chain are claimed to slow the deposition of asphalt from emulsions in contact with electronegative mineral surfaces (63). Polyethoxylated derivatives of quaternary ammonium

8. ASPHALT EMULSIONS   397

salts, such as the halides or sulfates, have been used to improve asphalt-emulsion stability (64). Combinations of quaternaries and nonionics are claimed to give superior CSS emulsions (65).

Amine-salt derivatives of lignin are useful for making CSS emulsions (66). A reaction product of tetraethylenepentamine and pinewood-resin acid has also been used as a CSS-type emulsifier (67).

D. Amphoteric Emulsifiers

Amphoteric compounds are excellent asphalt emulsifiers, but have not been widely used because of their generally higher cost in comparison with other types. Improved adhesion and extent of coating of both electronegative and electropositive aggregates are claimed (68) for amphoteric emulsions over conventional anionic or cationic types due to the presence of both types of ionic groups in the same emulsifier.

E. Nonionic Emulsifiers

Although sometimes used as additives in minor amounts in anionic and cationic emulsions, nonionics have not been used to any extent as primary emulsifiers for asphalt. The absence of ionic character may give poor adhesion to aggregates, and the cost is generally higher than that of the common anionic emulsifiers. However, such an absence of ionic interaction between nonionic-emulsified asphalt and mineral surfaces is claimed as an advantage in making stable slurry mixtures (69). A wide variety of nonionics, including block copolymers of poly(ethylene oxide) and poly(propylene oxide), have been successfully used in making such slurries. Polyoxyethylated octyl alcohols derived from the Oxo process have also been used as asphalt emulsifiers (70).

F. Clay Emulsifiers

A useful property of certain clays is the ability to concentrate at an asphalt-water interface and thus function as a stabilizer for dispersed asphalt. There are many varieties of clay, varying in chemical composition and ability to form colloidal dispersions. The presence of electrolytes can have very pronounced effects on such surface activity and on the properties of the resulting asphalt emulsions (71). Bentonite, a clay of the montmorillonite group, is most widely used for making asphalt clay emulsions. Wyoming bentonite, which has a relatively high ratio of sodium to calcium ions in its structure, has an exceptional ability to hydrate and swell, and is widely used in emulsion manufacture. Since dispersed clay particles carry a negative charge, asphalt-clay emulsions are of the anionic type. The pH of the aqueous phase is especially important in developing maximum emulsion stability. The optimum pH depends on the particular clay, asphalt source, and the presence of electrolytes and other emulsion additives. Thus optimum pH must be determined experimentally for each particular emulsion system, but generally is in the range 5 to 8. Alum has been shown to be a particularly useful additive for the adjustment of pH in clay emulsions (72).

A disadvantage of clay emulsions is their relatively high viscosity compared with other asphalt emulsions. Generally the asphalt content must be limited to 50 to 55 wt % in order to obtain a product with satisfactory application characteristics. Viewed under the microscope, clay emulsions have "pickle"-shaped particles rather than the spheres obtained with other emulsifier types.

Clay has been used in combination with other asphalt emulsifiers. These include fatty-acid soaps, such as oleates (73), alkali-metal salts of carboxymethylcellulose (74), and various surfactant amines (75).

V. IMPROVEMENT OF ASPHALT EMULSIFIABILITY

When the content of naturally occurring organic acids in asphalt is too low to make RS-type emulsions by contacting with caustic solutions, surface-active acids may be added to asphalt to achieve emulsification. "Asphaltic acids" obtained by caustic extraction of still bottoms, petroleum residua, or oxidized lubricating oils have been used for this purpose (76). Blends of low- and high-acid-number asphalts can be used to improve emulsifiability (77).

Surface-active materials can also be formed in asphalt by treating it with certain reagents. Heating asphalt with maleic anhydride (78) or sulfuric acid (79) can improve emulsifiability. Air blowing (high-temperature air oxidation) of heavy oils may give asphalts with better emulsification characteristics than those produced by steam distillation or vacuum distillation alone (80).

High asphaltic acid concentrations (acid numbers of 2 or more) may cause difficulty in the manufacture of RS emulsions, which must begin to coalesce immediately when brought into contact with a mineral surface. Excessive emulsion stability can be reduced by adding an amino compound capable of reacting with part of the asphaltic acids to the asphalt prior to emulsification and thus limiting the amount of emulsifier soap that can be formed (81). Another approach is the use of lithium hydroxide, which forms less stable asphaltic acid soaps than does sodium or potassium hydroxide, as the emulsifying caustic (82).

VI. ADDITIVES IN EMULSIONS

Frequently it is necessary to modify one or more properties of an asphalt emulsion, such as viscosity, adhesion to mineral surfaces, or stability, by including additives to enhance or diminish

A. Viscosity Control

Either an increase or decrease in viscosity may be desired. The viscosity of anionic emulsions made with fatty-acid soaps can be increased by including starch, sugars, gelatin, glues, or glycogen, whereas phenols, creosotes, wood, and lignite tars reduce viscosity (83). Poly(ethylene glycol) esters of fatty acids also reduce the viscosity of RS emulsions (84). Guar gum may be used to increase asphalt-emulsion viscosity if stabilized with $Na_2S_2O_3$ (85). The condensation products of chlorinated paraffin oil and amines like propylenediamine or hexamethylenetetramine can be used to increase anionic emulsion viscosity (86). Combinations of oil-soluble sulfonates in the asphalt phase with water-soluble sulfonates in the aqueous phase will increase emulsion viscosity (87).

Inorganic salts, such as calcium chloride, are frequently used to reduce the viscosity of cationic emulsions. The viscosity of cationics made with quaternary ammonium halide emulsifiers can be increased by adding high-molecular-weight unsaturated monohydric alcohols, such as cholesterol (60).

Lignosulfonates (88) and bis(2-ethylhexyl) esters of phosphoric acid (89) reduce the viscosity of bentonite-clay emulsions.

B. Adhesion Improvement

Poor adhesion of asphalt to mineral aggregates occasionally is a problem in road building. This may occur when the mixture is first made or after the road is in service. In the latter case the phenomenon is called "stripping" and is caused by the displacement

8. ASPHALT EMULSIONS

of the asphalt film from the mineral surface by the action of water. Any displacement of the road structure by traffic greatly accelerates the stripping action. Adhesion problems can occur with either hot-mix or emulsified-asphalt road construction. Such adhesion problems are usually regarded as related primarily to the nature of the aggregate surface and are usually relatively insensitive to asphalts from different sources.

Fatty amines, such as tallow propylenediamine or tall-oil-substituted imidazoline, in amounts of a few tenths to several percent can often eliminate adhesion problems with hot asphalt mixes. This perhaps explains why cationic emulsions, which often contain such amines as emulsifiers, are generally regarded as giving better adhesion to a wider variety of aggregates than do anionic emulsions. Gastmans (90) and Mouton (91) have discussed the various types of amine adhesion agents and some adhesion test procedures.

The coating of mineral aggregates by cationic emulsions can be further improved by the presence of minor amounts of nitrogen bases obtained by the distillation of Gilsonite (92) or of polyamides made by the reaction of polymerized linoleic acid with polyalkylene polyamines, such as those sold commercially under the trade name Versamid (93). The adhesion of cationic emulsions made with alkyl quaternary ammonium halide emulsifiers can be improved by the addition of alkali-metal acetates (94).

Inorganic salts, such as ammonium chloride, nitrite, or sulfate (95); sodium dichromate (96); potassium thiocyanate (97); or ammonium thiosulfate (98), are useful for improving the adhesion of anionic emulsions. Alkali-metal phenates (99), lead naphthenate (100), or additives made by reacting phosphorus pentasulfide with olefins followed by reaction with C_8 to C_{24} aliphatic amines (101) have also been used for this function. The reaction products of aliphatic polyamines with chlorinated paraffins are claimed to improve the adhesion of anionic asphalt emulsions (102).

C. Stability Improvement

The type of emulsion-stability enhancement needed may concern storage life, ability to withstand mechanical shearing action like that encountered in pumping, or the ability to mix with aggregates without coalescence. Often a combination of two or more emulsifiers increases stability more than the equivalent amount of a single emulsifier (103). Thus sulfuric acid esters of secondary aliphatic alcohols of eight or more carbon atoms improve the stability of emulsions made with fatty-diamine mineral acid salt emulsifiers (104). Proteins (105), small amounts of clay (106), or polycyclic aromatic hydrocarbons having a maximum of three rings, such as anthracene oil (107), are useful in reducing settlement in emulsions during storage. Sorbitol (108), gelatin (109) and sodium silicate (110) are claimed to reduce the freezing points of asphalt emulsions.

D. Workability Enhancement

Hydrocarbon solvents, usually in the kerosene boiling-point range, are frequently added to asphalt emulsions to keep mixtures with aggregates workable during construction or while the mixture is stored in a stockpile. The addition of such hydrocarbons, often referred to as "cutter stock," may be necessary to prevent premature coalescence when the emulsion is mixed with dry aggregates. Thus incorporation of 10 to 15% by volume of hydrocarbon solvent in certain CMS emulsions often may give uniformly coated, easy to handle mixes with sands and fine aggregates. Microscopic examination of such mixes will usually show that the emulsion has coalesced. The solvent plasticizes the asphalt and aids the spreading of thin asphalt films over the aggregate.

Hydrocarbons mixed into a preformed asphalt emulsion are dispersed first as a separate phase. Thus only emulsions made with emulsifiers that easily disperse kerosene-type hydrocarbons will

8. ASPHALT EMULSIONS

accept cutter-stock addition. The fatty-diamine salts of mineral acids are examples of such emulsifiers. By contrast, lignin sulfonates are poor emulsifiers for cutter stock; hydrocarbons mixed with such emulsions will separate on the surface. Hydrocarbons can also be incorporated in emulsions by mixing with the asphalt prior to emulsification. Thus liquid or cutback asphalts may be emulsified.

Addition of hydrocarbon solvents to emulsions has the disadvantage of increasing the curing time of the aggregate mixtures. Road bases made with such mixtures may require months to attain their ultimate strength, depending on temperature, handling procedure prior to placement, and pavement thickness. Proper construction practices can minimize this problem.

VII. MANUFACTURE OF EMULSIONS

A. Emulsification

The function of the basic equipment required is to bring the asphalt and water phases together in a manner conducive to the formation of the desired type of emulsion. This equipment may be a colloid mill, homogenizer, or simple mechanical mixer, depending on the amount of shearing force required. Since it is generally desired to make asphalt-in-water emulsions, the asphalt phase is usually introduced in a manner to promote its fragmentation. This can be accomplished by first starting the flow of the water phase through the shearing device (e.g., propeller mixer mounted in a tank) and then injecting the asphalt phase into the water feed line so that a coarse dispersion is obtained prior to passage through the shear field.

The amount of shear required to give the desired particle size will depend greatly on the particular emulsifier employed. Anionic emulsions often can be made with low-power shearing equipment, such as propeller or turbine mixers. Cationic emulsions almost always require colloid-mill manufacture. Asphalt type and viscosity, manufacturing temperature, and colloid-mill pressure drop also affect the particle-size distribution obtained.

Colloid mills are usually so constructed that the spacing between rotor and stator plates can be changed to give the shear intensity desired. Furthermore, since they can be operated under pressure, colloid mills can be used to emulsify high-melting asphalts that require manufacturing temperatures above the normal water boiling point.

Any method of producing a shear force is of potential interest for making asphalt emulsions. Emulsification systems demonstrated to be useful range from a closed mixing circuit containing a centrifugal pump (111) to cavitation dispersion by ultrasonic energy (112). The operation of a colloid mill within a strong magnetic field is claimed to improve asphalt emulsification (113).

B. Plant Design

The complete emulsion plant consists of tanks for hot storage asphalt, batch preparation tanks for the emulsifying water, feed pumps, colloid mill, or other emulsifying equipment, a heat exchanger for cooling the emulsion after it leaves the mill, emulsion storage tanks, and transfer pumps for loading emulsion into trucks or tank cars.

The asphalt storage tanks, transfer lines, and pumps are generally heated with high-pressure steam in order to bring the asphalt to the mill at temperatures of about 250 to 270°F. The exact temperature employed will depend on the viscosity of the particular asphalt used.

Several tanks are generally provided for the preparation of the emulsifying water in order that the mill may be operated continuously. In some cases smaller tanks are used to prepare concentrated

8. ASPHALT EMULSIONS

solutions of emulsifying agents; these are then metered into the larger emulsifying-water tanks for dilution to the final concentration. The emulsifying-water tanks, lines, and pumps must be constructed of materials able to resist corrosion due to acids or alkalies. Either mechanical mixers or air spargers can be used in the emulsifying-water tanks. Steam coils can also be provided in these tanks. Emulsifying-water temperatures vary from ambient to about 140°F, depending on the type of emulsion being made and asphalt viscosity. Provision is generally made for measuring the pH of the emulsifying water and adjusting it to the desired range by adding acid or caustic.

A heat exchanger to cool the emulsion product after it leaves the colloid mill is required if emulsification is conducted at a temperature above the normal boiling point of water. Generally in the manufacture of paving emulsions this exit temperature is in the range of 180 to 200°F. Thus the heat exchanger may not be required in some cases. However, some formulations require rapid cooling to between 140 and 160°F to prevent development of undesirably large particles.

C. Storage and Handling

Asphalt emulsions are usually stored in mild-steel tanks with as high a ratio of height to diameter as practical in order to expose the least surface area and thus minimize evaporation. Carbon steel is a satisfactory material of construction, even for storing the acidic cationic emulsions, due to a protective film of asphalt that deposits on the metal. Most emulsion formulations have maximum storage stability at temperatures in the range 140 to 160°F. However, in some cases longer emulsion life is obtained at 70 to 100°F.

There is presently no way of predicting storage behavior, which must be determined from experience. Emulsions must not be subjected to freezing temperatures in any case. Also storage in uninsulated or unheated tanks subject to large changes in temperature between day and night is undesirable, as formation of large asphalt particles, commonly referred to as "shot," may be greatly accelerated. These particles are usually spherical and may range up to several millimeters in diameter. The heating of emulsion storage tanks should always be accomplished with hot-water coils. Even low-pressure steam in heating coils may cause local overheating, resulting in asphalt phase separation. Agitation in the tank with either a propeller mixer or an air sparger is desirable to prevent settling of asphalt particles and keep the product in a homogenous condition. Air spargers should be used only intermittently to avoid excessive evaporation. A fine-mist water spray is often used in the top of the tank to reduce evaporation and to prevent the formation of an asphalt skin at the surface.

Asphalt emulsions stored in drums or tanks without agitation often form deposits consisting of closely packed asphalt particles. Such settlement may, in some cases, easily be remixed to give an emulsion unchanged from its original state; in other cases it may form a coagulated mass that must be discarded. This behavior depends mostly on the particular emulsifier used. If the asphalt is lighter than water, the emulsion particles will concentrate at the top; this is called "creaming." Generally an emulsion that creams can be restored to its original condition by agitation. If the storage stability is poor, emulsions may invert, so that the water phase becomes dispersed in a continuous asphalt phase.

The mixing of anionic and cationic emulsions in tanks and lines must always be avoided to prevent coagulation. Such equipment must be thoroughly cleaned before switching from one type of emulsion to another.

Gear-type pumps are usually employed for transfering emulsions. These must have sufficient clearance between their moving parts to avoid excessive shearing of the emulsion. In the case of a new pump a few thousandths of an inch must be machined off gear surfaces. A pump somewhat worn from some other type of service may

8. ASPHALT EMULSIONS

be excellent for pumping asphalt emulsions. It is also possible to prepare a new pump for emulsion service by pumping a slurry of abrasive, but this must be done very carefully to avoid excessive wear. Pumps that give rise to high shear forces, such as nylon-roller pumps, are sure to fail quickly in emulsion service.

D. Waste Disposal

Equipment cleaning, such as line washing, is certain to produce dilute asphalt-emulsion waste, which cannot be discharged into sewers without danger of pollution. Good management can reduce such waste to a relatively small volume, which can usually be disposed of by recycling in the emulsifying water of the next emulsion batch. Segregation of wastes according to product type may be necessary.

VIII. EMULSION PROPERTIES AND TESTING

A. Total Asphalt and Solids Content

The American Society for Testing and Materials (ASTM) uses a distillation method for determining the water content of asphalt emulsions using xylene or other petroleum distillates; the procedure is described in ASTM Designation D 244-69 (114). An alternative evaporation procedure, also described in this ASTM designation, allows determination of total volatiles, including any hydrocarbon solvent in the emulsion. The residue from either of these methods will contain emulsifiers and any clay or other mineral fillers present in the emulsion in addition to the asphalt. The asphalt content of commercial paving emulsions generally ranges from 55 to 70%.

B. Viscosity

The industry standard for determining asphalt-emulsion viscosity is the Saybolt Furol method; this is also described in ASTM D 244-69. Although this test is the basis of present-day specifications, it leaves much to be desired, as its reproducibility is poor. Capillary viscometers do not work well with emulsions, which tend to foam due to their surface-active components. However, rotational viscometers, such as the Brookfield or Ferranti, do work well with emulsions. Moreover, they have the advantage that shear rate can be varied so as to obtain a more complete rheological description of the emulsion. The Ferranti machine has the further advantage that shear rate can be varied over a wide range and can be assigned an absolute value. This is important because asphalt emulsions are typically non-Newtonian. Both thixotropic and dilatant behaviors are observed for various emulsion types.

The asphalt content and the nature of the emulsifiers are very important factors in determining emulsion viscosity. As would be expected, the particle-size distribution also has a large influence on this property. High-viscosity emulsions observed under the microscope usually exhibit "chaining" or "clustering" between the particles.

C. Particle Size and Distribution

Asphalt-emulsion particle size is usually determined with the optical microscope. An emulsion of normal asphalt content must, of course, be diluted. This is best done with the emulsifying water used in the preparation of the particular emulsion in order to avoid coagulation. If the emulsifying water is not available, a dilute aqueous solution of a nonionic emulsifier like Triton X-100 can be used to increase stability.

A new, more rapid, and accurate method for determining asphalt-emulsion particle size and distribution employs the Coulter Counter

8. ASPHALT EMULSIONS

(Coulter Electronics, Inc., Hialeah, Florida). With this apparatus particle size is related to changes in the electrical conductivity of an aqueous circuit whose limiting portion is a precisely formed orifice through which a portion of the emulsion sample is passed. The emulsion must be highly diluted with an aqueous electrolyte and stabilized by adding an excess of surfactant for Coulter Counter examination. An extensive study of the use of the Coulter Counter with cationic asphalt emulsions has been made (115). The use of this instrument with other types of emulsions has also been thoroughly described (116). The orifice size selected depends on the emulsion particle size; the 30- or 70-μ orifices are satisfactory for most asphalt emulsions.

Spectrophotometric methods have also been used to determine the mean surface area, and thus the mean particle size, of asphalt emulsions (117).

D. Particle Charge

Estimations of the ζ potential of asphalt-emulsion particles are satisfactorily made by the usual electrophoretic-mobility-measurement techniques employed for other emulsion and dispersion systems. Use of the emulsifying water employed in making the emulsion as a diluent is desirable. A "particle-charge test" has been demonstrated (118) as a useful means of identifying certain asphalt emulsions by electrolytic deposition of asphalt on a stainless-steel electrode. This method has been adopted by the American Society for Testing and Materials as a standard test method, ASTM D 244-69, for identifying cationic emulsions (114) and is included in the Asphalt Institute specifications (119) for CRS and CMS emulsions. However, the test is not definitive for all CSS emulsions nor the various anionic emulsion types.

E. Storage Stability

Storage stability is determined by observing the formation of shot, settlement, creaming, or changes in the viscosity of emulsion samples over a period of time. If the samples are retained in the laboratory, the conditions of storage, such as temperature and container materials, should be the same as for storage in tankage. Well-formulated emulsions should show little or no change in properties after 6 months of storage and are often found to be little changed after several years of storage if protected from evaporation. Settlement that can be easily remixed is not considered a serious problem.

F. Mechanical Stability

There are no standard tests for determining mechanical stability, but many laboratories have procedures for simulating commercial handling operations. These usually involve pumping and recylcing an emulsion sample and noting either an increase in particle size or the time required to cause gross coagulation. The value of such a test depends on the correlation with full-scale pumping conditions.

G. Dehydration Rate

The rate of dehydration is also a property of great practical importance, but no standard tests for determining it have been established. The maintenance of uniform humidity over the entire sample in such a manner as to simulate actual use conditions requires great care. The presence of fillers and aggregates mixed with the emulsion can greatly influence evaporation rates. Emulsifier type is also a factor in determining dehydration tendencies.

H. Stability in the Presence of Mineral Aggregates

Depending on their intended end use, the various grades of paving emulsions are designed to have different stabilities on contact with mineral aggregates. The RS and CRS emulsions should coalesce almost immediately when brought into contact with a mineral surface. The rate of coalescence will depend mainly on the particular emulsifier used and its concentration, the nature of the aggregate, temperature, and the amount of shear developed by the application technique. The SS emulsions have such great stability that they can be mixed with the aggregate for a prolonged time without any coalescence. Dehydration is required to coalesce these emulsions. The MS types are of intermediate stability. The effect of emulsifier-aggregate adsorption on coalescence rates has been discussed by Gaestel (120).

A wide variety of laboratory tests have been devised to determine emulsion stability in the presence of aggregates. The most useful of these involve mixing procedures simulating use under actual construction conditions. The ASTM Coating Ability and Water-Resistance Test, D 244-69, is a good example (114). Although a standard aggregate from a specific source (i.e., limestone from the Monon Stone Company of Monon, Indiana, mixed with about 1% chemically pure precipitated calcium carbonate dust) is suggested, results will be more meaningful if the actual field-job aggregate is used instead. This test is intended for evaluating emulsions to be mixed with coarse-graded calcareous aggregates. The ASTM Cement Mixing Test, D 244-69 (114), is intended to indicate the capability of stable-type emulsions (SS) to mix with aggregates containing a large amount of fine material (dense-graded aggregates). Type III portland cement is used as the aggregate in this test. Although the Cement Mixing Test is fairly good for its intended purpose, the author has encountered numerous emulsions that fail the test but mix satisfactorily with dense-graded aggregates.

A further improvement in the testing of emulsions with dense-graded aggregates is the use of mechanized mixing test equipment so that the conditions are more reproducible. The ASTM test employs hand mixing. However, no standard, industry-wide mechanical mixing test has been adopted.

Attempts have been made to eliminate the problem of obtaining suitable reference aggregates for mixing tests by replacing them with aqueous salt solutions. The best example is the ASTM Demulsibility Test, D 244-69 (114), which uses solutions of calcium chloride to measure the stability of anionic emulsions. A similar test that uses a solution of sodium dioctyl sulfosuccinate has been proposed for cationic emulsions (11). The ability of these tests to correlate with actual performance in road building is very questionable; the only sure indication they give is the degree of compatibility with the particular chemical reagent used.

IX. THE TECHNOLOGY OF PAVING WITH EMULSIONS

A. Seal Coating

Seal coating is a surface treatment usually performed by spraying an asphalt emulsion on a weathered road, followed by the immediate application of a suitable aggregate. Generally the final surface is rolled to ensure good contact between the covering aggregate and the substrate. The usual construction procedure is first to clean the pavement with a rotatory power broom, apply the emulsion from a tank truck equipped with a spray distributor, and immediately spread the aggregate from a dump truck equipped with a spreader box. The truck moves in reverse gear, so that its wheels pass over the freshly spread aggregate. Hard, dust-free, crushed rock chips about 1/4 to 3/8 in. in size are the best aggregate. Coarse sands have been used successfully in areas where rock is not

8. ASPHALT EMULSIONS

readily available. Emulsion-application rates vary with the nature of the surface and are generally in the range 0.25 to 0.35 gal/yard^2. Several seal coats can be applied successively if it is desired to build up the pavement thickness.

Seal coats have the advantage of relatively low cost and of almost immediate use by traffic after laydown. A disadvantage is the hazard of flying rock chips when the surface is new. This can be minimized by brooming off excess rock chips as soon as construction is finished. The seal coat can give optimum maintenance value only if applied before excessive deterioration of the pavement has taken place. Cracks and faults that develop in the base will almost always in time reflect through a new seal coat. The value of maintaining an intact seal coat to prevent entrance of water into the road base and the resulting damage to the base cannot be overemphasized.

Emulsions for seal coating are of the RS or CRS type. These should be formulated to give good wetting and spreading on the particular aggregate as soon as it is applied in order to avoid runoff loss to the side of the road. The asphalt used in making the emulsions must have high enough viscosity to hold the aggregate chips during hot weather. Thus emulsions are climatized in accord with the weather conditions in the areas in which they are to be used. Asphalts with a penetration grade of 200 to 250 are typically used in cold-weather climates, whereas the harder asphalts, with penetration grades of 100 to 200, are used in warm areas. The superior performance of cationic emulsions in comparison with anionics for seal coating was recognized soon after their commercial introduction (7, 8); most seal coating today is done with cationics. The main advantages cationics have over anionics are faster set, early rain resistance, and better adhesion to a wider variety of aggregates (121).

A seal-coating technique that utilizes a dense-graded aggregate is described by Bower (122). A cationic asphalt emulsion containing a minor amount of petroleum distillate is mixed with the aggregate at a temperature of about 350°F. The resulting mix is sufficiently fluid to be spread on the road surface and compacted at a temperature of about 200°F. This gives a more dense and durable surface than can be obtained with many rock-chip seals. It is difficult to satisfactorily spread ordinary hot asphalt paving mixtures in the thin overlays required for seal coating because of the 350 to 400°F temperatures required for workability.

Synthetic rubber latices have been used with asphalt emulsions for seal coating with varying degrees of success for a number of years. Benefits claimed include improved pavement durability, better skid resistance because of elimination of asphalt bleeding, and quicker opening of new construction to traffic (123). There is no question that the addition of a suitable latex can greatly increase the strength and durability of a cured asphalt emulsion-aggregate mixture if used in sufficient amount. However, the low levels used in some tests have given insignificant improvements. Probably a minimum of about 5% of polymer solids based on asphalt solids is needed to give measurable improvement in durability. The styrene-butadiene latices have proved particularly promising for this use. Formulations for latex-asphalt emulsion compositions will be discussed in Section X.A.

B. Pothole Patching

Pothole patching must frequently be made prior to seal coating or as a general maintenance procedure on rural roads where it is not economically feasible to bring in hot asphalt mixes. Sometimes road patches made with stockpile cutback asphalt cure so slowly that failure results from traffic action. At best, such cutback repairs usually leave a depression after the mix finally cures and is compacted. Preliminary tests in the author's laboratory have demonstrated that some of the new Quick-Set slurry-seal mixes (10) show promise for making improved patches of this type. Another new

8. ASPHALT EMULSIONS

patching technique that gives a strong, durable repair uses a polypropylene mat impregnated with hot asphalt or an asphalt emulsion (124); this is applied over the surface of a conventional patching mixture to prevent displacement from the hole by traffic.

C. Emulsion Paving

Mixtures of aggregates and asphalt emulsions are finding increasing use for road-base and final pavement construction. The aggregates may range from sandy silts to graded rock for base construction. Construction techniques range from simple mixing in place with a motorized blade to plant-prepared batches of paving mixture laid down with specially designed machines. Mixing may be with wet or dry aggregates. A variety of emulsion tupes are required to handle this broad range of conditions. Cationic emulsions are gaining favor for making aggregate paving mixtures, although anionic emulsions are also used. The SS and CSS emulsions are used where maximum emulsion stability is required, such as when mixing with soils or dense-graded aggregates. These emulsions must be used with wet aggregates to prevent coalescence. This can present a curing-rate problem, particularly with anionics, if dehydration is slow. The CMS emulsions containing 10 to 15% hydrocarbon solvent mix with either wet or dry aggregates. These mixes have the disadvantage of sometimes remaining tender to traffic for a long time while the hydrocarbon solvent they contain vaporizes. Coarse-graded aggregates are mixed with emulsions of intermediate stability of the MS or CMS type. The advantages of cationic mixing-grade emulsions over anionics for a variety of road-construction uses have been reported by Borgfeldt and Ferm (125).

Increased use of asphalt emulsions for base stabilization is expected. A series of case histories of such construction show excellent performance under a variety of conditions (126). A successful large-scale base-construction project with blow-sand and asphalt emulsion (127) illustrates the versatility that can be obtained in aggregate usage with proper engineering laboratory design of the emulsion-aggregate mixture.

D. Soil Stabilization

Subbases for roads, airfields, building construction, reservoirs, or other hydraulic works needing erosion protection frequently require the stabilization of soil. Soil differs from the aggregates normally used in road construction by its clay content. Asphalt emulsions, as well as portland cement, lime, and other chemicals, have been used for soil stabilization. However, ordinary paving emulsions work well for stabilization only when the clay content of the soil is low and does not exceed 10 to 20%, depending on the particular soil. The maximum soil-plasticity index for satisfactory asphalt stabilization is about 6. Cement or lime is presently preferred for high-clay soils. Subgrade stabilization beneath railroad ballast in Japan with cationic asphalt emulsion was highly successful (128).

Emulsion additives used to improve soil stabilization include sodium silicate or silica hydrogel (44), sulfate liquor from the Kraft wood-pulping process (129), calcium chloride (130), ferric or aluminum chloride (131), phosphates, carbonates, or oxalates (132-134), as well as oil-soluble surfactant amines (135). A process of solidifying soils by penetration first employs a very stable emulsion, which is percolated through the soil without coagulation; this is followed by the application of a less stable emulsion that coagulates in a uniform manner throughout the soil (136). Hard asphalts (penetration number less than 25 at room temperature) are claimed to be superior to the softer grade asphalts for soil stabilization, but must be diluted with a volatile petroleum solvent prior to emulsification (137). Surface stabilization of soils can be attained by spraying with a mixture of a rubber

8. ASPHALT EMULSIONS

latex and asphalt emulsion (138). Such a treatment would be used to prevent soil erosion. A hydraulic method of mixing sandy soils with asphalt emulsions using a fluid jet to give a protective soil coating of moderate depth has been devised (139).

The emulsifiers themselves in soil-stabilizing emulsions can have an important effect on soil properties. Shirley (140) found that anionic and nonionic surfactants are more effective in increasing the density of clay soils than cationic or amphoteric types.

E. Slurry Sealing

Slurry sealing is a surface treatment conducted by spreading a fluid mixture of asphalt emulsion and fine aggregate on the pavement surface. It combines the advantages of speed, economy, and a smooth surface, free of rock chips that can be loosened by traffic. This last property makes slurry sealing particularly desirable for airfields where flying rocks could be sucked into jet engines. Asphalt slurry sealing also is an effective treatment for spalled concrete surfaces. Although slurry sealing was used in the 1930s in Germany, commercialization did not take place until suitable equipment was developed in recent years. Mobile slurry-seal units now combine aggregate, emulsion and water storage, metering devices, a pug-mill mixer, and an adjustable spreader box. In another less used technique the slurry is mixed in a batch plant and then transported to the job site in a concrete transit truck to which a spreader box is attached.

Slurry-seal mixtures usually contain 15 to 25% wt % of asphalt emulsion (asphalt content 55 to 65%) and sufficient additional water to give maximum fluidity to the slurry without segregation of the fine aggregate. A special aggregate grading is required to prevent streaking, sanding, or pavement embrittlement (9). The effect of operating variables on slurry sealing has been summarized by Clifton and Hiscock (141). Portland cement is generally added from a separate feeder to provide sufficient fines if the aggregate is lacking in these. Such addition seldom amounts to more than 1 or 2% of the total aggregate. The addition of asbestos fiber is claimed to increase pavement strength (142).

The advent of the new Quick-Set slurry-seal formulations is expected to greatly increase the use of this paving method. Slurry-seal coatings of this type can be used by traffic within a fraction of an hour after application, amking possible the repair of high-traffic-density roads that cannot be closed without great inconvenience. Both anionic and cationic Quick-Set slurry emulsions have been developed. The anionic types utilize sulfonate emulsifiers in combination with other additives that react with alkaline components in the aggregate to cause emulsion coalescence (10, 143). Such alkalinity can be provided by the addition of lime, portland cement, or naturally occurring minerals in the aggregate. A slurry-seal emulsion made with either a sodium alkylbenzene or a lignin sulfonate emulsifier and containing a reaction product of tall oil and Vinsol resin with maleic anhydride is claimed to have improved thixotropic properties and consequently better slurry-handling characteristics (144).

The cationic Quick-Set emulsions use polyalkoxylated quaternary ammonium halide emulsifiers (145) or combinations of emulsifier types normally used to make rapid-setting cationic emulsions with emulsifiers that give stable or mixing-grade emulsions. An example of such a combination is a mixture of the reaction product of tall-oil fatty acid and diethylenetriamine with the reaction product of split-palm-kernel-oil fatty acid and pentaethylenehexamine, used as the hydrochloride salt (146). Another Quick-Set cationic slurry-seal formulation employs a combination of three emulsifier mineral acid salts: the products of fatty-acid reaction with polyethylene polyamines, oxyethylated fatty alkyl polyamines, and fatty-alkyl-substituted ammonium quaternary salts (147).

Once the slurry seal is laid down, its main economic benefit to the taxpayer is wear resistance. A simple laboratory test that

8. ASPHALT EMULSIONS

correlates well with field life has been developed. This is the Wet Track Abrasion Test (148), which involves determination of weight loss from a slurry-mix sample cured under standard conditions and abraded with a piece of rubber hose attached to a Hobart laboratory mixer. Aggregate type, asphalt, and emulsifier are all important factors in slurry-seal performance.

X. INDUSTRIAL EMULSIONS

The adhesive, waterproof, and inert character of asphalt has prompted the development of a wide variety of emulsion products for the construction, manufacturing, and mining industries. Patent reviews are available on such products (149).

A. Roof and Surfacing Coatings

Asphalt emulsions for these uses generally contain mineral fillers, fibers, pigments, and sometimes a polymeric latex. The absence of flammable solvents is a distinct advantage over conventional liquid-asphalt coatings.

1. FILLED COATINGS

Clay emulsions have been successfully used for roof coatings for many years. A performance comparison with products of the asphalt-solvent type is given by Dickson (150). Clay asphalt emulsions dry from the bottom up and thus do not skin over as do some of the asphalt emulsions made with other emulsifier types (151). Polymer latex has been added to clay emulsions (152).

A fire-resistant asphalt-emulsion coating composition contains asbestos fiber, pulverized oyster shells, mica, metal salts of carboxymethylcellulose, and ammonium hydrogen phosphate (153).

Most present-day coating emulsions are of the anionic type. Cationic types should be advantageous because of faster set and early rain resistance. However, the incorporation of inorganic fillers like clays or asbestos, which become negatively charged when moist, is a problem with cationic emulsions. Pretreating such negatively charged materials with an aqueous solution of the nitrate, nitrite, chloride, or sulfate salts of sodium, potassium, or ammonium allows a stable dispersion to be formed in a cationic emulsion (154). Alkyl-substituted quaternary ammonium compounds are also useful as pretreatemnt agents for such fillers. Organic fillers, such as cellulosic fibers, wood flour, or coal dust, can also be dispersed in cationic emulsions by this technique. The use of cationic emulsifiers that interact with negatively charged aggregates at a very slow rate is another possibility for making stable compositions. Alkyl quaternary ammonium emulsifiers or diamine salts containing an aryl group attached to the aliphatic side chain have this property (155).

2. COMBINATIONS WITH POLYMERS

Compatibility of the polymer with the asphalt used in coating compositions is important if a durable coating is to be obtained. This can be determined by examining dehydrated films from emulsion-latex mixtures under a microscope. Styrene-butadiene and neoprene are examples of asphalt-compatible polymers. The difference in water content between typical asphalt emulsions, usually 40 to 30%, and that of polymer latices, frequently about 50%, can result in a stability problem when these two types of materials are mixed. Additional water can be added, but this is undesirable from the standpoint of keeping the solids content of the product high. Addition of about 0.01 to 0.05% of an amine, such as ehtanolamine, to the asphalt emulsion will sometimes permit mixing with a concentrated polymer latex without coagulation (156). Hydrocarbon solvents are included in some rubber latex-asphalt emulsion compositions (157).

8. ASPHALT EMULSIONS

Polymers can be incorporated into the asphalt phase prior to emulsification. Asphalt emulsions containing polyepoxides have been made in this fashion (158). Another preparation method involves directly combining the asphalt and a latex-surfactant aqueous phase in a colloid mill (159).

Mixtures of asphalt emulsion and urea-formaldehyde resins have been sprayed with an acid catalyst in such a manner as to produce a foamed reaction mass useful as an insulation material (160). Asphalt-polymer coatings can be formed by simultaneously spraying an emulsion and a latex, so that coagulation takes place on the substrate (161).

B. Paper and Building-Board Sizes

Anionic asphalt emulsions have been used extensively to waterproof kraft paper, corrugating stock, jute liners, insulation board, and other building boards (162). As little as 1 or 2% of asphalt solids based on fiber weight may be effective. Clay or pinewood resin-type emulsifiers are generally used in such emulsions. An emulsion formulation consisting of a monocarboxylic acid soap in combination with organic sulfonate metallic salts has also been used for sizing (163). An alkali sulfonate of a polyaryl condensation product has been used as an emulsifier for an asphalt-wax emulsion employed as an additive in gypsum board (164). Wax, in combination with polyisobutene, has been incorporated into asphalt emulsions used for waterproofing paperboard (165). The wet strength of paper sized with an asphalt-wax emulsion is improved if a copolymer of ethylene and vinyl acetate is included in the formulation (166).

Cationic asphalt paper-sizing emulsions utilize the electronegativity of moist cellulose fibers to obtain better asphalt retention and adhesion (167).

A sheet material for sound-deadening purposes that contains not less than 80 wt % of finely divided mineral matter uses an asphalt-emulsion binder (168).

The coating of fiber-glass materials with asphalt emulsion is improved by the presence of phosphated-alcohol and sulfonated-polyester additives in the emulsion (169).

C. Adhesives

The water resistance of paper-box adhesives, particularly those used for corrugating, has been improved by the incorporation of asphalt emulsions. Combinations with starch-urea-formaldehyde adhesives are especially useful (170). Asphalt-emulsion mixtures with natural rubber or synthetic latices also give water-resistant adhesives for paperboard (171). A water-in-asphalt emulsion containing aqueous lignin sulfonate is reported to be particularly useful as an adhesive for porous material (172). Asphalt emulsions mixed with polypropylene tow have been used as sealers for cracks and joints (173). Floor-tile cements frequently contain emulsified asphalts.

D. Metal-Coating Uses

Asphalt-emulsion products for automobile undercoating have an advantage over products containing asphalt and a hydrocarbon solvent in that they can be applied to moist surfaces, which may result from the washing operation needed to clean the metal surface before application. An emulsified asphalt with a softening point of 150 to 220°F, in combination with a minor amount of a volatile petroleum solvent, gives a hard, rust-preventing coating (174). The sodium salt of N-cocoaminobutyric acid is used as the emulsifier in this case. A clay asphalt emulsion has also been used for undercoating (175).

8. ASPHALT EMULSIONS

Buried lead cables have been successfully protected from corrosion by coating with an asphalt emulsion containing sodium silicate and sodium dichromate (176). Processes for the electrodeposition of asphalt from emulsions onto pipes and other metal objects have been reported (177).

E. Applications in the Mineral Industries

Coal can be protected from oxidation and dusting by spraying with an anionic asphalt emulsion (178). Ores are sometimes pelletized with the aid of emulsified asphalt as a binder (179). Asphalt emulsions diluted with water are frequently used in open-pit mining to control dust.

F. Concrete Additives

The addition of asphalt emulsion to portland-cement concrete or mortar reduces permeability to water and imparts corrosion resistance to steel reinforcing materials; it does, however, diminish strength somewhat. Hydropel (Chevron Asphalt Company) is used for this purpose at rates of about 0.5 to 1.5 gal per sack of cement. A high concentration, 20%, of asphalt emulsion in concrete-floor mortar containing 10 to 30% portland cement can impart resistance to attack by dilute sulfuric acid and lactic acid (180). This type of floor surfacing is useful for dairy and milk-processing buildings. Mixtures of asphalt emulsion made with an asphalt cutback with a penetration grade of 85 to 100 asphalt, petroleum wax (mp 175 to 185°F), and cork (150- to 200-mesh grade) are useful for sealing joints in concrete (181). Portland-cement concrete with reduced efflorescence and improved water resistance is made by mixing with the cement-aggregate mixture a dispersion of soft asphalt (250 to 500 penetration grade) with rosin in an aqueous solution of pine-pitch soaps and a protective colloid (182).

G. Soil Building Blocks

Asphalt materials from oil seeps were used by the ancients of biblical times for making waterproof clay soil building blocks. This technology was rediscovered about 40 years ago and with further refinement has provided stabilized soil building materials for constructing many fine homes in the western United States, particularly in California. Asphalt emulsion has been used as the stabilizing agent. Since a large percentage of the world's population lives in mud-brick houses, this technology has a large potential in developing countries. Mud-brick stabilization with asphalt is presently used commercially only in the United States. The manufacture of stabilized bricks, which involves mixing asphalt emulsion with a selected soil, molding, and sundrying the bricks, has been described in detail (183). A mobile brick-molding machine that permits economical mass production of the bricks has been developed (184). About 5 wt%, based on the dry soil, of a stable, mixing-grade asphalt emulsion, such as an SS type, is required with average soils.

Improved compressive strength and greater density for asphalt soil bricks is obtained by molding under a pressure of 1000 psi or more followed by heating in air at about 400°F or higher for about 16 h (185). Dimensionally stable articles of complex shape can be made by this process. This new material, called BMX, is a development of the Esso Research and Engineering Company. Asphalt emulsion, as well as hot or cutback asphalts, can be used.

H. Oil-Well Treatment

Asphalt-emulsion drilling muds are reported to speed drilling, reduce filter loss, and reduce the breakup of cuttings (186). The permeability of a porous rock formation may be reduced by injecting simultaneously, but through separate tubing, a cationic emulsion and an aqueous solution of a metal hydroxide (187). Asphalt emulsions containing bentonite clay have been used as drilling muds (188). The electrodeposition of asphalt coatings from emulsions onto metals has already been mentioned. This technique has been successfully used to deposit asphalt coatings on oil-well casings to reduce corrosion (189).

XI. AGRICULTURAL AND WATER-CONSERVATION USES

The low cost and versatility of emulsified asphalt make it a useful material for soil coatings, sealants, and soil stabilizers that can be used for environmental improvement. Some of the techniques discussed in this section can be used in combinations to make new ecosystems for the improvement of arid-land productivity.

A. Agricultural Mulches

Asphalt emulsions have been used for many years in some areas as soil mulches along roadsides to promote grass growth and thus control erosion. For this use, the emulsion is generally applied in combination with chopped straw with an especially designed machine. Sometimes grass seed is also blown on the soil with the mulch with these machines. The potential value of asphalt mulches in agriculture was not recognized until the early 1960s. Asphalt mulch is presently used to a limited extent on certain high-value row crops in the western United States.

Increased soil temperature is a major benefit from asphalt mulch. This accelerates early plant growth and may increase the availability of certain minerals from the soil. This and other benefits have been studied in detail (190, 191). Under some conditions substantial increases in crop yields can be obtained (192), but where climatic and moisture conditions are ideal, little improvement results. The ability of asphalt mulch to prevent soil crusting and thus ensure seedling emergence is now regarded as its greatest general economic value. Lettuce and directly seeded tomatoes are crops that profit greatly from this property. Earlier and more uniform ripening with mulch is particularly beneficial for melon crops. The potential usefulness of asphalt mulch for a variety of crops has been estimated (193).

A number of patents relate to the application of asphalt mulch consisting of conventional asphalt emulsions (194). Although cationic emulsions have been claimed to be superior for mulching, the author has found that some anionic mulches work equally well. The consistency of the asphalt in the mulch is particularly important. The asphalt should be as soft as possible without obtaining soil penetration. Mulches made with hard asphalts can hinder seedling emergence. Soil moisture content at the time of application is also an important factor. If the soil is too moist, the mulch will penetrate without forming a satisfactory surface film. It is also desirable to break up large clods during seedbed preparation in order to obtain a uniform, continuous mulch film. Rubber rollers added to the planter can often accomplish this, although care must be taken not to cause too much soil compaction.

In the usual farming operations the mulch is broken up and buried by cultivation as the crop matures. This does not reduce the value of the mulch, as its maximum benefits are obtained during the first few weeks of plant growth. Once incorporated into the soil, the mulch disappears quickly, probably due to bacterial attack (193).

A variety of additives have been proposed for asphalt mulches, primarily to increase their ability to retain moisture in the soil. These include petroleum waxes (195), clays (196), and polymeric materials (197). Herbicides or growth stimulants can also be incorporated in asphalt mulch (198).

A foamed asphalt mulch has been developed (199). This is a

foamed asphalt emulsion stabilized with poly(vinyl alcohol). It can give improved coverage of irregular soil surfaces with a continuous film.

B. Soil-Erosion Control

The first extensive study of asphalt emulsion-derived soil coatings for wind-erosion control was made by Chepil (200). Emulsions of asphalt or other petroleum resinous residue, when highly diluted with water and applied to soils, penetrate to form a consolidated soil layer beneath any film deposited on the surface. This aids in controlling erosion (201). Many areas of the world have soils susceptible to wind-erosion damage. These may profit from the use of erosion-control agents if their use for agricultural purposes is to be expanded. Additional research is needed to develop economical systems for accomplishing this. In other regions moving sand dunes encroach on arable land and need to be controlled. The possibility of using petroleum-derived soil coatings for the reclamation of dunes has been demonstrated (202).

C. Canal and Reservoir Sealants

Unlined canals and reservoirs account for enormous water losses in the United States and throughout the world. The search for lower cost lining materials and sealants is a continuing effort. Synthetic rubber membranes are widely used to line small reservoirs, but are too costly and difficult to install for general use in canals. Asphalt emulsions have been mixed with canal-bank soil to provide stabilization (203). Cationic asphalt emulsions added to water-filled canals or reservoirs can greatly reduce seepage (131, 133, 204). Results with cationic emulsions made with Redicote TXO emulsifiers (Armour Company) have been particularly promising (205).

Although many cationic emulsions give good initial results as waterborne sealants, their performance is not long lasting in many cases because they are deposited mainly as surface films. These are usually easily washed away. Certain very-small-particle-size asphalt emulsions (median particle diameter 1 μ or less) can be percolated into soils to form discrete, continuous membranes at depths of up to and exceeding several feet below the surface (206). Both cationic and anionic fine-particle asphalt emulsions have been used to make subsurface seals in this manner.

A method has been developed for reducing seepage from water bodies by removing the soil underlying the water, contacting the soil with diluted asphalt emulsion, and redepositing the treated soil beneath the water body (207). An underwater grouting procedure with cationic asphalt emulsions for permeable natural or manmade hydraulic structures uses a dilute solution of lime as an emulsion coagulant (208). Leaks in pipelines, particularly around packed joints, can be sealed by filling with asphalt emulsion under pressure (209).

D. Water Harvesting

The term "water harvesting" describes the collection of rainfall runoff by catchments. Although the ancients demonstrated the feasibility of such systems for agricultural purposes in the Negev desert by covering watershed areas with gravel (210), only recently have membrane soil coatings been used to collect water for livestock in arid regions (211). Asphalt emulsion was used to construct a water-harvesting catchment near Kona, Hawaii, in 1959 (212). This installation had a decline in collection efficiency from 93 to 78% over a 3-year period mainly due to the growth of vegetation through the membrane. Construction details and short-term observations on surface conditions are reported by Myers, Frasier, and Griggs (213) for a number of catchments that were constructed in the southwestern

8. ASPHALT EMULSIONS

United States with emulsified and liquid asphalts. The life of asphalt emulsion-derived soil coatings is greatly extended by application of an asphalt-leafing aluminum pigment composition (214).

E. Underground Sealing of Sandy Fields

Some sandy soils cannot be used for agriculture because water loss by seepage would exceed available supplies. Crop yields in some presently producing sandy areas could be increased if soil permeability could be reduced in an economical manner. The feasibility of creating an asphalt subsurface membrane by spraying asphalt emulsion into a cavity created by a moving subsoiler-type plow has been demonstrated (215). A cationic emulsion is coagulated by injecting ammonia gas into the cavity during the spraying operation. The membranes thus formed about 2 ft below the surface gave substantial yield increases for several crops. A technique for installing subsoil membranes consists of lifting the soil overburden with a scoop-and-conveyer device as asphalt emulsion is sprayed on the newly uncovered surface and then redepositing the soil overburden as the machine progresses (216).

F. Climate Modification

Coating large earth surface areas with asphalt coatings has been proposed by Black and Tarmy (217,219) as a potential means of increasing rainfall. Heated air masses may produce a "thermal mountain" effect and trigger precipitation in moist air. Additional meteorological analyses by Malkus (218) support the theoretical basis for this suggestion.

G. Fertilizer Component

Asphalt emulsion mixed with aqueous ammonia, ammonium nitrate, or urea reduces nitrogen loss when used on sandy soils (220). Diethylene glycol in a minor amount reduces the viscosity of such mixtures.

REFERENCES

1. A. J. Day and E. C. Herbert, in *Bituminous Materials: Asphalts, Tars, and Pitches*, Interscience, New York, 1965, p. 333.
2. E. W. Mertens and M. J. Borgfeldt, ibid., p. 359.
3. J. J. Drukker, ibid., p. 391.
4. C. N. Forrest, *Proc. Assoc. Asphalt Paving Tech.*, 2, 22 (1929).
5. E. S. Ross, *Proc. Assoc. Asphalt Paving Tech.*, 2, 23 (1929).
6. U.S. Department of the Interior, Bureau of Mines, *Mineral Industries Survey, Asphalt Shipments, Annual 1969*, Washington, D.C., 1970, pp. 3-6.
7. Anon., *Chem. Eng. News*, April 27, 1959, p. 54.
8. Anon., *Eng. News Record*, July 2, 1959, p. 44.
9. International Slurry Seal Association, *Report on Slurry Seal Applications*, No. 1, Rev. ed., 1968; R. J. Province, *Bituminous Slurry Surfaces Handbook*, Slurry Seal Inc., Waco, Texas, 1966; W. J. Kari and L. D. Coyne, *Proc. Assoc. Asphalt Paving Tech.*, 33, 502 (1964).
10. R. L. Ferm and C. C. Latif, French Pat. 1,589,212 (1970); W. Hoff, *International Slurry Seal Assoc., 7th Annual Convention Transcript*, Miami, Fla., January 1969.
11. American Society for Testing and Materials, Tentative Specification Designations D 977-68T and D 2397-68T, Philadelphia, 1968.

12. J. H. Caro, *Erdöl-Zeit.*, No. 7 (1962); through *Bitumen, Teere, Asphalte, Peche*, *13*, 526 (1962).
13. H. J. Sommer, U.S. Pat. 2,730,506 (1956).
14. N. V. de Bataafsche Petroleum Maatschappij, Dutch Pat. 72,781 (1953); through *C.A.*, *50*, 6038 (1956).
15. W. K. Seifert and R. M. Teeter, *Chem. Ind. (London)*, 1464 (1969).
16. N. H. Greatorex and K. N. Shaw, Canadian Pat. 812,658 (1969).
17. Shell Internationale Research Maatschappij, British Pat. 864,102 (1961).
18. L. C. Bradshaw and J. M. Durst, British Pat. 1,149,257 (1969).
19. R. Tourret and J. Valayer, British Pat. 1,165,517 (1969).
20. N. E. Lemmon and F. W. Schuessler, U.S. Pat. 2,436,046 (1948).
21. K. E. McConnaughay, U.S. Pat. 2,855,319 (1958).
22. P. E. McCoy, U.S. Pat. 2,512,580 (1950).
23. W. F. Jense, Dutch Pat. 63,008 (1949); through *C.A.*, *43*, 5186 (1949).
24. K. Ichihara, Japanese Pat. 12,771 (1961); through *C.A.*, *56*, 5027 (1962).
25. O. N. Sola, *Rev. Fac. Cienc. Quim.* (Univ. Nacl. La Plata), *25*, 111(1950); through *C.A.*, *48*, 4798 (1954).
26. D. N. Myers, U.S. Pat. 1,957,031 (1934).
27. J. N. Borglin, U.S. Pat. 2,351,912 (1944).
28. C. Maters and R. J. Riemersma, U.S. Pat. 2,155,141 (1939); W. W. DeLaney, U.S. Pat. 2,350,548 (1944); R. W. Martin, U.S. Pat. 2,376,498 (1945).
29. W. A. McIntosh, *Ind. Eng. Chem.*, *44*, 1656 (1952).
30. P. E. McCoy, U.S. Pat. 2,481,322 (1949).
31. H. J. Sommer and C. C. Evans, South African Pat. Appl. 69/2968 (1969).
32. W. D. Buckley and E. O. Bly, U.S. Pat. 2,372,658 (1945).
33. I. Tomita, Japanese Pat. 54 (1952); through *C.A.*, *49*, 1320 (1955).
34. L. Kirschbraum and H. L. Levin, U.S. Pat. 1,960,112 (1934).

35. J. Parker, British Pat. 333,303 (1929).
36. A. W. Hixson and J. M. Fain, *Ind. Eng. Chem.*, **24**, 1336 (1932).
37. R. L. Ferm, U.S. Pat. 3,340,203 (1967).
38. A. Schwieger, U.S. Pat. 2,020,662 (1936).
39. Hanseatische Mühlenwerke A.G., British Pat. 382,432 (1932).
40. A. H. Michelsen, Danish Pat. 38,178 (1927); through *C.A.*, **22**, 3042 (1928).
41. J. W. Redman, British Pat. 1,109,917 (1968).
42. L. Cohen, U.S. Pat. 2,972,588 (1961).
43. M. Asfazadourian and P. Bernard, French Pat. 1,541,748 (1967).
44. H. J. Sommer and R. L. Griffin, U.S. Pat. 2,706,688 (1955).
45. Esso Standard Societé Anonyme Francaise, British Pat. 792,648 (1958).
46. P. L. DuBrow, J. N. Dybalski, and E. Fischer, German Pat. Appl. 1,594,724 (1969).
47. S. D. Turk, A. M. Schnitzer, and B. D. Simpson, U.S. Pat. 3,389,090 (1968).
48. Kao Soap Co. Ltd., French Pat. 1,576,697 (1968).
49. R. A. Rick, J. S. Stalioraitis, and J. N. Dybalski, German Pat. 1,292,572 (1969).
50. P. L. DuBrow, J. N. Dybalski, and E. Fischer, Canadian Pat. 849,080 (1960).
51. Esso Research and Engineering Co., British Pat. 849,080 (1960).
52. Imperial Chemical Industries Ltd., Australian Pat. 250,441 (1964).
53. E. W. Mertens, U.S. Pat. 3,096,292 (1963).
54. E. W. Mertens, U.S. Pat. 3,097,174 (1963).
55. C. W. Falkenberg, R. A. Paley, and J. J. Pati, U.S. Pat. 3,230,104 (1966).
56. J. N. Dybalski, P. L. DuBrow, and A. S. Michaels, U.S. Pat. 3,236,671 (1966).
57. D. F. Levy and D. W. Gagle, U.S. Pat. 3,093,595 (1963).

58. A. C. Pitchford, U.S. Pat. 3,276,887 (1966).
59. J. R. Wright, U.S. Pat. 3,050,468 (1962).
60. J. R. Wright, U.S. Pat. 3,032,507 (1962).
61. O. Dobozy, *Fette, Seifen, Anstrichm.*, **68**, 982 (1966).
62. A. C. Pitchford and J. T. Gragson, U.S. Pat. 3,276,888 (1966).
63. E. J. Miller, Jr., and H. E. Tiefenthal, U.S. Pat. 3,467,708 (1969).
64. S. Ohtsuka and T. Dori, U.S. Pat. 3,466,247 (1969).
65. J. N. Dybalski, South African Pat. 63/4551.
66. M. J. Borgfeldt, U.S. Pat. 3,126,350 (1964).
67. M. J. Borgfeldt, U.S. Pat. 3,432,221 (1969).
68. A. C. Pitchford, Canadian Pat. 813,241 (1969); J. R. Wright, U.S. Pat. 3,433,026 (1969); L. Havestadt, German Pat. 1,180,300 (1964).
69. A. C. Pitchford, U.S. Pat. 3,432,320 (1969).
70. M. E. Conn and A. H. Popkin, U.S. Pat. 3,511,676 (1970).
71. L. Worson, U.S. Pat. 2,468,533 (1949).
72. Flintkote Co., British Pat. 32,721 (1928).
73. N. V. de Bataafsche Petroleum Maatschappij, French Pat. 690,242 (1929); through *C.A.*, **25**, 1071 (1931); International Bitumen Emulsions Corp., British Pat. 400,409 (1933); through *C.A.*, **28**, 2178 (1934).
74. W. E. Heinz, Australian Pat. 61,809 (1961).
75. J. Conort, Canadian Pat. 803,914 (1969); I. B. Chang, U.S. Pat. 3,497,371 (1970); J. J. Christie, U.S. Pat. 2,889,230 (1959).
76. J. E. Fratis and E. H. Oakley, U.S. Pat. 2,406,823 (1946).
77. E. Bergel, Austrian Pat. 134,983 (1933); through *C.A.*, **28**, 875 (1934).
78. A. Champagnat, French Pat. 1,053,174 (1954).
79. Y. Kishi and M. Fiyita, Japanese Pat. 16,990 (1963); through *C.A.*, **59**, 11174 (1963).
80. H. Bouis, German Pat. 1,494,420 (1969).
81. E. W. Mertens and J. R. Wright, U.S. Pat. 2,974,107 (1961).

82. E. W. Mertens, U.S. Pat. 3,240,716 (1966).
83. E. Roualt, French Pat. 748,886 (1933).
84. J. R. Wright and E. W. Mertens, U.S. Pat. 3,052,639 (1962); R. W. Farris, U.S. Pat. 2,701,777 (1955).
85. A. M. Goldstein and E. N. Alter, U.S. Pat. 3,146,200 (1964).
86. K. Keller, K. Keller, and P. Koppe, German Pat. 1,163,726 (1964).
87. Shell Internationale Research Maatschappij, British Pat. 864,102 (1967).
88. E. C. Brown, W. H. Driesen, and J. S. Gupet, U.S. Pat. 2,782,169 (1957).
89. R. W. Farris, U.S. Pat. 2,712,506 (1955).
90. A. Gastmans, *Bitumen, Teere, Asphalte, Peche*, 19, 441 (1968).
91. Y. Mouton, *Bitumen, Teere, Asphalte, Peche*, 20, 19 (1969).
92. R. L. Ferm and R. F. Boynton, U.S. Pat. 3,445,258 (1969).
93. E. W. Mertens and J. R. Wright, U.S. Pat. 3,026,266 (1962).
94. M. J. Borgfeldt, U.S. Pat. 3,220,953 (1965).
95. D. N. Manzer, U.S. Pat. 2,615,851 (1952).
96. P. E. McCoy, U.S. Pat. 2,412,526 (1946).
97. H. F. Hardman and E. D. Wells, U.S. Pat. 2,789,918 (1957).
98. R. L. Ferm, U.S. Pat. 3,305,379 (1967).
99. D. Wagner and F. Hugel, U.S. Pat. 2,726,163 (1955).
100. E. D. Wells and R. B. Faris, Jr., U.S. Pat. 2,786,775 (1957).
101. H. F. Hardman and R. F. Jenkins, U.S. Pat. 2,780,557 (1957).
102. W. Hilgers Co. and Cassella Farbwerke Mainkur A. A., French Pat. 1,337,523 (1963).
103. N. V. deBataaffsche Petroleum Maatschappij, Dutch Pat. 64,891 (1949); through *C.A.*, 44, 5622 (1950).
104. D. J. Dwyer, Canadian Pat. 804,500 (1969).
105. M. Ceintrey, German Pat. 1,297,519 (1969).
106. L. G. Thompson, U.S. Pat. 1,884,919 (1933).
107. K. Meyer, F. Eisenhunt, and A. Siegel, U.S. Pat. 2,774,676 (1956).
108. T. G. Swift, U.S. Pat. 3,085,889 (1963).

109. Y. Itikawa and S. Maki, Japanese Pat. 96,313 (1932); through C.A., *27*, 3812 (1933).
110. A. Tsuchiya and T. Furuta, Japanese Pat. 2485 (1957); through C.A., *52*, 5811 (1958).
111. Etablissements Brunel Freres, British Pat. 966,336 (1964).
112. Sonic Engineering Corp., Bulletin 640, Norwalk, Conn.
113. F. Fukushima, U.S. Pat. 3,236,503 (1966).
114. Anon., *Annual Book of ASTM Standards,* American Society for Testing and Materials, Philadelphia, 1969, Designation D 244.
115. R. Sauterey, Comision Permanente del Asfalto, XII Reunion Anual, Buenos Aires, 1964, p. 138.
116. B. A. Matthews and C. T. Rhodes, *J. Colloid Interface Sci.*, *32*, 339 (1970); P. Walstra and H. Oortwijn, ibid., *29*, 424 (1969); E. Sholton and S. S. Davis, *J. Pharm. Pharmacol.*, 430 (1968).
117. J. C. Vogt, *Rev. Gen. Routes Aerodromes*, 430 (1968).
118. E. W. Mertens, L. D. Coyne, and E. D. Rogers, ASTM Special Technical Publication No. 294, 1960, p. 68.
119. The Asphalt Institute, *Asphalts--Paving and Liquid,* Berkeley, Calif., 1970, p. 19.
120. C. Gaestel, *Chem. Ind. (London)*, 221 (1967).
121. E. W. Mertens and J. R. Wright, *Highway Res. Board Proc.*, *38*, 386 (1959).
122. H. C. Bower, U.S. Pat. 3,270,631 (1966).
123. E. A. Sinclair and K. E. Bristol, *Rubber World*, No. 12, 67 (1969).
124. G. N. Woodruff, U.S. Pat. 3,505,260 (1970).
125. M. J. Borgfeldt and R. L. Ferm, *Highway Res. Board Proc.*, *41*, 195 (1962).
126. Chevron Asphalt Co., *Bitumuls Base Treatment Manual*, San Francisco, 1969.
127. Anon., *Roads and Streets*, *108(9),* 83 (1965).
128. Anon., *Asphalt, Asphalt Inst. Quart.*, October 1965, p. 12.
129. A. C. Pitchford, Canadian Pat. 753m572 (1967).

130. J. Oberbach, British Pat. 990,189 (1963).
131. J. N. Dybalski, Canadian Pat. 766,138 (1967).
132. Shell Internationale Research Maatschappij N.V., Australian Pat. 267,506 (1963).
133. E. Nasallah, British Pat., 1,099,475 (1966).
134. C. Duruy and G. Samman, French Pat. 1,327,277 (1963).
135. T. K. Miles, U.S. Pat. 2,378,235 (1945).
136. J. van Hulst, U.S. Pat. 2,051,505 (1936).
137. J. R. Benson, U.S. Pat. 3,399,608 (1968).
138. D. A. Bennett, Canadian Pat. 795,041 (1968).
139. Shell Internationale Research Maatschappij N.V., British Pat. 935,818 (1963).
140. H. G. Shirley, *AD624659*, U.S. Department of Commerce, Washington, D.C., 1965.
141. D. J. Clifton, *Roads and Road Construction*, March 1967, p. 60; W. J. Hiscock, ibid., p. 66.
142. F. Hardisty and J. A. Sinclair, British Pat. 1,124,498 (1968).
143. L. C. Brasshaw and W. J. M. Durst, British Pat. 1,130,332 (1968).
144. J. F. T. Blott and L. Bolgar, British Pat. 1,024,588 (1966).
145. J. N. Dybalski, Australian Pat. 277,872 (1965); Canadian Pat. 741,715 (1966).
146. N. H. Greatorex and K. N. Shaw, South African Pat. Appl. 66/6702 (1966).
147. Etablissements Lassailly and Bichebois, French Pat. 1,418,824 (1966).
148. P. E. McCoy, *Transcript of Principal Speeches*, International Slurry Seal Association 6th Annual Convention, San Francisco, 1968, p. 18.
149. F. Rosendahl, *Bitumen, Teere, Asphalte, Peche*, *20(2)*, 52 (1969); M. Boettcher, ibid., *16(2)*, 54 (1965).
150. W. J. Dickson, *Ind. Eng. Chem.*, *58*, 28 (1966).
151. R. N. Traxler, *Roads and Streets*, *103*, 158 (1960).

152. J. J. Drukker, U.S. Pat. 3,493,408 (1970).
153. W. E. Skelton and C. E. Wilkinson, U.S. Pat. 3,224,890 (1965).
154. Armour and Co., British Pat. 1,125,518 (1968).
155. Armour and Co., British Pat. 1,190,808 (1970).
156. I. B. Chang, Canadian Pat. 785,827 (1968).
157. H. J. Sarrett, Jr., and W. N. Axe, Canadian Pat. 716,627 (1965).
158. Shell Internationale Research Maatschappij N.V., British Pat. 960,911 (1964).
159. R. D. Timmons and L. M. Harkness, German Pat. 1,800,995 (1969).
160. G. Giesemann, German Pat. 1,273,185 (1968).
161. Farbenfabriken Bayer A.G., French Pat. 1,517,381 (1967).
162. E. M. Lorenzini, *Paper Trade J.*, *124(23)*, 43 (1947); J. J. Perot, *Fibre Contianers*, *37(10)*, 74 (1952); J. J. Perot, *Tappi*, *36*, 69 (1953).
163. E. E. Woodward, U.S. Pat. 3,036,015 (1962).
164. R. L. Selbe, U.S. Pat. 2,699,414 (1955).
165. R. H. Cubberley and W. Thomas, U.S. Pat. 3,007,825 (1956).
166. A. A. Goldstein, U.S. Pat. 3,525,668 (1970).
167. W. J. Kari, E. M. Lorenzini, and L. C. Wingerd, U.S. Pat. 3,271,240 (1966).
168. E. O. Groskopf, U.S. Pat. 2,742,373 (1956).
169. R. W. Farris, U.S. Pat. 2,676,155 (1954).
170. L. C. Wooster and E. E. Gardiner, U.S. Pat. 2,489,170 (1949).
171. H. E. Tarbell and D. W. Mogg, Canadian Pat. 795,225 (1968).
172. W. D. Buckley, P. E. McCoy, and L. G. Thompson, U.S. Pat. 2,506,339 (1950).
173. D. W. Gagle and H. L. Draper, U.S. Pat. 3,503,311 (1970).
174. J. C. Roediger, E. E. Tompkins, and G. H. Schoenbaum, U.S. Pat. 3,427,172 (1969).
175. British Petroleum Co., British Pat. 1,111,650 (1968).
176. W. D. Sanderson, *Corrosion*, *7(11)*, 1 (1951).

177. P. E. McCoy, U.S. Pat. 3,159,558 (1964); J. E. Putnam, U.S. Pat. 1,897,011 (1933).
178. C. R. Rosencranse, U.S. Pat. 2,431,891 (1947); Reissue 23,462 (1952).
179. W. E. Heinz and R. E. Dodd, Canadian Pat. 686,198 (1964).
180. O. Matsuda, T. Kawano, H. Nishi, and S. Matsuyoshi, Semento Konkuruto, 2o2, 7 (1966); through C.A., 62, 314 (1965).
181. F. T. Helm and T. D. Just, U.S. Pat. 2,748,012 (1956).
182. G. H. Eick, R. P. T. Young, and J. N. Stone, Canadian Pat. 779,172 (1968).
183. G. L. Kimmons, R. L. Ferm, and R. Matteson, Ind. Eng. Chem. Prod. Res. Develop., 8, 250 (1969).
184. H. C. Sumpf, U.S. Pat. 2,524,683 (1950).
185. D. T. Rogers and J. C. Munday, Ind. Eng. Chem. Prod. Res. Develop., 8, 241 (1969).
186. Anon., Oil Gas J., May 9, 1966, p. 86.
187. C. T. Brandt and W. R. Bowles, U.S. Pat. 3,159,976 (1964).
188. W. M. Dobson, A. L. Frye, and T. A. Gruggs, U.S. Pat. 2,380,156 (1945).
189. L. C. Hill and L. C. Cronberger, Materials Protection, No. 9, 42 (1963).
190. N. Alpert and R. A. Louis, paper presented at the Sixth World Petroleum Conference, Frankfurt/Main, June 1963, Section VI.
191. F. H. Takatori, L. F. Lippert, and F. L. Whiting, Calif. Agriculture, 17(6), 2 (1963).
192. Anon., Chem. Eng., No. 5, 90 (1962).
193. L. F. Lippert, ed., Proc. Petroleum Mulch Conf., University of California, Riverside, December 1966.
194. R. A. Louis and I. F. Wagner, U.S. Pat. 3,061,974 (1962); H. J. Hibshman, R. A. Louis, and I. F. Wagner, U.S. Pat. 3,061,975 (1962).
195. J. F. T. Blott and D. R. Lamb, South African Pat. Appl. 64/2658, (1964).
196. L. Henschel, U.S. Pat. 3,336,146 (1967).

197. R. H. Salvesen, U.S. Pat. 3,274,138 (1966); L. W. Blanken, U.S. Pat. 3,323,254 (1967).
198. L. F. Wagner, U.S. Pat. 3,207,594 (1965); A. G. Harshman, U.S. Pat. 3,210,174 (1965); R. H. Salvesen, U.S. Pat. 3,252,784 (1966); T. M. Mozell, U.S. Pat. 3,210,173 (1965).
199. J. M. Iwasyk and B. C. Lawes, U.S. Pat. 3,387,405 (1968).
200. W. S. Chepil, *Soil Sci. Soc. Proc.*, *19*, 125 (1955).
201. M. B. Goren and B. G. Marquardt, U.S. Pat. 2,927,402 (1960).
202. G. P. Richard and Tl L Les, paper presented at the Sixth World Petroleum Conference, Frankfurt/Main, June 1963, Section VI.
203. M. E. Hickey, paper presented at the Conference Asphalt in Hydraulic Construction, Asphalt Institute, Bakersvield, Calif., May 1961.
204. K. Blair and T. L. Lee, British Pat. 932,726 (1963); D. H. Labon, G. Whitehouse, and E. Nasrallah, British Pat. 1,003,323 (1965); M. E. Hickey and P. F. Enger, *Lower Cost Canal Lining Program*, General Report No. 32, U.S. Department of Interior, Bureau of Reclamation, Washington, D.C., 1963
205. Anon., *Chem. Week,*, November 4, 1967, p. 44.
206. R. L. Ferm, unpublished results; U.S. Pat. Appl. (1970).
207. J. N. Dybalski, U.S. Pat. 3,359,738 (1967).
208. D. W. Gagle and D. F. Levy, U.S. Pat. 3,252,290 (1966).
209. J. P. Harmsworth and N. R. Bangert, British Pat. 1,062,278 (1967).
210. M. Evenari, L. Shanan, N. Tadmore, and Y. Aharoni, *Science.* *133*, 979 (1961).
211. C. W. Lauritzen, *Crops Soils*, August-September 1961, p. 7.
212. S. S. W. Chinn, *Geological Survey Water-Supply Paper 1809-P*, U.S. Department of Interior, Washington, D.C., 1965.
213. L. E. Myers, G. W. Frasier, and J. R. Griggs, *J. Irrigation Drainage Div.* (Amer. Soc. Civil Engr.) *93*, 79, No. IR3, Paper 5413 (1967).
214. R. L. Ferm, U.S. Pat. 3,394,550 (1968).

215. C. M. Hansen and A. E. Erickson, *Ind. Eng. Chem. Prod. Res. Develop.*, *8*, 256 (1969); J. A. Bolt, U.S. Pat. 3,276,208 (1966).
216. W. R. Thompson and B. Adams, U.S. Pat. 3,394,551 (1968).
217. J. F. Black and B. L. Tarmy, *J. Appl. Meteorol.*, *2*, 557 (1963).
218. J. S. Malkus, *J. Appl. Meteorol.*, *2*, 547 (1963).
219. J. F. Black, U.S. Pat. 3,409,220 (1968); British Pat. 988,109 (1965).
220. A. G. Harshman and R. W. Sage, U.S. Pat. 3,164,925 (1965).